(*continued on back*)

(*continued from front*)

Model Selection

Model Selection

H. LINHART

Universität Göttingen

W. ZUCCHINI

University of Cape Town

JOHN WILEY & SONS

New York • Chichester • Brisbane • Toronto • Singapore

Copyright © 1986 by John Wiley & Sons, Inc.

All rights reserved. Published simultaneously in Canada.

Reproduction or translation of any part of this work
beyond that permitted by Section 107 or 108 of the
1976 United States Copyright Act without the permission
of the copyright owner is unlawful. Requests for
permission or further information should be addressed to
the Permissions Department, John Wiley & Sons, Inc.

Library of Congress Cataloging in Publication Data :

Linhart, H.
  Model selection.

  (Wiley series in probability and mathematical
statistics, ISSN 0271-6038. Applied probability and
statistics)
  Bibliography: p.
  Includes index.
  1. Mathematical statistics. I. Zucchini, W.
II. Title.   III. Series.

QA276.L5486   1986   519.5   86-7763
ISBN 0-471-83722-9

Printed in the United States of America

10 9 8 7 6 5 4 3 2 1

TO OUR FAMILIES

Gertrud, Hans, Eva, Heinz
Sue, Leon, Laura

Preface

This book describes a systematic way of selecting between competing statistical models. We outline a general strategy and then apply it to develop methods to solve specific selection problems. Examples of application are used to illustrate the methods.

Naturally, the choice of topics and examples is based in favor of our own research interests, but we have tried to cover many of the typical situations that arise in practice. These include the selection of univariate distributions, simple and multiple regression, the analysis of variance and covariance, the analysis of proportions and contingency tables, and time series analysis.

The mathematical level is quite modest. The somewhat more demanding asymptotic theory is relegated to the Appendix. A relatively broad statistical knowledge, however, is desirable. We assume that the reader is familiar with standard statistical terminology and techniques.

We have not attempted an exhaustive survey of model selection methods—this would have required a book for each of the present chapters. The list of references is also far from complete. In particular, we write only about classical statistics; Bayesian methods are not discussed. Also omitted are selection methods based on standard tests of null hypotheses that certain simple models hold.

Although we deal exclusively with the selection of parametric models we do not necessarily assume that the operating model has a simple parametric form. In other words, we wish to select a simple model for what could possibly be a complex situation.

The selection methods that we describe are purely data based. As with any other automatic method in statistics they should be uses critically; they are meant to supplement but not to replace sound judgment.

This research was partially supported by the Water Research Commision and the Council of Scientific and Industrial Research, South Africa. It gives us pleasure to acknowledge the help of many colleagues. We thank Professor A. Linder, Dr. P. Volkers (who wrote Section A.2.7 and as coauthor of relevant papers had a strong influence on Sections A.1, A.2.5, and A.2.6), Dr. F. Böker, Mr. R. Bust, Miss M. Hochhaus, Professor J. Juritz, Dr. G. Rubinstein, Dr. R. S. Sparks, Professor T. Stewart, and Mr. S. Witzel. Mr. R. Ederhof helped with the computing problems. Finally, we wish to thank Mrs. E. Wegener and Mrs. I. Biedekarken for their patient and expert typing.

H. LINHART
W. ZUCCHINI

Göttingen, West Germany
Cape Town, South Africa
July 1986

Contents

Model Selection

CHAPTER 1

The Basic Ideas

One of the main tasks of applied statisticians is to construct probability models for observations. One begins by representing the observations in terms of random variables. To "fit a model" means to provide an estimate of the probability distribution of these random variables.

A probability model is an abstract mechanism which one can imagine as having generated the data. Naturally, one does not believe that the data were actually generated in this way. The model is merely a convenient conceptual representation of the observed phenomenon. Tukey (1961) wrote: "In a single sentence the moral is: admit that complexity always increases, first from the model you fit to the data, thence to the model you use to think and plan about the experiment and its analysis, and thence to the true situation."

1.1. FITTING A MODEL TO OBSERVATIONS

1.1.1. The Operating Model and the Operating Family

We call the "model you use to think about the data" the *operating model*. It constitutes the nearest representation of the "true situation" which it is possible to construct by means of a probability model and, ideally, it is on the operating model that interpretations, predictions, and decisions would be based.

Information on the operating model is obtained from knowledge about the subject matter under investigation. One may know that the random variables must be non-negative, that certain events can be assumed to be equally probable, certain others statistically independent, or that under certain conditions the probability of a given event is infinitesimal compared to

1

that of a second event, and so on. A further source of information is Probability Theory, in particular, asymptotic results such as the central limit theorem.

However, it is only in exceptional cases that sufficient information is available to fully specify the operating model, the distribution of the random variables of interest. As a rule one can only circumscribe the family of models to which the operating model belongs. We call this the *operating family*. The more one knows about the given application the "smaller" the operating family will be.

1.1.2. Approximating Families

Conventionally, the "size" of a family of models is indicated by the number of its independent parameters, that is, the number of quantities which must be specified to identify a specific member of the family. The parameters have to be estimated from observations. The "accuracy" with which this can be done depends essentially on the amount of data available relative to the number of parameters to be estimated; it improves as the sample size increases or the number of parameters decreases.

If it were always possible on the basis of well-grounded assumptions alone to arrive at an operating family having an appropriately small number of parameters, then model fitting would, at least in principle, be straightforward. What often happens is that there are more parameters than can be reasonably estimated from the data. Attempting to estimate too many parameters is referred to as overfitting. The disadvantage of overfitting is instability; repeated samples collected under comparable conditions lead to widely different models, each of which conveys more about the particularities of its corresponding data set than about the underlying phenomenon. The hazards of using such models for interpretation or prediction are obvious.

The above discussion suggests that successful model fitting is simply a matter of ensuring that there are sufficient data to achieve the required level of accuracy. In practice, the sample size is more often dictated by physical and economic constraints than by statisticians; one has to make do with what is available. To proceed at all one is often obliged to compromise by using a simpler *approximating family* of models. This family from which "the model you fit to the data" is selected may not even contain the operating model.

1.1.3. An Example

Table 1.1 gives the annual flow of the Vaal River at Standerton in the years

Table 1.1. Annual Flow ($10^6 m^3$) of the Vaal River at Standerton

Year	Flow	Year	Flow	Year	Flow
1905	222	1927	235	1949	534
1906	1094	1928	346	1950	129
1907	452	1929	778	1951	317
1908	1298	1930	95	1952	640
1909	882	1931	111	1953	291
1910	988	1932	78	1954	1461
1911	276	1933	554	1955	611
1912	216	1934	364	1956	809
1913	103	1935	460	1957	637
1914	490	1936	1151	1958	336
1915	446	1937	286	1959	245
1916	386	1938	1401	1960	683
1917	2580	1939	651	1961	319
1918	408	1940	746	1962	365
1919	258	1941	224	1963	309
1920	606	1942	567	1964	479
1921	715	1943	1593	1965	147
1922	1539	1944	217	1966	683
1923	183	1945	496	1967	250
1924	696	1946	256	1968	324
1925	110	1947	295	1969	556
1926	193	1948	274		

1905–1969. The flow is in millions of cubic meters measured over "water years," which run from October to September.

The following three questions are of particular interest to water resource planners, Zucchini and Adamson (1984b).

1.   *What is the distribution of the annual flow?*

The flow varies so much from year to year that it is inadequate to base any plans solely on the average flow—the *distribution* of flows is required. It can be assumed that conditions on the river catchment have remained unchanged over the period of observation and that they are likely to remain unchanged in the foreseeable future. In other words, we may regard the flows as being identically distributed. They are also obviously non-negative. Whether they can be taken as independently distributed is not clear. In general, annual river flow is serially correlated, but data from other rivers in comparable regions suggest that the serial correlations for this river are negligible. The estimated first-order serial correlation coefficient is small

(-0.046) and is not significantly different from zero. Suppose tentatively that the flows are independently distributed.

With these assumptions one can conclude that the distribution of flows can be described by some univariate distribution function which vanishes for negative arguments. The operating family has an infinite number of parameters and there can be no question of estimating all of them, even if one were prepared to wait for more data. In this example, one has no choice but to use some simpler approximating family of models, for example, the lognormal or gamma families or possibly a histogram-type family.

2. *What is the probability of a drought?*

The second question of interest has to do with the frequency with which abnormally low flows (droughts) can be expected. Suppose that a drought is said to occur when the annual flow is less than 200 million cubic meters. Here it is not necessary to fit a distribution to the annual flows; it is sufficient to fit a distribution to the discrete variable "drought." The operating family is the one-parameter family of Bernoulli distributions which is sufficiently simple for one to proceed without having to consider any approximating families.

3. *What is the probability of three or more successive droughts?*

The lowest recorded flow occurred in 1932. The fact that this drought immediately followed the droughts of 1930 and 1931 compounded its severity. The probability of the event that a run of three or more droughts occurs can be derived in terms of the simple model discussed above, but the derivation is based on the assumption that droughts occur *independently*, which is a consequence of the tentative assumption that the flows are independently distributed.

If one relaxes this assumption a little and assumes instead that the operating model is a simple Markov chain (with two parameters) then one obtains a different estimate of the probability of the event. The question then arises as to whether it is preferable to use the estimate based on the operating family or that based on the approximating (Bernoulli) family. The former family may be closer to the "true situation" but the latter has one less parameter to be estimated and consequently is less subject to sampling variation.

1.2. STRATEGIES FOR SELECTING AN APPROXIMATING FAMILY OF MODELS

The question which arises from our discussion in Section 1.1 is how one should go about selecting an approximating family of models. We assume that a number of such families are available and that the operating family itself may also be a candidate for selection. In this section we discuss general strategies on which one can base the selection.

1.2.1. Selection Based on Tests of Null Hypotheses

A common approach to fitting a model is to *select the simplest approximating family which is not inconsistent with the data.* From a catalog of families of models one selects a family which matches the general features evident in the data. The parameters are estimated under the assumption that the family holds, that it contains a member which conforms to the true situation. The next step is to verify, usually by means of a test of a hypothesis, that the model is not inconsistent with some specific aspects of the data. If inconsistencies are detected then an alternative family is selected.

The compromise which one makes by using this strategy is that all interpretations and predictions have to be prefaced with the remark: "Assuming that the selected family of models holds (and there is no evidence to suggest that it does not)." It is under this assumption that optimal estimators are derived, hypotheses are tested, and decisions are made.

This strategy does have several important advantages. As a rule, one only has to consider relatively simple models. There is also a good deal of literature available on methods of estimation for different circumstances and much reported experience has been accumulated from its frequent application.

In practice, however, there has been a tendency to forget that one is making what are really ad hoc assumptions. Lack of evidence of inconsistency with the data is often tacitly taken to mean that the "true" family of models is known. Under this assumption the parameters, being the only unknowns, become the focus of interest. The problem of fitting a model is reduced to the problem of fitting parameters. But parameters per se are only of interest if they can be naturally and usefully interpreted in the context of the phenomenon under observation, if they describe something real.

An improvement on the above approach is to use estimators which are robust to deviations from the assumption that the family used for fitting holds. Such methods of estimation are discussed in the literature. With this approach the cost of insuring against possible deviations is a loss of efficiency if the envisaged deviations do not arise. The qualifying preface necessary if one uses the traditional strategy, however, can be relaxed somewhat; the selected family of models need only contain an approximation to the operating model.

1.2.2. Selection Based on Discrepancies

An alternative strategy is the one discussed in this book. The idea is to select the family of models which is *estimated to be the "most appropriate" in the circumstances*, namely, the background assumptions, the sample size, and the specific requirements of the user. Briefly, one begins by specifying in which sense the fitted model is required to best conform to the operating model, that is one specifies a discrepancy which measures the lack of fit. The approximating family which minimizes an estimate of the expected discrepancy is selected. It is not assumed that this family contains the operating model; in fact, one may know a priori that it does not.

This approach is fairly flexible; the user can determine which aspects of the model are of importance and then arrange matters so that an appropriate fitting procedure is used. One can to a large extent avoid making ad hoc assumptions. In cases where such assumptions are essential the strategy can nevertheless still be used, but with the qualifications mentioned in Section 1.2.1.

To implement this strategy one has to estimate the expected discrepancy. Although this is the only unknown that depends on the operating model which has to be estimated, it is nevertheless an aspect of a model whose properties can be estimated relatively imprecisely. This estimation problem is the weakest link in the selection procedure based on discrepancies.

1.3. DISCREPANCIES

Whatever strategy is employed to select and fit a model, there will be as a rule a number of aspects in which the operating model and the model which is ultimately fitted differ. Each aspect of the "lack of fit" can be measured by some discrepancy and the relative importance of the different discrepancies will differ according to the purpose of the envisaged statistical analysis. Consequently, it is proposed that the user should decide which discrepancy is to be minimized. In what follows, we suppose that this decision has been taken and we will refer to the selected discrepancy as *the discrepancy*.

A discrepancy allocates a real number to any ordered pair of fully specified models. The discrepancy between the operating model and that member of the approximating family which is "nearest" to it is called the *discrepancy due to approximation*. It does not depend in any way on the data, the sample size, or the method of estimation employed.

In contrast, the *discrepancy due to estimation* is a random variable and does depend on the data and the method of estimation used. It is the discrepancy between the model which is fitted and the (unknown) "nearest" approximating model.

The *overall discrepancy*, which is in some (but not all) cases the sum of the above two components, is the discrepancy between the fitted model and the operating model.

1.3.1. An Example

Figure 1.1 gives the age distribution of the inhabitants of the Federal Republic of Germany in 1974 (*Statistisches Bundesamt*, 1976, p. 58). Table 1.2 gives a random sample of size $n = 100$ from this population.

In this situation the operating model, the complete age distribution, is described by a probability density function f, which is known. Suppose, however, for the purpose of illustration, that we wished to estimate f by

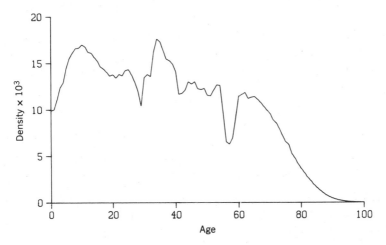

Figure 1.1. Age distribution in the Federal Republic of Germany, 1974.

Table 1.2. A Random Sample of 100 Ages from the Distribution in Figure 1.1

77	17	55	39	63	4	82	41	68	26
1	40	34	11	14	13	7	27	24	10
39	68	20	68	15	4	50	66	3	67
34	48	46	62	32	61	46	11	35	43
48	50	34	4	36	40	52	19	75	3
10	23	70	12	56	34	79	74	8	7
54	67	65	64	66	67	28	42	9	63
59	3	48	11	2	7	39	15	18	10
63	16	79	44	13	37	5	13	6	32
37	39	20	15	8	24	17	26	7	26

means of a histogram-type probability density function $g_\theta^{(I)}$ with I intervals of equal length between 0 and 100. The parameters $\theta = (\theta_1, \theta_2, \ldots, \theta_I)^\mathsf{T}$ represent the heights of the rectangles which constitute $g_\theta^{(I)}$. For a given I there only are $p = I - 1$ free parameters because the area under the density is equal to one and so θ_I is determined by the other parameters. Each I leads to a different approximating family.

Suppose that we wished to obtain an estimate of the general shape of $f(x)$, with no particular emphasis on any one age group. A suitable discrepancy for this is

$$\Delta(\theta) = \int_0^{100}(f(x) - g_\theta^{(I)}(x))^2 dx.$$

Figures 1.2 and 1.3 give the best approximating densities, $g_{\theta_0}^{(10)}$ and $g_{\theta_0}^{(50)}$, the densities which minimize the discrepancy $\Delta(\theta)$ for $I = 10$ and $I = 50$. The corresponding *discrepancies due to approximation* can be calculated:

Discrepancy due to approximation $= 217 \times 10^{-6}$ for $I = 10$

$$= 20 \times 10^{-6} \quad \text{for } I = 50.$$

As would be expected, the discrepancy due to approximation is smaller for $I = 50$ than for $I = 10$ and clearly it will continue to decrease as I increases. Note also that this discrepancy does not depend in any way on the sample size or the sample.

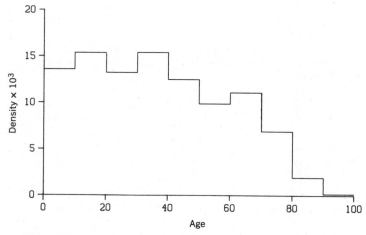

Figure 1.2. Age distribution, best approximating density for $I = 10$.

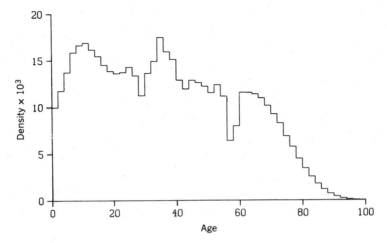

Figure 1.3. Age distribution, best approximating density for $I = 50$.

The *discrepancy due to estimation* does depend on the particular values of the random sample and so differs from sample to sample; that is, it is a random variable. In general, its expectation increases with increasing I because as I increases the number of parameters to be estimated also increases. For the sample given here and using $\hat{\theta}_i = n_i I/100n$, $i = 1, 2, \ldots, I$, where n_i is the frequency in the ith interval, one obtains:

$$\text{Discrepancy due to estimation} = 65 \times 10^{-5} \quad \text{for } I = 10$$

$$= 313 \times 10^{-5} \quad \text{for } I = 50.$$

The fitted models for $I = 10$ and $I = 50$ are given in Figures 1.4 and 1.5, respectively, together with the corresponding best approximating models.

The *overall discrepancy* is in this case simply the sum of the discrepancy due to approximation and the discrepancy due to estimation. One obtains:

$$\text{Overall discrepancy} = 87 \times 10^{-5} \quad \text{for } I = 10$$

$$= 315 \times 10^{-5} \quad \text{for } I = 50.$$

In this example and for the given discrepancy, we have shown that the approximating family with $I = 10$ leads to a smaller overall discrepancy than the family with $I = 50$. One could even find the value of I which leads to the smallest overall discrepancy, but that is not the purpose of the example. What does emerge is that as the number of parameters increases so the discrepancy due to approximation decreases, but (on average) the discrepancy

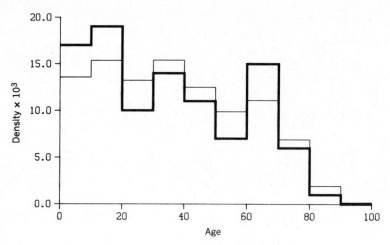

Figure 1.4.   Age distribution, best approximating and fitted (bold) densities ($I = 10, n = 100$).

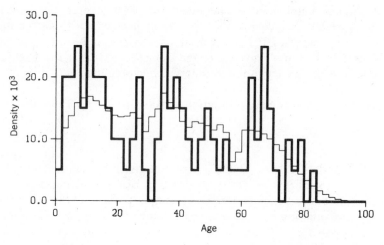

Figure 1.5.   Age distribution, best approximating and fitted (bold) densities ($I = 50, n = 100$).

due to estimation increases. The overall discrepancy tends to decrease and then to increase again.

In practice one can not, as we did in this example, compute the overall discrepancy because the operating model is not known. However, one can use an estimate of its expectation as the basis of a method for model selection. This will be discussed in Section 1.4.

1.3.2. Some Definitions Relating to Discrepancies

To formalize the ideas outlined above, suppose that we have n independent observations on k variables and that each observation can be regarded as a realization of a k-dimensional random vector having distribution function F. Let M be the set of all k-dimensional distribution functions. Each member of M is a fully specified model.

A *family of models*, G_θ, $\theta \in \Theta$, is a subset of M whose individual members are identified by the vector of parameters $\theta = (\theta_1, \theta_2, \ldots, \theta_p)^\mathsf{T}$.

A *fitted model*, $G_{\hat\theta}$, is a member of a family of models G_θ, $\theta \in \Theta$, which has been selected by estimating the parameters using the observations.

A *discrepancy* is a functional, Δ, on $M \times M$ which has the property

$$\Delta(G, F) \geqslant \Delta(F, F) \quad \text{for } G, F \in M.$$

This functional which expresses some aspect of the dissimilarity between two models should be such that its value increases if G and F are considered to become "less similar." The discrepancy between a member G_θ of an approximating family of models and the operating F will be denoted by the abbreviated notation

$$\Delta(\theta) = \Delta(\theta, F) = \Delta(G_\theta, F).$$

The *discrepancy due to approximation* between an approximating family, G_θ, $\theta \in \Theta$, and an operating model, F, is given by $\Delta(\theta_0)$, where

$$\theta_0 = \arg\min\{\Delta(\theta) : \theta \in \Theta\}.$$

We will usually assume that θ_0 exists and is unique. Clearly, if $F \in \{G_\theta, \theta \in \Theta\}$ then by definition θ_0 exists. The model G_{θ_0} is called the *best approximating model* for the family G_θ, $\theta \in \Theta$, and the discrepancy Δ.

The *discrepancy due to estimation* is defined as $\Delta(G_{\hat\theta}, G_{\theta_0})$. It expresses the magnitude of the lack of fit due to sampling variation.

The *overall discrepancy* is defined as $\Delta(\hat\theta) = \Delta(G_{\hat\theta}, F)$. In many of the cases considered here the overall discrepancy is simply the sum of the two component discrepancies, but even where it is not the two "effects" still act in opposition to each other. To minimize the overall discrepancy, the respective contributions of these two opposing component discrepancies must be balanced. It is the overall discrepancy which, ideally, one wishes to minimize.

There are two stages in fitting a model to data. First, an approximating family must be selected. This determines θ_0 and the discrepancy due to approximation, $\Delta(\theta_0)$. Second, some method of estimation is required to obtain $\hat{\theta}$, an estimator of θ_0. By a *fitting procedure* we mean the pair consisting of the *approximating family* and an associated *method of estimation*. Although the primary interest in model selection is often to select between competing approximating families, one can in fact select between different fitting procedures.

The term *discrepancy* is not new in the statistical literature and appears, for example, in Haldane (1951), Sahler (1970), Robertson (1972), Durbin and Knott (1972), Durbin (1973, p. 26), Geisser (1974, p. 106), Parr and Schucany (1980), Parr and De Wet (1981), Ponnapalli (1976), Sakamoto and Akaike (1978a, p. 186), and Serfling (1980, p. 164). The more general term "loss function" is inappropriate in this context where we are dealing with a particular class of loss functions for which "loss" is really a misnomer. The term distance would also be inaccurate. In analysis, "distance" is endowed with properties which we do not want to demand. The word "discrepancy" is free of any specific mathematical, technical, or economic connotations and the sense in which we use it corresponds closely to the sense in which it is commonly used.

1.3.3.   Empirical Discrepancies and Minimum Discrepancy Estimators

A fitting procedure is determined by an approximating family and an associated method of estimation. Obviously, for a given family of models, one wants to choose an estimation method which leads to small expected discrepancies. The expected discrepancy depends on the operating model, which is not known in a given practical situation. Estimators which should produce good results whatever the operating model are the *minimum discrepancy estimators*.

A consistent estimator of $\Delta(\theta)$ is called an *empirical discrepancy* and is denoted by $\Delta_n(\theta)$. If

$$\hat{\theta} = \arg\min\{\Delta_n(\theta): \theta \in \Theta\}$$

exists almost surely we call it a minimum discrepancy estimator of θ_0 [corresponding to $\Delta(\theta)$].

A suitable $\Delta_n(\theta)$ for $\Delta(\theta) = \Delta(\theta, F)$ is usually $\Delta(\theta, F_n)$, where F_n is the empirical distribution function, that is, the defining function of the distribution with mass $1/n$ at the observed points x_i, $i = 1, 2, \ldots, n$. In the following, we shall mostly use minimum discrepancy estimators. They are often referred to as minimum distance estimators in the literature and were first studied by Wolfowitz (1953). A bibliography on this method of estimation is given in Parr (1981). Some recent work is by Boos (1981, 1982).

1.4. CRITERIA

1.4.1. Selection Based on Estimated Expected Discrepancies

The overall discrepancy is a random variable. Its distribution under the operating model determines the quality of a given procedure and thus constitutes a basis for comparing different fitting procedures. It is not always possible to estimate the complete distribution of the overall discrepancy, but only some characteristic of it such as the expectation. Clearly, one would prefer procedures which, on average, result in low overall discrepancies and so it is reasonable to judge a procedure by the expected overall discrepancy, $E_F \Delta(\hat{\theta})$. For simplicity, this will be referred to as the *expected discrepancy*. It is important to note that the expectation is taken under the operating model.

In what follows, we will declare one fitting procedure to be better than another if it has a smaller expected discrepancy.

The expected discrepancy depends on the operating model which is of course unknown, but it can in many cases be estimated. An estimator of the expected discrepancy is called a *criterion*.

To select between competing fitting procedures, it is proposed that one uses the one which leads to the smallest value of the criterion. In cases where the same method of estimation is used, one is in effect selecting between competing approximating families.

Some expected discrepancies (see Example 1.4.1) can be expressed as the sum of a number of terms, some of which do not depend on the fitting procedure used. For the purposes of comparing different fitting procedures, such terms may therefore be omitted. An estimator of the remaining terms (which we refer to as the *essential part of the expected discrepancy*) will also be called a criterion.

EXAMPLE 1.4.1.   The expected discrepancy in the example in 1.3.1 is given by

$$E_F \Delta(\hat{\theta}) = E_F \int_0^{100} (f(x) - g_{\hat{\theta}}^{(I)}(x))^2 dx,$$

where

$$g_{\hat{\theta}}^{(I)}(x) = \frac{n_i I}{100 n} \quad \text{for } \frac{100(i-1)}{I} < x \leqslant \frac{100i}{I}, i = 1, 2, \ldots, I,$$

and n_i is the frequency in the ith interval. After suitably subdividing the integral and taking the expectation one obtains

$$E_F\Delta(\hat{\theta}) = \int_0^{100} f(x)^2 dx + \frac{I}{100n}\left(1 - (n+1)\sum_{i=1}^{I}\pi_i^2\right),$$

where

$$\pi_i = \int_{100(i-1)/I}^{100i/I} f(x)dx.$$

The first term does not depend on the approximating family used, and so the essential part of the expected discrepancy is given by the second term. An (unbiased) estimator of this term, a criterion, is

$$\frac{I}{100n}\left[1 - \frac{n+1}{n-1}\left(\sum_{i=1}^{I}\frac{n_i^2}{n} - 1\right)\right].$$

The criterion was evaluated for different values of I, that is, for different approximating families, for the random sample given in Table 1.2 and the results are illustrated in Figure 1.6. The approximating family corresponding to $I = 6$ leads to the smallest value of the criterion and is therefore estimated to be the most appropriate.

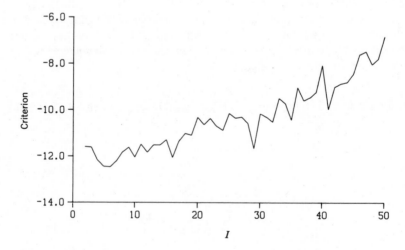

Figure 1.6. Criterion for choice of the number of intervals in a histogram for the age distribution in Germany.

1.4.2. Selection Based on Tests of Hypotheses About Expected Discrepancies

The selection procedure outlined above does not especially favor any particular approximating family of models except on the basis of their criteria. In some situations there are reasons to prefer a particular approximating family of models and to bias the selection in its favor. For example, one may want to use a simple model unless there is strong evidence that it is inferior to a more complex one.

For some of the applications which we consider, it is possible to construct a test of the hypothesis that a given approximating family has a smaller expected discrepancy than a rival family. The required strength of evidence can be achieved by selecting the significance level of the test. It should be noted that such tests are based on the distribution of the test statistic under the operating model. The hypothesis tested is not the usual one, namely, that the simpler model holds.

In many cases tests are difficult to construct and one must then rely on a simple comparison between estimated expected discrepancies (criteria). This corresponds roughly to a test with a significance level of 50%.

1.4.3. Selection Based on Discrepancies Due to Approximation

The expected discrepancy depends strongly on the sample size. This dependence is an essential aspect of the argument which recommends its use. However, in some applications one wants to select the approximating family which *potentially* gives the best fit, irrespective of whether this potential can be achieved with the given sample. In pilot studies to larger experiments, for example, one may want to select on this basis. In such cases the appropriate basis for model selection is the discrepancy due to approximation.

It is often easy to determine which model leads to the smallest discrepancy due to approximation. In the example in 1.3.1 it is clear that $\Delta(\theta_0)$ can be decreased by simply increasing p, the number of parameters. Trivially, the operating family is always best in this sense but one may want to select among simpler families only. In some cases it is not obvious which family leads to the smallest $\Delta(\theta_0)$ and so this discrepancy has to be estimated. An estimator of $\Delta(\theta_0)$ will also be called a criterion.

In most applications the appropriate basis for selection is the expected discrepancy. It takes into account the disparity due to misspecification and the potential sampling variation. Approximating families which, on average, result in a good fit for the *given sample size* are selected.

CHAPTER 2

An Outline
Of General Methods

2.1. SELECTING A DISCREPANCY

The first two steps required to implement the method described in Chapter 1 are to identify an operating family of models and then to construct a discrepancy.

The first step is seldom difficult because the operating family is usually determined by the type of analysis which one intends to carry out (e.g., regression analysis, contingency table analysis, time series analysis). There may be special features in the situation which can be used to reduce the complexity of the models. These should be utilized, but ad hoc assumptions should be avoided at this stage.

The second step is to choose a discrepancy. Here one has to consider the use to which the fitted model is to be put and so decide which aspect of it is required to conform particularly to the operating model.

In theory, any functional on the product space of distribution functions satisfying the simple condition of Section 1.3.2 can be used as a discrepancy and the only difficulty is to select one that is appropriate for the problem in hand.

EXAMPLE 2.1. Suppose that we have IJ pairs of observations (y_{ij}, x_{ij}), $i = 1, 2, \ldots, I, j = 1, 2, \ldots, J$, whose operating model can be described by means of a simple analysis of covariance model:

$$y_{ij} = \alpha_i^* + \beta^* x_{ij} + e_{ij},$$

where y_{ij} are the responses, x_{ij} the (fixed) covariates, α_i^* the treatment means, β^* the covariate regression coefficient, and the e_{ij} are independently distributed with expectation zero and variance σ_i^2.

We think initially in terms of a fully specified operating model, not in terms of an operating family of models. The α_i^* and β^* are fixed constants (unknown to us) and the e_{ij} have some joint distribution which we need not describe in any further detail.

One of the problems encountered when fitting models to such data is to decide whether the covariate should be included in the fitted model. In particular, if β^* is close to zero and the sample size is small, then it can happen that the conventional estimate of β^* leads to a worse fit than if one had simply set the estimate to zero. In other words, one wants to select one of the two approximating families

$$G^{(1)}: y_{ij} = \alpha_i + \beta x_{ij} + e_{ij},$$
$$G^{(2)}: y_{ij} = \alpha_i + e_{ij}, \qquad i = 1, 2, \ldots, I,$$
$$j = 1, 2, \ldots, J.$$

The first of these two is of course the operating family.

Now if one wants to select that family which yields the better estimate of the expected response then it would be reasonable to select the family which leads to the smaller value of $\sum_{ij} (\alpha_i^* + \beta^* x_{ij} - \hat{E}_G y_{ij})^2$. This quantity is simply the overall discrepancy which emerges if the original discrepancy is $\sum_{ij} (\alpha_i^* + \beta^* x_{ij} - E_G y_{ij})^2$.

Here the approximating family $G^{(1)}$ has a discrepancy of approximation equal to zero but the expectation of its discrepancy due to estimation is greater than that of $G^{(2)}$ because it has one more parameter than $G^{(2)}$ to be estimated, namely, β.

In other applications the *treatment means*, rather than the expected responses, are the quantities of primary interest. The covariate should be included in the model only if it leads to better estimates of the α_i^*, $i = 1, 2, \ldots, I$. One would then prefer the family of models which leads to a smaller value of, for example, $\sum_i (\alpha_i^* - \hat{\alpha}_i)^2$, where $\hat{\alpha}_i$ is an estimator of the treatment mean under the approximating family. This overall discrepancy belongs to the discrepancy $\sum_i (\alpha_i^* - \alpha_i)^2$. Again the model $G^{(1)}$ has a discrepancy due to approximation equal to zero but a higher expected discrepancy due to estimation than $G^{(2)}$.

For this particular example it turns out (see Section 9.4.1) that the two discrepancies lead to equivalent criteria. However, for the case of two or more covariates they lead to different criteria.

2.2. SOME IMPORTANT DISCREPANCIES

The *Kullback–Leibler discrepancy* (Kullback and Leibler, 1951) is based on the likelihood ratio and as such is almost universally applicable. It is defined as

$$\Delta_{K-L}(\theta) = -E_F \log g_\theta(x),$$

where g_θ is the probability density function (or probability function) which characterizes the approximating family of models.

The minimum discrepancy estimator associated with this discrepancy is the maximum likelihood estimator. As well as being associated with this well-documented method of estimation the discrepancy has the additional advantage that it often leads to particularly simple criteria (Akaike, 1973).

Other possible discrepancies are the *Kolmogorov discrepancy*

$$\Delta_K(\theta) = \sup_x |F(x) - G_\theta(x)|$$

and the *Cramér–von Mises discrepancy*

$$\Delta_{C-M}(\theta) = E_{G_\theta}(F(x) - G_\theta(x))^2.$$

The *Pearson chi-squared* and *Neyman chi-squared* discrepancies, which are suitable for discrete data or grouped data, are given by

$$\Delta_P(\theta) = \sum_x (f(x) - g_\theta(x))^2/g_\theta(x), \qquad g_\theta(x) \neq 0,$$

and

$$\Delta_N(\theta) = \sum_x (f(x) - g_\theta(x))^2/f(x), \qquad f(x) \neq 0,$$

respectively, where f and g_θ are the probability functions which characterize the operating model and approximating family.

Another simple discrepancy for this type of data which we will call the *Gauss discrepancy* is

$$\Delta_G(\theta) = \sum_x (f(x) - g_\theta(x))^2.$$

Analogous discrepancies suitable for observations on continuous random variables are obtained by replacing the sums by integrals and probability functions by probability density functions.

Discrepancies may be based on some specific aspect of the participating distributions; they need not depend on every single detail of these distributions. For example, in regression analysis it is often the case that only certain *expectations* are of interest. A discrepancy such as Δ_G (or even Δ_P and Δ_N) but where f and g_θ are replaced by expectations under the operating and the approximating models, respectively, is also suitable for model selection.

Some of the discrepancies mentioned above are also known by different names. The Kolmogorov discrepancy is related to the supremum norm and the Gauss discrepancy to the L_2 norm. The Kullback–Leibler discrepancy is the essential part of the expected log-likelihood ratio and is related to the entropy.

2.3. DERIVATION OF CRITERIA

Each pair (operating model, discrepancy) requires its own method of analysis. In the following chapters we will derive methods for a number of the typical operating families which arise in practice given typical assumptions and data sets. Where appropriate, we consider alternative discrepancies for the same operating family. For nonstandard situations—when some of the usual assumptions do not hold, when additional assumptions can be made, or when a special discrepancy is required—new methods will have to be developed. The remainder of this chapter describes a number of approaches to deriving such methods.

Given an operating model, a discrepancy, and a number of approximating families together with their associated methods of estimation, one has to construct a criterion, an estimator of the expected discrepancy $E_F\Delta(\hat{\theta})$. This expectation is taken with respect to the operating model, F.

The derivation of the expected discrepancy can be straightforward in some situations (see Example 1.4.1) but can lead to insurmountable difficulties in others. In particular, it is hopeless to try to derive $E_F\Delta(\hat{\theta})$ if the properties of the estimator $\hat{\theta}$ can not be obtained. This is the case for a number of minimum discrepancy estimators. So although such estimators can be recommended as being the "natural" estimators they can lead to this difficulty.

Once one has obtained an expression for the expected discrepancy the next problem is to give an estimator for it. This too is not always easy. The problem here is that one has to estimate a function of some of the parameters of the operating family. Whether or not a reasonable estimator can be obtained depends on the precise form of this function and the complexity of the operating family.

These remarks apply if one tries to derive *finite sample methods*. In view of the difficulties pointed out it is perhaps surprising that in many applications such methods can be derived. Where finite sample methods are unobtainable one can resort to *asymptotic methods* which are discussed in the next section. *Bootstrap methods*, discussed in Section 2.5, are another possibility. These provide a means of directly estimating the expected discrepancy without the necessity of deriving an expression for it. They have the additional advantage of being easy to implement for a wide variety of discrepancies. A third alternative is to use *cross-validatory methods*, discussed in Section 2.6.

2.4. ASYMPTOTIC METHODS

In situations where the expected discrepancy, $E_F\Delta(\hat{\theta})$, is too difficult to derive, or the result is too complicated to be usefully estimated for the finite sample case, asymptotic methods will sometimes work. In this section we will simply state the relevant results without proving them. The proofs are given in the Appendix.

Essentially, one approximates $\Delta(\hat{\theta})$ by the first two terms of its Taylor expansion about the point θ_0 and then calculates its expectation. To do this the asymptotic distribution of $\hat{\theta}$ is required.

The results are particularly convenient for minimum discrepancy estimators. Let $\Delta_n(\theta)$ be an empirical discrepancy, and let $\hat{\theta}$ be the corresponding minimum discrepancy estimator, that is,

$$\hat{\theta} = \arg\min\{\Delta_n(\theta):\theta \in \Theta\}.$$

Suppose that θ is real, p-dimensional, and contains only independent parameters, that is, no parameter in θ can be expressed in terms of the others.

Under certain regularity conditions which are discussed in the Appendix the following two propositions hold:

Proposition 1

$$E_F\Delta(\hat{\theta}) \approx \Delta(\theta_0) + \operatorname{tr}\Omega^{-1}\Sigma/2n$$

and

$$\operatorname{Var}_F\Delta(\hat{\theta}) \approx \operatorname{tr}\Omega^{-1}\Sigma\Omega^{-1}\Sigma/2n^2,$$

where

$$\Omega = \{\partial^2\Delta(\theta_0)/\partial\theta_i\partial\theta_j: i, j = 1, 2, \ldots, p\}$$

and Σ is the asymptotic covariance matrix of

$$\{\sqrt{n}\partial\Delta_n(\theta_0)/\partial\theta_i : i = 1, 2, \ldots, p\}.$$

Proposition 2

$$E_F\Delta(\hat{\theta}) \approx E_F\Delta_n(\hat{\theta}) + \text{tr } \Omega^{-1}\Sigma/n,$$

and

$$\Delta(\theta_0) \approx E_F\Delta_n(\hat{\theta}) + \text{tr } \Omega^{-1}\Sigma/2n.$$

Remark. The regularity conditions under which the two propositions hold are B1 to B8. (See Appendix, A.1.1.) For Proposition 2 it is also assumed that the bias of $\Delta_n(\theta)$ as estimator of $\Delta(\theta)$ is $o(1/n)$. The approximations are obtained by replacing the covariance matrix of $\sqrt{n}(\hat{\theta} - \theta_0)$ by its asymptotic covariance matrix and by omitting an $o_p(1/n)$ term. Under additional conditions (which are given in the Appendix) the approximation is $o(1/n)$, and

$$\Sigma = \{\lim_{n\to\infty} E_F n(\partial\Delta_n(\theta_0)/\partial\theta_i)(\partial\Delta_n(\theta_0)/\partial\theta_j) : i, j = 1, 2, \ldots, p\}.$$

As usual $o(\ )$ and $O(\ )$ are Landau's symbols; they characterize the limiting behavior of sequences of real numbers. For sequences of random variables and convergence in probability, $o_p(\ )$ and $O_p(\ )$ are used. (See, for example, Bishop et al., 1975, Sections 14.2 and 14.4.) ∎

Proposition 2 provides a means of estimating the expected discrepancy, $E_F\Delta(\hat{\theta})$, and the discrepancy due to approximation, $\Delta(\theta_0)$; it can be used to derive criteria. It also provides a means of estimating $\text{Var}_F\Delta(\hat{\theta})$.

If $\text{tr } \Omega^{-1}\Sigma$ is not known it is estimated by $\text{tr } \Omega_n^{-1}\Sigma_n$, where Ω_n and Σ_n are estimators of Ω and Σ, and one uses the following:

(I) $\Delta_n(\hat{\theta}) + \text{tr } \Omega_n^{-1}\Sigma_n/n$ as criterion based on $E_F\Delta(\hat{\theta})$.

(II) $\Delta_n(\hat{\theta}) + \text{tr } \Omega_n^{-1}\Sigma_n/2n$ as criterion based on $\Delta(\theta_0)$.

(III) $\text{tr}(\Omega_n^{-1}\Sigma_n)^2/2n^2$ as estimator of $\text{Var}_F\Delta(\hat{\theta})$.

For some discrepancies it can be shown that if the approximating family *contains* the operating model then $\Omega^{-1}\Sigma$ depends only on $p = \dim(\theta)$ and so simpler criteria can be given. For the Pearson chi-squared and Neyman chi-squared discrepancies this trace reduces to $2p$. For the Kullback–Leibler discrepancy it reduces to p and thus to the simpler criteria given by

(I)* $\Delta_n(\hat{\theta}) + p/n.$

(II)* $\Delta_n(\hat{\theta}) + p/2n.$

(III)* $p/2n^2.$

Now although one does not necessarily want to make the assumption that the approximating family contains the operating model one can still use (I)*, (II)*, and (III)* as estimators for cases where the operating model is not very different from the nearest approximating model.

It has been our experience that unless the operating and fitted models are obviously grossly different these simpler criteria perform as well (and often better) than those which use $\text{tr } \Omega_n^{-1}\Sigma_n$. If the approximating family is sufficiently flexible to accommodate a large variety of models then there is a good chance that the discrepancy of approximation is small and hence the approximation which leads to the simpler criteria is relatively accurate. As a rough rule the simpler criteria should be used unless the number of parameters in the approximating family (and hence its flexibility) is very small, for example, $p = 1, 2$.

For the case of the Kullback–Leibler discrepancy, (I)* is simply the well-known Akaike Information Criterion (Akaike, 1973). The selection of models by this criterion is treated in detail in the book edited by Kitagawa (1983).

2.5.   BOOTSTRAP METHODS

Bootstrap methods, introduced by Efron (1979), provide a simple and effective means of circumventing the technical problems sometimes encountered in deriving the expected discrepancy and an estimator for it. An outline of the method in our context is as follows:

Let F be the operating model, F_n the empirical distribution function, and $\hat{\theta}$ an estimator for an approximating family of models with members G_θ and use

$$\Delta_n(\theta) = \Delta(G_\theta, F_n).$$

The expected discrepancy is in fact $E_F\Delta(G_{\hat{\theta}}, F)$ and a natural estimator of this is

$$E_{F_n}\Delta(G_{\hat{\theta}}, F_n) = E_{F_n}\Delta_n(\hat{\theta}).$$

Note that Δ_n and F_n are now fixed, and the only random variable is $\hat{\theta}$.

This last expectation is difficult to evaluate theoretically but it can be evaluated to an arbitrary degree of accuracy by means of Monte Carlo methods.

One generates repeated (Bootstrap) samples of size n using the now given and fixed distribution function F_n. Each sample leads to another (Bootstrap) estimate $\hat{\theta}^*$ and a corresponding $\Delta_n(\hat{\theta}^*)$. The average of the generated $\Delta_n(\hat{\theta}^*)$ converges to $E_{F_n}\Delta_n(\hat{\theta})$. Furthermore, the observed distribution of the generated $\Delta_n(\hat{\theta}^*)$ converges to the distribution of $\Delta_n(\hat{\theta})$ under F_n (Δ_n is fixed, only $\hat{\theta}$ is a random variable), which is an estimate of the distribution of $\Delta(\hat{\theta})$ under the operating model F.

In other words, by means of Bootstrap methods one can not only estimate the expectation of the overall discrepancy but also its distribution.

To estimate a functional $S(F)$ of the distribution function F by $S(F_n)$, the same functional taken at the empirical distribution function F_n, is well-established practice in statistics. (See, e.g., von Mises, 1947.) Many standard estimators, for example, $\bar{x}$ as estimator of Ex, are of this form. New—and extremely important—is the idea to use Monte Carlo methods to calculate $S(F_n)$ in certain cases in which analytical methods are not available.

The properties of estimators $S(F_n)$ depend entirely on S and on F. In the context of model selection $E_F\Delta(G_{\hat{\theta}}, F) = E_F\Delta(\hat{\theta})$ plays the rôle of $S(F)$ and $E_{F_n}\Delta(G_{\hat{\theta}}, F_n) = E_{F_n}\Delta_n(\hat{\theta})$ the rôle of $S(F_n)$. Although the expectation $E_{F_n}\Delta_n(\hat{\theta})$ can be computed by Bootstrap methods to any desired accuracy, this does not of course guarantee that it is a good estimator of $E_F\Delta(\hat{\theta})$. Bootstrap estimators are still relatively new and much theoretical work needs to be carried out to determine their precise properties. In the context of model selection such properties are particularly difficult to derive.

An additional advantage of the Bootstrap method is the ease with which it is possible to switch from one discrepancy to another without having to carry out tedious derivations for each new case. One does of course need to use a computer to carry out the computations, but for the problems which we will consider the computations can in most cases be comfortably carried out on a typical desk-top microcomputer.

2.6. CROSS-VALIDATORY METHODS

Cross-validation is a statistical technique to assess the "validity" of statistical analyses. The idea is an old one; it was explicitly stated in the psychometric literature (in the 1930s and later) in connection with prediction studies.

The data are subdivided into two parts: the first part is used to fit the model, and the second part to validate it. Stone (1974a) gives a short historical introduction in which Mosteller and Tukey (1968) are credited with the first general statement of the refinement in which the validation sample consists of a single observation. The whole procedure of fitting and validating is repeated n times, one for each of the possible subdivisions. The method is referred to as "one item out" cross-validation.

So far emphasis was on *assessment* of a method, but Stone (1974a) suggested using cross-validation for the *choice* of a method. It is exactly this aspect which makes cross-validation interesting for model selection. In principle, cross-validation could be applied to almost all problems considered in this book.

There are cases in which one must leave out at least *two* items to form reasonable criteria. We therefore introduce rules for the construction of "*m* items out" criteria, $m = 1, 2, \ldots$, and state their fundamental property. Subsequently, we explicitly give the criteria for some important discrepancies and data situations.

To fix the terminology, a convention is needed on what constitutes an item. We use the following. Let the available data consist of n observations (scalars or k-dimensional vectors) which under the operating model are considered as realizations of *independently* distributed random variables or random vectors. The observations constitute n items, unless—in the case of vectors—the elements are independently distributed. In the latter case the nk elements of the observed vectors are taken as items.

We consider initially the case in which the items belong to *identically* distributed random variables or random vectors. The modifications which are necessary if this is not so are explained later.

Identically Distributed Items

Let $\Delta(G_\theta, F)$ be the discrepancy on which the model selection is to be based and $\hat{\theta}$ be the estimator of θ_0 in the envisaged fitting procedure. A cross-validatory criterion for m items out based on the items $y(1), y(2), \ldots, y(n)$ is then constructed as follows:

(I)   Form, for fixed θ, an unbiased estimator of $\Delta(G_\theta, F)$ based on the m items $y(v_1), \ldots, y(v_m)$ which are set aside.

(II) Substitute $\hat{\theta}^{(v_1, \ldots, v_m)}$, the estimator of θ based on the remaining $n - m$ items, for θ in the estimator obtained under (I).

(III) Repeat (I) and (II) for all possible m-tupels. The average, over these repetitions, of the estimates obtained in (II) is the cross-validatory criterion.

The use of the term *criterion* is justified by the following fundamental result.

Proposition.   For *identically* distributed items the m items out cross-validatory criterion is an unbiased estimator of the expected discrepancy belonging to a sample of size $n - m$.

If Δ is *linear* in the distribution function F or the corresponding density f, an estimator of $\Delta(G_\theta, F)$ can be found on the basis of a *single* item $y(v)$ set aside. In the *discrete* case one gets this estimator by estimating $f(i)$ by $\delta_{y(v)i}$ (Kronecker delta) and in the continuous case by estimating $f(y)$ by $\delta(y - y(v))$ (Dirac delta function). The estimator of F is obtained by summing or integrating the estimators of f.

If f^m or F^m appears in $\Delta(G_\theta, F)$ one needs at least m items, $y(v_1), \ldots, y(v_m)$, for its estimation. An unbiased estimator of $f^r(i)$, $r = 1, 2, \ldots, m$, is then

$$\frac{1}{\binom{m}{r}} \Sigma_{j_1 < j_2 < \cdots < j_r \in (v_1, v_2, \ldots, v_m)} \delta_{y(j_1)i} \delta_{y(j_2)i} \cdots \delta_{y(j_r)i}.$$

In the continuous case $f^r(y)$ and $F^r(y)$ are estimated by

$$\frac{1}{\binom{m}{r}} \Sigma_{j_1 < j_2 < \cdots < j_r \in (v_1, v_2, \ldots, v_m)} \delta(y - y(j_1)) \delta(y - y(j_2)) \cdots \delta(y - y(j_r)),$$

and

$$\frac{1}{\binom{m}{r}} \Sigma_{j_1 < j_2 < \cdots < j_r \in (v_1, v_2, \ldots, v_m)} H(y - y(j_1)) H(y - y(j_2)) \cdots H(y - y(j_r)),$$

where H is the Heavyside function.

With these rules cross-validatory criteria are usually easily constructed. For convenience we give these criteria explicitly for a few important cases. Inessential factors like $1/n$ or $1/n(n - 1)$ are omitted.

Discrete Random Variables

Gauss

Discrepancy	$-2 \sum_i f(i) g_\theta(i) + \sum_i g_\theta^2(i)$
Criterion	$-2 \sum_i n_i g_{\hat{\theta}^{(i)}}(i) + \sum_i n_i \sum_j g_{\hat{\theta}^{(i)}}^2(j)$

Kullback–Leibler

Discrepancy	$-E_F \log g_\theta(i)$
Criterion	$-\sum_i n_i \log g_{\hat{\theta}^{(i)}}(i)$

Pearson

Discrepancy
$$\sum_{i=1}^{I} \frac{f^2(i)}{g_\theta(i)}$$

Criterion
(two items out)
$$\sum_{i=1}^{I} \frac{n_i(n_i - 1)}{g_{\hat\theta^{(i,i)}}(i)}$$

Here n_i is the frequency of the value i and $\hat\theta^{(i)}$ (and $\hat\theta^{(i,i)}$) are estimators of θ if one item (two items) with value i is (are) set aside.

Continuous Random Variables

Gauss

Discrepancy
$$-2\int f(y)g_\theta(y)dy + \int g_\theta^2(y)dy$$

Criterion
$$\sum_{v=1}^{n} \left[-2g_{\hat\theta^{(v)}}(y(v)) + \int g_{\hat\theta^{(v)}}^2(y)dy\right]$$

Kullback–Leibler

Discrepancy
$$-E_F \log g_\theta(y)$$

Criterion
$$-\sum_{v=1}^{n} \log g_{\hat\theta^{(v)}}(y(v))$$

Cramér–von Mises
Discrepancy
$$\int F^2(y)dG_\theta(y) - 2\int F(y)G_\theta(y)dG_\theta(y)$$

Criterion
(two items out)
$$\sum_{v_1 \neq v_2 = 1}^{n} \{G_{\hat\theta^{(v_1,v_2)}}^2(y(v_1)) + G_{\hat\theta^{(v_1,v_2)}}^2(y(v_2))$$
$$- 2G_{\hat\theta^{(v_1,v_2)}}[\max(y(v_1), y(v_2))]\}$$

Here $y(v)$ are the values of the observed items and $\hat\theta^{(v)}$ (and $\hat\theta^{(v_1,v_2)}$) are estimators of θ if the item $y(v)$ (the items $y(v_1)$ and $y(v_2)$) is (are) set aside.

Not Identically Distributed Items

If the items are not identically distributed in the operating model the method of constructing a criterion has to be modified. Assume that the discrepancy is the sum of components belonging to operating model F_i:

$$\Delta(\theta) = \sum_i \Delta(F_i, G_{i,\theta}) = \sum_i \Delta_i(\theta),$$

where θ does not depend on i. The criterion is then obtained by constructing a component for each i and summing over i. The construction follows the steps (I), (II), (III) above with the following modification. For the estimation of $\Delta_i(\theta)$ for fixed θ in step (I) only (set aside) items with operating model F_i are used, but for the estimation of θ in step (II) *all* (remaining, not set aside) items are used, whatever their operating model. The fundamental result becomes then:

Proposition. Let $\eta(v_1, v_2, \ldots, v_m)$ be the expected discrepancy belonging to the design generated by the $n - m$ items which were not set aside, then the cross-validatory criterion is an unbiased estimator of

$$\frac{1}{n(n - 1) \cdots (n - m + 1)} \Sigma_{v_1, v_2, \ldots, v_m, v_i \neq v_j \text{ for } i \neq j} \eta(v_1, v_2, \ldots, v_m),$$

the average of these expected discrepancies.

Key references to cross-validation in general are by Stone (1974a, 1978). The above summary of cross-validation concerns its use for model selection. A special case of the property described in the first proposition was noticed by Hall (1983a, p. 1006). The view of cross-validatory criteria as estimators of expected discrepancies, and their property described in the above propositions is implicit in Efron (1985). Closely related is cross-validation for the choice of smoothing parameters which can, for example, be applied in kernel estimation of probabilities or densities. Some recent studies relating to the properties of estimators obtained by cross-validation are described in Hall (1982, 1983b), Chow et al. (1983), Bowman (1984) and Bowman et al. (1984). These papers are mainly concerned with questions of consistency. Titterington (1985) gives a unified treatment of smoothing techniques and many further references on cross-validation.

CHAPTER 3

Histograms

The problem which we considered in Section 1.3.1 and in Example 1.4.1 has some practical significance and we deal with it here in more detail.

Traditionally, a histogram is constructed to summarize a set of data graphically. If the observations are regarded as realizations of some random variable then the histogram constitutes an estimate of the underlying probability density function. It can be used either as a final estimate or as an intermediate estimate in the search for some smoother model. The properties of the histogram as an estimator of the probability density function depend strongly on the sample size and on the number of intervals used in its construction. This was illustrated in Section 1.3.1 and in Example 1.4.1. If, for a fixed sample size, the number of intervals (resolution) is increased then the estimate tends to display more variation. If this number continues to be increased, it becomes doubtful whether the individual dips and bumps evident in the histogram represent population rather than sample characteristics.

As was pointed out in Chapter 1 the problem of selecting an appropriate resolution for histograms is a problem of model selection. The unknown probability density function is the operating model. The histogram is the fitted model belonging to the family of histograms having a given number of intervals (at preassigned positions). By changing the number or positions of the intervals we change the approximating family.

Although we derive a criterion to select the optimal number of equally spaced intervals there is no difficulty in deriving a criterion to select between any two (or more) histograms with arbitrary preassigned intervals.

3.1. APPROXIMATING MODELS AND DISCREPANCY

Suppose that n observations are available, which can be regarded as realizations of n independently and identically distributed random variables,

28

and that we wish to estimate the common probability density function using histogram-type density functions having I equally spaced intervals. We denote these densities by $g_\theta^{(I)}(x)$, where $\theta = (\theta_1, \theta_2, \ldots, \theta_I)$ represents the heights of the rectangles, only $p = I - 1$ of which are free to vary because the total area under the density must equal one.

When constructing an ordinary histogram one begins by specifying a convenient interval, say $(a, a + L)$, which contains all the observed values. The approximating densities are then given by

$$g_\theta^{(I)}(x) = 0, \qquad x - a \leqslant 0 \quad \text{or} \quad L < x - a$$

$$= \theta_i, \qquad \frac{(i-1)L}{I} < x - a \leqslant \frac{iL}{I}, \quad i = 1, \ldots, I,$$

where

$$\theta_i \geqslant 0, i = 1, 2, \ldots, I, \qquad \theta_I = \frac{I}{L} - \sum_{i=1}^{I-1} \theta_i.$$

The discrepancy which we considered in Example 1.3.1 is

$$\int_{-\infty}^{\infty} (f(x) - g_\theta(x))^2 dx,$$

where f is the density of the operating model. Since all approximating models have zero densities outside $(a, a + L)$, it is sufficient for the comparison of different models to integrate over this interval only. Now $\int f^2(x)dx$ does not depend on the approximating family, so for the purposes of comparing models the following discrepancy is equivalent to that given above:

$$\Delta(\theta) = -2 \int_a^{a+L} f(x)g_\theta(x)dx + \int_a^{a+L} g_\theta^2(x)dx.$$

Since g_θ is piecewise constant this expression reduces to

$$\Delta(\theta) = -2 \sum_{i=1}^{I} \pi_i \theta_i + \sum_{i=1}^{I} \frac{\theta_i^2 L}{I},$$

where

$$\pi_i = F\left(a + \frac{iL}{I}\right) - F\left(a + \frac{(i-1)L}{I}\right)$$

is the probability of the ith interval under the operating model with distribution function F. Clearly, this discrepancy can also be used in cases where the operating distribution function F does not have a density. The random variable need not be continuous.

3.2. ESTIMATION, EXPECTED DISCREPANCY, AND CRITERION

The maximum likelihood estimator of θ_i is the relative frequency in the ith interval divided by the length of the interval:

$$\hat{\theta}_i = \frac{n_i I}{nL}, \, i = 1, 2, \ldots, I.$$

It is easy to check that this is also a minimum discrepancy estimator for the following empirical discrepancy:

$$\Delta_n(\theta) = -2 \sum_{i=1}^{I} \frac{n_i \theta_i}{n} + \sum_{i=1}^{I} \frac{\theta_i^2 L}{I}.$$

Now $E_F n_i / n = \pi_i$ and

$$E_F \left(\frac{n_i}{n} \right)^2 = \frac{\pi_i + (n-1)\pi_i^2}{n}.$$

The expected discrepancy is thus

$$E_F \Delta(\hat{\theta}) = -2 \sum \pi_i E \hat{\theta}_i + \sum \frac{E \hat{\theta}_i^2 L}{I}$$

$$= \frac{I(1 - (n+1)\sum \pi_i^2)}{nL}.$$

Since $(\sum n_i^2 / n - 1)/(n - 1)$ is an unbiased estimator of $\sum \pi_i^2$ a criterion is

$$\frac{I}{nL} \left[1 - \frac{n+1}{n-1} \left(\sum_{i=1}^{I} \frac{n_i^2}{n} - 1 \right) \right].$$

Remarks. The criterion can be generalized to deal with the more general case in which the intervals are not necessarily all of equal length. For example one can use fixed intervals at the extremes of the distribution and then apply the above method to determine how many intervals of equal length to use in the middle.

The methods of this chapter are the subject of a thesis by Krahl (1980). Essentially, the same methods were recently published by Rudemo (1982), in whose paper a review of the literature on this problem can be found. Interesting is a suggestion by Scott (1979) to use the integer nearest to $Ln^{1/3}/3.49s$ as the number of intervals, I, where s is an estimator of the standard deviation. This last result is based on asymptotic considerations using the discrepancy described in this chapter. ∎

3.3. EXAMPLES OF APPLICATION

EXAMPLE 3.3.1.   Table 3.1 gives observed barley yields in grams in 400 plots (Rayner, 1967). The purpose of collecting these data was to estimate the distribution of the yields in small plots (1 square meter) for barley cultivated under specific conditions in this type of region.

The method described in this chapter was applied to select a histogram

Table 3.1.   Barley Yields of 400 Plots

```
162 136 157 141 130 129 176 171 190 157 147 176 126 175 134 169 129 180 128
205 129 117 144 125 165 170 153 186 164 123 165 203 156 182 164 176 176 150
154 184 203 166 155 215 190 164 204 194 148 162 146 174 185 171 181 158 147
157 180 165 127 186 133 170 134 177 109 169 128 152 165 139 146 144 178 188
128 161 160 167 156 125 162 128 103 116  87 123 143 130 119 141 174 157 168
180 158 139 139 168 145 166 118 171 143 132 126 171 176 115 165 147 186 157
174 172 191 155 169 139 144 130 146 159 164 160 122 175 156 119 135 116 134
182 209 136 153 160 142 179 125 149 171 186 196 175 189 214 169 166 164 195
108 188 149 178 171 151 192 127 148 158 174 191 134 188 248 164 206 185 192
178 189 141 173 187 167 128 139 152 167 131 203 231 214 177 161 194 141 161
130 112 122 192 155 196 179 166 156 131 179 201 122 207 189 164 131 211 172
140 156 199 181 181 150 184 154 200 187 169 155 107 143 145 190 176 162 123
194 146  22 160 107  70  84 112 162 124 156 138 101 138 141 143 135 163 183
118 150 151  83 136 171 191 155 164  98 136 115 168 130 111 136 129 122 120
172 192 171 151 142 193 174 146 180 140 137 138 194 109 120 124 126 126 147
148 195 154 149 139 163 118 126 127 139 174 167 175 179 172 174 167 142 169
163 144 147 123 160 137 161 122 101 158 103 119 164 112  57  94 106 132 122
142 155 147 115 143  68 184 183 167 160 138 191 133 160 156 122 111 153 148
131 180 142 191 175 146 181 111 110 154 176 168 175 175 146 148 167 106 123
154 148  91  93  74 113  79 131 119  96  80  97  98 106 107  69  86  94 129
```

over the interval (20, 260] to estimate the yield distribution. Figure 3.1 shows the values of the criterion which were computed for different values of I. The smallest value occurs when $I = 12$. Examining Figure 3.1 as a whole, that is, ignoring the high-frequency irregularities, one can see that the criterion first decreases sharply, reaches a minimum for I between about 12 and 20, and then increases gradually. The histograms for $I = 12, I = 15$, and $I = 20$, which give essentially the same description of the data, are shown in Figure 3.2.

EXAMPLE 3.3.2.   We give the marks scored by 63 students in an examination following an introductory course in statistics at the University of Göttingen in June 1980:

11	11	13	15	19	20	21	24	24	24	24	25	26
28	28	29	32	32	32	33	33	34	34	35	35	35
36	36	36	37	37	37	39	39	40	40	40	41	41
42	42	43	43	43	43	44	44	45	45	47	47	47
47	47	48	48	50	50	50	50	52	59	66		

We wish to assess the effectiveness of the given course–examination combination by means of the mark distribution. The distribution which we want to estimate is not that for the particular 63 students who attended the course, but rather for a larger population of students who might have attended the course or who may attend the course in the future. Clearly,

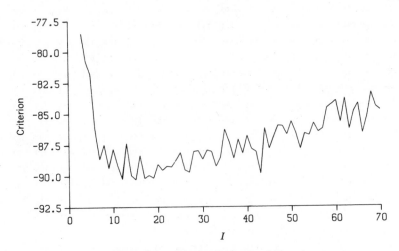

Figure 3.1.   Criterion for barley yields.

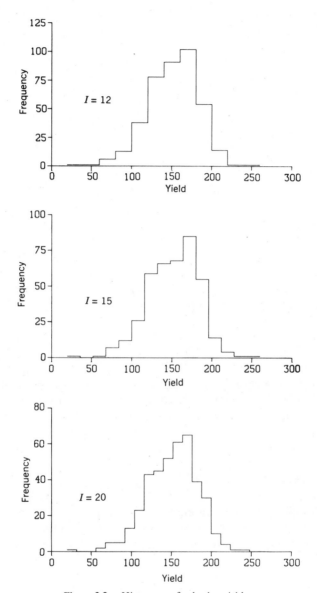

Figure 3.2. Histograms for barley yields.

we need to exercise care to ensure as far as possible that the above sample is representative of the population which we have in mind.

In this example it is convenient to work with the basic interval (0, 80]. Then $L = 80$ and, to remain with convenient class intervals, we restrict our

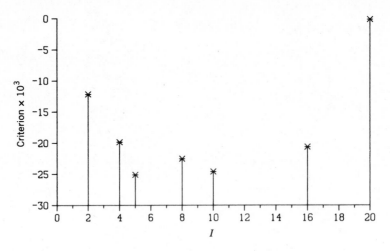

Figure 3.3. Criterion for examination marks.

attention to $I = 2, 4, 5, 8, 10, 16, 20$. We take the histogram intervals as being closed on the left and open on the right except for the last interval which is also closed on the right.

The values of the criterion corresponding to the given values of I are illustrated in Figure 3.3. The lowest value of the criterion occurs at $I = 5$ and the second lowest at $I = 10$. The criterion at $I = 8$ is a little higher, but looking at Figure 3.3 as a whole and taking into account that for this I the class intervals are particularly convenient, the model with $I = 8$ would seem to be a reasonable choice. The estimates for $I = 5, 8,$ and 10 are in Figure 3.4.

3.4. BIVARIATE HISTOGRAMS

There is no difficulty in extending the above methods to the bivariate case. Two-dimensional intervals can be generated by selecting I intervals for one of the variables and J for the other. The criterion is a function of both I and J, namely,

$$\frac{IJ}{nL_1L_2}\left[1 - \frac{n+1}{n-1}\left(\sum_{i=1}^{I}\sum_{j=1}^{J}\frac{n_{ij}^2}{n} - 1\right)\right],$$

where n_{ij} is the frequency in the (i, j)th interval, $n = \sum_{ij} n_{ij}$ is the number of observations, and the domain of the histogram has been specified to be of the form $(a_1, a_1 + L_1] \times (a_2, a_2 + L_2]$.

Considerably less computation is required if selection is restricted to

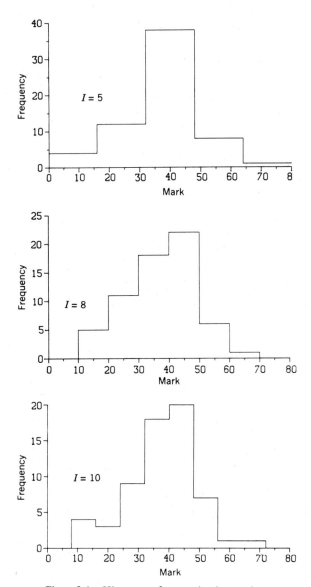

Figure 3.4. Histograms for examination marks.

histograms having $I = J$. By applying this restriction one is in effect selecting from a smaller set of approximating families.

EXAMPLE 3.4. Ekhart (1949) observed temperature inversions near Klagenfurt over a period of two years. He measured the thickness of the

Table 3.2. Thickness, x_1, and Magnitude, x_2, of Ground Inversions

x_1	x_2	x_1	x_2	x_1	x_2	x_1	x_2	x_1	x_2	x_1	x_2
1	14	15	19	25	28	37	12	55	20	78	41
2	5	15	27	25	35	38	4	55	35	78	71
2	10	15	30	25	45	38	8	55	37	79	40
2	14	16	16	26	29	38	22	55	78	79	47
2	17	16	25	26	35	38	24	55	95	79	92
3	2	16	27	26	41	38	31	56	2	80	43
3	3	16	132	26	44	38	63	56	15	80	110
3	5	17	1	27	12	39	12	56	16	81	30
3	10	17	15	27	16	39	17	56	20	81	105
3	17	17	15	27	18	40	50	56	31	82	40
3	31	17	20	27	22	40	91	57	16	82	42
4	8	17	20	27	24	40	122	57	16	84	142
4	13	17	20	27	24	41	20	57	27	86	9
4	18	17	27	28	2	41	38	57	32	86	32
5	10	17	28	28	4	41	51	57	41	86	78
6	7	17	30	28	7	41	54	57	45	88	30
6	8	18	4	28	10	41	120	57	77	88	45
7	3	18	9	28	15	42	9	58	10	88	77
7	3	18	18	28	18	42	11	58	15	89	95
7	6	19	4	28	10	42	19	58	31	90	21
7	6	19	16	28	22	42	21	59	10	90	43
7	6	19	18	28	30	42	30	59	68	91	44
7	7	20	3	28	71	42	37	59	96	92	25
7	17	20	4	29	30	42	39	60	2	92	46
8	1	20	5	29	71	42	44	60	20	92	98
8	5	20	24	30	1	42	53	60	26	93	60
8	6	20	25	30	15	42	54	60	35	94	41
8	22	20	28	30	17	42	123	61	15	95	10
10	1	20	31	30	22	44	41	63	11	95	39
10	5	21	10	30	31	45	61	63	13	95	43
10	6	21	11	31	10	45	72	63	27	95	141
10	7	21	11	31	26	46	6	63	40	96	24
10	9	21	12	31	32	46	11	63	41	97	85
10	10	21	20	32	3	46	12	63	51	98	52
10	7	21	32	32	12	46	35	63	73	99	132
11	1	21	34	32	17	46	41	64	20	101	4
11	25	21	60	32	22	46	56	64	44	101	28
12	1	22	2	32	24	46	65	64	82	102	46
12	2	22	8	32	32	47	18	65	66	102	83
12	5	22	13	32	35	47	20	66	29	102	123
12	5	22	16	32	37	47	66	66	74	104	103
12	7	22	18	32	40	48	10	67	32	104	123
12	7	22	20	32	42	48	76	67	43	105	62
12	7	22	22	32	45	49	19	69	29	106	112
12	9	22	22	32	47	50	13	69	36	112	100
12	10	22	24	32	48	50	32	70	84	114	22
12	12	22	37	32	56	50	43	70	115	115	100
12	12	22	42	33	6	51	2	71	14	118	111
12	13	22	59	33	8	51	10	71	19	118	116
12	15	22	60	33	8	51	41	72	9	123	62
12	16	22	62	33	9	51	68	72	10	125	92
12	18	23	5	33	34	52	3	72	12	125	100
12	19	23	10	33	48	52	5	72	13	126	95
12	38	23	14	34	11	52	30	72	51	130	92
12	40	23	17	35	3	52	31	73	30	132	37
12	42	23	18	35	6	52	55	73	39	132	106
13	13	23	56	36	9	52	60	73	67	136	79
13	21	23	98	36	14	52	63	73	73	137	35
13	27	24	4	36	28	53	1	74	5	141	48
14	5	24	13	36	31	53	31	75	15	148	40
14	9	24	21	36	32	54	11	75	40	165	58
14	16	24	21	36	34	54	18	76	82	175	95
15	4	24	28	36	39	54	28	77	35	179	71
15	5	24	40	36	58	55	15	77	42	192	76
15	5	24	45	37	5	55	15	77	59		
15	15	25	8	37	11	55	19	78	15		

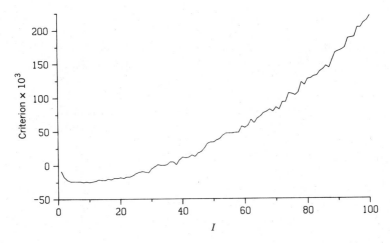

Figure 3.5. Criterion for joint distribution of thickness and magnitude of ground inversions.

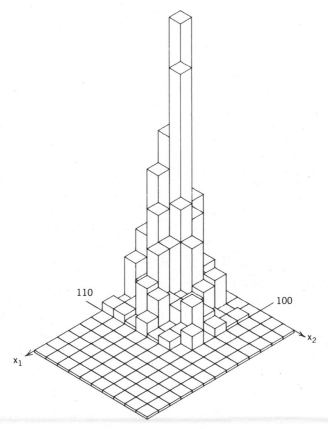

Figure 3.6. Histogram for thickness and magnitude of ground inversions.

layer (gdm/10), x_1, and the magnitude of the inversion (10 × °C), x_2. His observations for 394 "ground inversions" are given in Table 3.2.

We used as the basic intervals $0 \leqslant x_1 \leqslant 110$ and $0 \leqslant x_2 \leqslant 100$, that is, $L_1 = 110$ and $L_2 = 100$. There are altogether 33 observations outside the basic interval; they are accommodated in one large L-shaped region. This large region remained fixed (see the Remark in Section 3.2) and the criterion was calculated to determine the optimal subdivision of the basic interval. Selection was restricted to histograms with $I = J$; the obtained values of the criterion are illustrated in Figure 3.5.

The optimal value of I was 8 but it can be seen that there is hardly any change in the criterion for I between 5 and 10. In the end $I = 7$ was used since this emerged as optimal in a similar study of 271 "free inversions" and which is acceptable for ground inversions as well. The resulting histogram is given in Figure 3.6.

CHAPTER 4

Univariate Distributions

Suppose that the observations $x_1, x_2, \ldots, x_n$ can be regarded as realizations of n independently and identically distributed random variables and that we wish to estimate their common distribution. In some applications it is possible to deduce that the distribution belongs to a simple (operating) family, in which case there is no point in fitting approximating families. For example, the random variable in question, x, may be known to be non-negative and "without memory," that is, $P(x > a + b|x > a) = P(x > b)$ for all $a, b > 0$. This property implies that x is exponentially distributed. In other circumstances we can deduce that the operating distribution must belong, for example, to the Poisson, normal, or one of the extreme-type families.

This chapter is concerned with the many remaining applications where the operating family is too general to be of use, and consequently a model from some approximating family must be selected instead.

The histogram-type densities discussed in Chapter 3 are universally applicable, but we can often achieve lower discrepancies by fitting smoother approximating densities which depend on fewer parameters. Models such as the normal, lognormal, gamma, and so on provide more concise descriptions of the data and are also more convenient to work with than histograms. A further advantage of fitting smooth densities is that smoothness is a property which is associated with most operating densities. Consequently, histograms are mainly fitted as intermediate models, as aids in the preliminary selection of smoother approximating densities. They are also useful for detecting possible anomalies such as bimodality.

As a rule a number of different univariate models can be fitted to the data and the question arises as to which one should be selected. Traditionally, selection has been based on naive methods comparing the values of certain goodness-of-fit statistics. Cox (1961, 1962) considered the problem more systematically from the classical point of view. A bibliography of subsequent developments along these lines is given in Pereira (1977).

This chapter describes an alternative approach to the problem, based on the ideas outlined in Chapters 1 and 2. A well-known result which emerges in this framework is Akaike's (1973) Information Criterion. A second example, based on the Cramér–von Mises discrepancy, is given in Linhart and Zucchini (1984b).

It was mentioned in Section 2.3 that, unless the distributional properties of the estimator of relevant parameters are available, it is not possible to derive an exact expression for the expected discrepancy. The problem discussed in this chapter is a case in point. The finite sample distributions of the estimators are too difficult to derive for most of the discrepancies and approximating families of interest. One has to rely on asymptotic methods, Bootstrap methods, or cross-validatory methods. An exception to this is discussed in Section 4.2.

We begin (Section 4.1) by giving a number of asymptotic results. These relate primarily to applications where one is selecting one of the standard distributions, such as normal, lognormal, Poisson, and so on. Approximating models based on orthogonal systems of functions are discussed in Section 4.2. It is shown that, under certain suitably chosen discrepancies, finite sample methods can be derived. A typical Bootstrap algorithm for model selection is outlined in Section 4.3. Finally, in Section 4.4, a number of examples of application are used to illustrate the methods.

4.1. ASYMPTOTIC CRITERIA

All the criteria discussed in this section are based on the asymptotic results outlined in Section 2.4. Recall that to obtain asymptotic criteria it is necessary to estimate a trace term, tr $\Omega^{-1}\Sigma$, and that there are two ways to go about doing this. The trace term can be estimated from the data using an estimator of Ω and Σ, respectively. The criterion (based on the expected discrepancy) is then

$$\Delta_n(\hat{\theta}) + \text{tr } \Omega_n^{-1}\Sigma_n/n,$$

Both Ω and Σ have to be derived and estimated for each approximating family under consideration.

Alternatively, tr $\Omega^{-1}\Sigma$ can be approximated by the value which it would attain if there were no misspecification, that is, if the operating model were a member of the approximating family. Under this condition the trace term becomes significantly simpler. For a number of discrepancies it is simply a (known) multiple of p, the number of free parameters in the approximating family. For the Kullback–Leibler discrepancy it is equal to p which, being known, does not need to be estimated from the data. We refer to criteria derived in this way as *simpler criteria*.

The approximation on which the derivation of simpler criteria is based will be inaccurate whenever the discrepancy due to approximation is large. The families of models which are relevant to this section have typically only one or two parameters and are therefore relatively inflexible. Their associated discrepancy due to approximation can be relatively large, and consequently this method of estimating the expected discrepancy is inaccurate. Fortunately, it is not difficult to derive estimators for Ω and Σ for many of the cases of interest and the resulting estimates are easy to compute.

Perhaps the most important general purpose discrepancy is the Kullback–Leibler discrepancy. We now discuss this discrepancy and asymptotic criteria to which it gives rise for a number of the standard distributions.

4.1.1. Kullback–Leibler Discrepancy

The operating model is the unknown distribution function, F, of n identically and independently distributed random variables, $x_1, x_2, \ldots, x_n$. Let G_θ be the distribution function of an approximating family having p free parameters, $\theta = (\theta_1, \theta_2, \ldots, \theta_p)^\mathsf{T}$. We use g_θ to denote the corresponding probability density function if the random variables are continuous, and the probability function if they are discrete.

The Kullback–Leibler discrepancy is

$$\Delta(\theta) = \Delta(G_\theta, F) = -E_F \log g_\theta(x)$$

and the corresponding empirical discrepancy is

$$\Delta_n(\theta) = \Delta(G_\theta, F_n) = -\frac{1}{n} \sum_{i=1}^{n} \log g_\theta(x_i).$$

The minimum discrepancy estimator

$$\hat{\theta} = \arg \min \{\Delta_n(\theta) : \theta \in \Theta\}$$

is the *maximum likelihood estimator*.

In Appendix A.2.1 it is proved that the asymptotic criterion (based on the expected discrepancy) is

$$\Delta_n(\hat{\theta}) + \text{tr } \Omega_n^{-1}\Sigma_n/n$$

and the simpler criterion is

$$\Delta_n(\hat{\theta}) + p/n,$$

where

$$\Omega_n = \left\{ -\frac{1}{n} \sum_{i=1}^{n} \frac{\partial^2 \log g_{\hat{\theta}}(x_i)}{\partial\theta_r \partial\theta_s} : r, s = 1, 2, \ldots, p \right\}$$

$$\Sigma_n = \left\{ \frac{1}{n} \sum_{i=1}^{n} \left(\frac{\partial \log g_{\hat{\theta}}(x_i)}{\partial\theta_r} \right) \left(\frac{\partial \log g_{\hat{\theta}}(x_i)}{\partial\theta_s} \right) : r, s = 1, 2, \ldots, p \right\}.$$

We show as an example how the criterion is obtained for the lognormal distribution and later summarize the results for a number of other important distributions.

We use $m_h'[\ ]$ and $m_h[\ ]$, $h = 1, 2, \ldots$, to denote the sample moments and the sample moments about the sample mean of the variables in the bracket, respectively. For example,

$$m_2'[\log x] = \frac{1}{n} \sum_{i=1}^{n} \log^2 x_i.$$

For $m_h'[x]$ and $m_h[x]$ we write simply m_h' and m_h. The analogous notation is used for the two-dimensional case, for example,

$$m_{11}[x, \log x] = \frac{1}{n} \sum_{i=1}^{n} (x_i - m_1')(\log x_i - m_1'[\log x]).$$

Lognormal Distribution

For the lognormal distribution the logarithm of the density function is

$$\log g_{\theta}(x) = -\log x - \frac{\log 2\pi\lambda}{2} - \frac{(\log x - \mu)^2}{2\lambda}, \qquad 0 \leqslant x,$$

where, for convenience, we have used λ instead of the usual σ^2 and $\theta = (\lambda, \mu)^\mathsf{T}$.

One has

$$\frac{\partial \log g_\theta(x)}{\partial \lambda} = \frac{(\log x - \mu)^2}{2\lambda^2} - \frac{1}{2\lambda},$$

$$\frac{\partial \log g_\theta(x)}{\partial \mu} = \frac{(\log x - \mu)}{\lambda},$$

and the maximum likelihood estimators are $\hat{\mu} = m_1'[\log \ x]$ and $\hat{\lambda} = m_2[\log x]$. With the notation $\theta_0 = (\lambda_0, \mu_0)^\mathsf{T}$, it follows that

$$\Sigma_{11} = \frac{E(\log x - \mu_0)^4}{4\lambda_0^4} - \frac{E(\log x - \mu_0)^2}{2\lambda_0^3} + \frac{1}{4\lambda_0^2},$$

$$\Sigma_{12} = \Sigma_{21} = \frac{E(\log x - \mu_0)^3}{2\lambda_0^3},$$

$$\Sigma_{22} = \frac{E(\log x - \mu_0)^2}{\lambda_0^2},$$

$$\Sigma_{n\,11} = \frac{m_4[\log x] - m_2^2[\log x]}{4m_2^4[\log x]},$$

$$\Sigma_{n\,12} = \Sigma_{n\,21} = \frac{m_3[\log x]}{2m_2^3[\log x]},$$

$$\Sigma_{n\,22} = \frac{1}{m_2[\log x]}.$$

Also,

$$\frac{\partial^2 \log g_\theta(x)}{\partial \mu^2} = -\frac{1}{\lambda},$$

$$\frac{\partial^2 \log g_\theta(x)}{\partial \mu \partial \lambda} = -\frac{\log x - \mu}{\lambda^2},$$

$$\frac{\partial^2 \log g_\theta(x)}{\partial \lambda^2} = \frac{1}{2\lambda^2} - \frac{(\log x - \mu)^2}{\lambda^3}.$$

The elements of $-\Omega$ are the expectations of these derivatives with respect to the operating model taken at the values λ_0 and μ_0 and it follows immediately that

$$\Omega_{n\,11} = \frac{1}{2\, m_2^2[\log x]}$$

$$\Omega_{n\,12} = \Omega_{n\,21} = 0,$$

$$\Omega_{n\,22} = \frac{1}{m_2[\log x]}.$$

Thus tr $\Omega_n^{-1}\Sigma_n = (m_4[\log x] + m_2^2[\log x])/2m_2^2[\log x]$ and the criterion is

$$-\frac{1}{n} \sum_{i=1}^{n} \log g_\theta(x_i) + \frac{1}{n} \,\text{tr}\, \Omega_n^{-1}\Sigma_n$$

$$= m_1'[\log x] + \frac{1 + \log 2\pi + \log m_2[\log x]}{2} + \frac{m_4[\log x] + m_2^2[\log x]}{2nm_2^2[\log x]}.$$

If the trace term is replaced by $p/n = 2/n$ the criterion is

$$m_1'[\log x] + \frac{1 + \log 2\pi + \log m_2[\log x]}{2} + \frac{2}{n},$$

Akaike's Information Criterion (AIC).

Normal Distribution

Density: $\log g_\theta(x) = -\dfrac{\log 2\pi\lambda}{2} - \dfrac{(x - \mu)^2}{2\lambda},\qquad -\infty < x < \infty.$

Estimators: $\qquad\qquad \hat{\lambda} = m_2, \qquad \hat{\mu} = m_1'.$

Trace term: $\qquad\qquad \dfrac{m_4 + m_2^2}{2nm_2^2}.$

Criterion: $\qquad\qquad \dfrac{1 + \log 2\pi m_2}{2} + \dfrac{m_4 + m_2^2}{2nm_2^2}.$

AIC: replace the trace term by $2/n$.

Gamma Distribution

Density:
$$g_\theta(x) = \frac{\lambda^v x^{v-1} e^{-\lambda x}}{\Gamma(v)}, \qquad x \geqslant 0.$$

Estimation: $\hat{v}$ and $\hat{\lambda}$ are the solutions to the equations

$$\frac{\hat{v}}{\hat{\lambda}} = m_1'$$

and

$$\log \hat{\lambda} - \psi(\hat{v}) = -m_1'[\log x].$$

Here $\psi(z) = \partial \log \Gamma(z)/\partial z$ is Euler's ψ function.

Trace term: $\qquad\qquad\qquad\qquad$ tr $\Omega_n^{-1}\Sigma_n/n$,

where

$$\Omega_n = \begin{bmatrix} m_1'^2/\hat{v} & -m_1'/\hat{v} \\ -m_1'/\hat{v} & \psi'(\hat{v}) \end{bmatrix}$$

and

$$\Sigma_n = \begin{bmatrix} m_2 & -m_{11}[x, \log x] \\ -m_{11}[x, \log x] & m_2[\log x] \end{bmatrix}.$$

Criterion: $\log \Gamma(\hat{v}) - \hat{v}(\log \hat{\lambda} - 1) - (\hat{v} - 1)m_1'[\log x] + \text{tr } \Omega_n^{-1}\Sigma_n/n.$

AIC: replace the trace term by $2/n$.

Geometric Distribution

Probability function: $\quad g_\theta(x) = \theta(1 - \theta)^x, \qquad x = 0, 1, 2, \ldots.$

Estimator:
$$\hat{\theta} = \frac{1}{1 + m_1'}.$$

Trace term:
$$\frac{m_2}{nm_1'(1 + m_1')}.$$

Criterion: $(1 + m_1') \log (1 + m_1') - m_1' \log m_1' + \dfrac{m_2}{m_1'(1 + m_1')n}$.

AIC: replace the trace term by $1/n$.

Poisson Distribution

Probability function:

$$\log g_\theta(x) = - \theta + x \log \theta - \log x!, \qquad x = 0, 1, 2, \ldots.$$

Estimator: $\qquad\qquad\qquad\qquad \hat{\theta} = m_1'.$

Trace term: $\qquad\qquad\qquad\qquad \dfrac{m_2}{nm_1'}.$

Criterion: $\quad m_1'(1 - \log m_1') + \dfrac{1}{n} \sum_{i=1}^{n} \log \Gamma(x_i + 1) + \dfrac{m_2}{nm_1'}.$

AIC: replace the trace term by $1/n$.

Negative Binomial Distribution

Probability function: $\quad g_\theta(x) = \dfrac{\Gamma(x + \beta)(1 - \alpha)^\beta \alpha^x}{\Gamma(\beta)\Gamma(x + 1)}, x = 0, 1, 2, \ldots.$

Estimation: The maximum likelihood estimators $\hat{\alpha}$ and $\hat{\beta}$ are the solutions to the equations

$$\sum_{k=0}^{\infty} \dfrac{n(1 - F_n(k))}{\beta + k} + n \log \dfrac{\beta}{\beta + m_1'} = 0$$

and

$$\alpha = \dfrac{m_1'}{\beta + m_1'},$$

where F_n is the empirical distribution function.

Trace term: $\qquad\qquad\qquad \text{tr } \Omega_n^{-1} \Sigma_n / n,$

where

$$\Omega_n = \begin{bmatrix} \hat{\beta}/\hat{\alpha}(1 - \hat{\alpha})^2 & 1/(1 - \hat{\alpha}) \\ 1/(1 - \hat{\alpha}) & \psi'(\hat{\beta}) - m_1'[\psi'(\hat{\beta} + x)] \end{bmatrix}$$

and

$$\Sigma_n = \begin{bmatrix} m_2/\hat{\alpha}^2 & m_{11}[x, \psi(\hat{\beta} + x)]/\hat{\alpha} \\ m_{11}[x, \psi(\hat{\beta} + x)]/\hat{\alpha} & m_2[\psi(\hat{\beta} + x)] \end{bmatrix}.$$

Criterion:

$$\frac{1}{n} \sum_{i=1}^{n} \log \frac{\Gamma(x_i + 1)}{\Gamma(x_i + \hat{\beta})} - \hat{\beta} \log(1 - \hat{\alpha}) - m_1' \log \hat{\alpha} + \log \Gamma(\hat{\beta}) + \frac{1}{n} \operatorname{tr} \Omega_n^{-1} \Sigma_n.$$

AIC: replace the trace term by $2/n$.

4.1.2. Three Criteria for Grouped Data

In general, the operating family for grouped data, or for discrete data with a finite number of possible values, is the family of *multinomial distributions* with parameters n and π_i, $i = 1, 2, \ldots, I$. The π_i are only subject to the usual restrictions $(0 \leqslant \pi_i \leqslant 1, \sum_i \pi_i = 1)$ and are otherwise arbitrary. Apart from n there are thus $I - 1$ free parameters in the operating family. Exceptions to this can arise if special features exist which can be exploited to postulate an operating family having fewer parameters.

We consider two discrepancies for grouped data. They are based on classical goodness-of-fit statistics and can also be motivated using the notion of Mahalanobis' generalized distance. These are the Neyman chi-squared discrepancy,

$$\Delta(\theta) = \sum_{i=1}^{I} \frac{(\pi_i - h_i(\theta))^2}{\pi_i}, \quad \pi_i \neq 0$$

and the Pearson chi-squared discrepancy,

$$\Delta(\theta) = \sum_{i=1}^{I} \frac{(\pi_i - h_i(\theta))^2}{h_i(\theta)}, \qquad h_i(\theta) \neq 0,$$

where π_i represents the probability associated with the ith observed frequency, n_i, under the operating model and $h_i(\theta)$ that under the approximating model, $i = 1, 2, \ldots, I$. Apart from the restriction $\pi_i \neq 0$ in the Neyman chi-squared discrepancy, the π_i are only subject to the usual restrictions pertaining to the multinomial distribution.

The empirical discrepancies are (Neyman chi-squared)

$$\Delta_n(\theta) = \frac{1}{n}\chi_N^2(\theta) = \frac{1}{n}\sum_{i=1}^{I}\frac{(n_i - nh_i(\theta))^2}{n_i}$$

and (Pearson chi-squared)

$$\Delta_n(\theta) = \frac{1}{n}\chi_P^2(\theta) = \frac{1}{n}\sum_{i=1}^{I}\frac{(n_i - nh_i(\theta))^2}{nh_i(\theta)}.$$

If $n_i = 0$ one has to replace n_i by $nh_i(\theta)$ in the denominator of Neyman's chi-squared.

The minimum discrepancy estimators are simply the usual minimum (Neyman or Pearson) chi-squared estimators. Selection criteria, based on the expected discrepancy, are given by

$$\chi_N^2(\hat{\theta}) + C_N(\hat{\theta}) + \operatorname{tr} \Omega_n^{-1}\Sigma_n$$

and

$$\chi_P^2(\hat{\theta}) + C_P(\hat{\theta}) + \operatorname{tr} \Omega_n^{-1}\Sigma_n,$$

respectively, where

$$C_N(\hat{\theta}) = \sum_{i=1}^{I}\frac{nh_i^2(\hat{\theta})(n_i - n - 1)}{n_i^2}$$

and

$$C_P(\hat{\theta}) = \sum_{i=1}^{I}\frac{n_i(n_i - n)}{n(n-1)h_i(\hat{\theta})}.$$

The C_N and C_P are bias corrections. The matrices Ω_n and Σ_n are computed by replacing θ_0 by $\hat{\theta}$ and π_i by n_i/n in the corresponding Ω and Σ, which are given in Appendix A.2.2 and A.2.3. They are

$$\Omega_n = 2n\sum_{i=1}^{n}\frac{h_i'(\hat{\theta})h_i'(\hat{\theta})^\top + h_i(\hat{\theta})h_i''(\hat{\theta})}{n_i},$$

$$\Sigma_n = 4n^3\sum_{i=1}^{n}\frac{h_i^2(\hat{\theta})h_i'(\hat{\theta})h_i'(\hat{\theta})^\top}{n_i^3},$$

in the case of Neyman chi-squared [with the n_i in the denominator replaced by $nh_i(\hat{\theta})$ whenever $n_i = 0$], and

$$\Omega_n = \sum_{i=1}^{n} \frac{n_i^2(2h_i'(\hat{\theta})h_i'(\hat{\theta})^{\mathsf{T}} - h_i(\hat{\theta})h_i''(\hat{\theta}))}{n^2 h_i^3(\hat{\theta})},$$

$$\Sigma_n = 4 \sum_{i=1}^{n} \frac{n_i^3 h_i'(\hat{\theta})h_i'(\hat{\theta})^{\mathsf{T}}}{n^3 h_i^4(\hat{\theta})},$$

for Pearson chi-squared, where

$$h_i'(\theta) = (\partial h_i(\theta)/\partial \theta_1, \ldots, \partial h_i(\theta)/\partial \theta_p)^{\mathsf{T}}$$

and

$$h_i''(\theta) = \{\partial^2 h_i(\theta)/\partial \theta_r \partial \theta_s : r, s = 1, 2, \ldots, p\}.$$

It can be shown, for both cases, that if the operating model belongs to the approximating family, then $\Sigma = 2\Omega$ and consequently tr $\Omega^{-1}\Sigma = 2p$. Furthermore, under this condition, the bias is independent of the approximating family and thus the terms C_{N} and C_{P} can be omitted from the criteria. One then obtains the following simpler criteria:

$$\chi_{\mathrm{N}}^2(\hat{\theta}) + 2p \quad \text{and} \quad \chi_{\mathrm{P}}^2(\hat{\theta}) + 2p,$$

where $\chi_{\mathrm{N}}^2(\hat{\theta})$ and $\chi_{\mathrm{P}}^2(\hat{\theta})$ are the classical Neyman and Pearson chi-squared statistics. Although the above condition that the discrepancy due to approximation is exactly equal to zero will rarely be met, these simple criteria are useful for cases where this discrepancy is relatively small.

The discussion above also applies to the Kullback–Leibler discrepancy in the case of multinomial operating and approximating models. To obtain a criterion that is comparable with those resulting from the chi-squared discrepancies, we modify the simpler criterion (AIC) by adding $\sum_i (n_i/n) \log (n_i/n)$ and multiplying by $2n$. This is permitted, the modification does not depend on the approximating model. The resulting criterion is then

$$G^2(\hat{\theta}) + 2p,$$

where G^2 is the likelihood ratio statistic,

$$G^2(\hat{\theta}) = 2 \sum_i (n_i \log n_i - n_i \log nh_i(\hat{\theta})),$$

and $\hat{\theta}$ is the maximum likelihood estimator,

$$\hat{\theta} = \arg\min\{-\sum_i n_i \log h_i(\theta): \theta \in \Theta\}.$$

The operating model admits arbitrary probabilities π_i. We refer to this model which has $I - 1$ free parameters as the *saturated model*. It can also be used as the "approximating model." The relative frequencies n_i/n are then the minimum discrepancy estimators of π_i for each of the three discrepancies above. Since this family has zero discrepancy due to approximation, the criteria for the saturated model reduce to $2(I - 1)$ in all three cases.

It can be shown that arguments which lead to the derivation of the three simple criteria discussed in this section also apply if the three methods of estimation are interchanged. The criteria remain the same no matter which estimator, $\hat{\theta}$, of the three is substituted in $\chi_N^2(\hat{\theta})$, in $\chi_P^2(\hat{\theta})$, or in $G^2(\hat{\theta})$. In particular, the maximum likelihood estimator can be used in either of the two chi-squared criteria. In effect this provides a means of comparing two fitting procedures, for example, $(h^{(1)}(\theta)$ and maximum likelihood estimation) with $(h^{(2)}(\theta)$ and maximum likelihood estimation), where the quality of fit is judged on the basis of the required chi-squared discrepancy.

4.1.3. A Criterion Based on the Discrepancy of Cramér and von Mises

The discrepancy

$$\Delta(\theta) = \Delta(G_\theta, F) = \int (F(x) - G_\theta(x))^2 dG_\theta(x)$$

is the basis of yet another classical goodness-of-fit statistic. Let

$$\Delta_n(\theta) = \Delta(G_\theta, F_n)$$

$$= \frac{1}{n} \sum_{i=1}^{n} \left(G_\theta(x_{(i)}) - \frac{(2i-1)}{2n}\right)^2 + \frac{1}{12n^2},$$

where $x_{(1)}, x_{(2)}, \ldots, x_{(n)}$ are the order statistics of $x_1, x_2, \ldots, x_n$. Then $\Delta_n(\hat{\theta})$, where $\hat{\theta}$ is an estimator of θ, is the Cramér–von Mises statistic. We will restrict our attention to minimum discrepancy estimators:

$$\hat{\theta} = \arg\min\{\Delta_n(\theta): \theta \in \Theta\}.$$

The asymptotic distribution of $\hat{\theta}$, under certain regularity conditions, is derived in Appendix A.2.4. The resulting expression for the expected discrepancy contains, as in other cases, a term $\mathrm{tr}\,\Omega^{-1}\Sigma$, where the matrices Ω and Σ are given in the same appendix. The resulting criterion [which, for convenience, is an estimator of $n^2 E\Delta(\hat\theta)$ rather than $E\Delta(\hat\theta)$] is of the form

$$n^2\Delta_n(\hat\theta) + C_C + n\,\mathrm{tr}\,\Omega_n^{-1}\Sigma_n,$$

where

$$C_C = \frac{(n+1)\sum_i G_{\hat\theta}(x_{(i)}) - 2\sum_i iG_{\hat\theta}(x_{(i)})}{n-1}$$

is a bias correction which is needed because $\Delta_n(\theta)$ is not an unbiased estimator of $\Delta(\theta)$ (see Appendix A.2.4). The matrices Ω_n and Σ_n, consistent estimators of Ω and Σ, need to be derived for each competing approximating family.

The task of finding estimators for Ω and Σ, which are given by rather lengthy expressions, is onerous. However, a simpler criterion can be derived under the assumption that $F \equiv G_{\theta_0}$. One obtains

$$\Sigma = 4 \int_0^1 \int_0^1 (\min(s, t) - st)\psi(s, \theta_0)\psi^{\mathsf T}(t, \theta_0)ds\,dt$$

and

$$\Omega = 2 \int_0^1 \psi(t, \theta_0)\psi^{\mathsf T}(t, \theta_0)dt,$$

where

$$\psi(t, \theta) = G'_\theta(F^{-1}(t)) = (\partial G_\theta(F^{-1}(t))/\partial\theta_1, \ldots, \partial G_\theta(F^{-1}(t))/\partial\theta_p)^{\mathsf T}.$$

Estimators of Ω and Σ are obtained by replacing F by F_n and θ_0 by $\hat\theta$:

$$\Sigma_n = \frac{4}{n^2}\sum_{ij}\left(\min\left(\frac{i}{n}, \frac{j}{n}\right) - \frac{ij}{n^2}\right) G'_{\hat\theta}(x_{(i)})G'^{\mathsf T}_{\hat\theta}(x_{(j)}),$$

$$\Omega_n = \frac{2}{n}\sum_i G'_{\hat\theta}(x_{(i)})G'^{\mathsf T}_{\hat\theta}(x_{(i)}).$$

Under the assumption that there is no misspecification, (i.e., $F \equiv G_{\theta_0}$), the bias of $\Delta_n(\theta)$ reduces to $-n/6$, and this does not depend on the approximating family under consideration. The bias correction term, C_C, can therefore be omitted from the criterion, which then reduces to

$$n^2\Delta_n(\hat{\theta}) + n \operatorname{tr} \Omega_n^{-1}\Sigma_n.$$

4.2. APPROXIMATING MODELS BASED ON ORTHOGONAL SYSTEMS

The representation of functions in terms of orthogonal systems of functions is a standard mathematical technique which has been employed by statisticians for estimating probability distributions (e.g., see Kendall and Stuart, 1979, Vol. 1).

In the context of model selection orthogonality has the great advantage that the criteria can be decomposed into components belonging to the individual parameters in the model.

Consider, for example, approximating families

$$g_\theta(x) = \sum_{i \in A} \theta_i \psi_i(x),$$

where $\psi_i(x)$, $i = 0, 1, \ldots$, is a suitable sequence of functions. Here A, a finite set of non-negative integers, determines the approximating family and the number p of free parameters in the family is equal to the size of A. Assume that all families using $\psi_0, \psi_1, \ldots, \psi_9$ or subsets of these ten functions are considered as candidate families. There are then $2^{10} - 1 = 1023$ such families and in general one would have to calculate so many criteria if one wanted to select the most appropriate family. If the functions ψ_i are orthogonal (with respect to a very specific discrepancy) it is possible to calculate components of the criterion for each of the parameters θ_i belonging to ψ_i (in our example there are ten of them) and decide in this way on the most appropriate approximating family. Each parameter gives a contribution (its component) to the criterion of the approximating model. If the contribution is negative the parameter should be used, otherwise it should be eliminated.

The advantages of orthogonality are not so decisive if univariate distributions are fitted since one uses as a rule only a few of the first terms in the series. In other situations requiring many parameters, orthogonality is very important. This will become clear in later chapters: without orthogonality the scanning of all possible approximating families is often an impossible task.

An orthogonal system of functions is linked to a certain discrepancy and if one wants to use such a system with advantage in approximating models one is forced to base the analysis on this associated discrepancy. This is a difficulty. There are many orthogonal systems of functions known in applied mathematics but not all of them are ideal for fitting of distributions and the associated discrepancy is often not suitable for statistical purposes. As a result of this there are just a few systems which could be used for fitting univariate distributions, systems of functions with shapes which allow a reasonable fit with few parameters *and* which have a useful associated discrepancy.

We will restrict our attention to operating distributions that are determined by some probability density function, $f(x)$, which has positive support over the entire real line. However, the methods described below can be modified to deal with discrete distributions and with cases where the density has only positive support over a bounded interval. Extensions to multivariate distributions are also possible (Hall, 1983a).

The operating density is approximated using models of the form

$$g_\theta(x) = \sum_{i \in A} \theta_i \psi_i(x),$$

where the $\psi_i(x)$ are orthogonal with respect to some weight function $w(x) > 0$:

$$\int_{-\infty}^{\infty} \psi_i(x)\psi_j(x)w(x)dx = \delta_{ij}K(i),$$

where $K(i)$ is a constant and δ_{ij} is the Kronecker delta. For a given system $\{\psi_i(x)\}$, the index set A characterizes the approximating family whose parameters, θ_i, $i \in A$, are estimated from the data.

The question as to which family (which index set) should be selected depends, as usual, on the discrepancy which we wish to minimize. However, as is shown below, the advantages of orthogonality only carry through to the criterion if one uses the discrepancy

$$\int_{-\infty}^{\infty} (f(x) - g_\theta(x))^2 w(x)dx.$$

If the inessential $\int f^2(x)w(x)dx$ is omitted one arrives by virtue of the orthogonality property at

$$\Delta(\theta) = -\sum_{i \in A} \left(2\theta_i \int_{-\infty}^{\infty} f(x)\psi_i(x)w(x)dx - \theta_i^2 K(i) \right).$$

The minimizing values of the parameters (which we assume to exist) are given by

$$\theta_{0i} = \frac{1}{K(i)} \int_{-\infty}^{\infty} f(x)\psi_i(x)w(x)dx, \qquad i \in A,$$

and thus the discrepancy can be expressed as

$$\Delta(\theta) = \sum_{i \in A}((\theta_i - \theta_{0i})^2 - \theta_{0i}^2)K(i).$$

An unbiased estimator of θ_{0i} based on a random sample $x_1, x_2, \ldots, x_n$ is given by

$$\hat{\theta}_i = \frac{1}{nK(i)} \sum_{j=1}^{n} \psi_i(x_j)w(x_j), \qquad i \in A.$$

The overall discrepancy, which in this case is the sum of the discrepancies due to approximation and estimation, has expectation

$$E\Delta(\hat{\theta}) = \sum_{i \in A}(\text{Var } \hat{\theta}_i - \theta_{0i}^2)K(i).$$

The contribution of a parameter θ_i to the expected discrepancy is

$$(\text{Var } \hat{\theta}_i - \theta_{0i}^2)K(i)$$

and since $E(\hat{\theta}_i^2 - \text{Var } \hat{\theta}_i) = \theta_{0i}^2$, an estimator of this contribution, that is, the contribution to the criterion, is

$$(2 \widehat{\text{Var }} \hat{\theta}_i - \hat{\theta}_i^2)K(i),$$

where $\widehat{\text{Var }} \hat{\theta}_i$ is an unbiased estimator of $\text{Var } \hat{\theta}_i$. The parameter θ_i should be used in the approximating family (the index i should be put into A) if its contribution to the criterion is *negative*, that is, if

$$\frac{\hat{\theta}_i^2}{\widehat{\text{Var }} \hat{\theta}_i} > 2.$$

The derivation above is very similar to that of Kronmal and Tarter (1968), who examined the case where $\{\psi_i(x)\}$ is the trigonometric system associated with ordinary Fourier series in some detail. The corresponding discrepancy is the integrated squared difference, that is, the Gauss discrepancy.

For operating models which are most common in applications, there are a number of established alternatives which are likely to be more suitable than Fourier series. One such alternative is the system generated by successively differentiating the standard normal density. Another system, generated by differentiating the density function of a gamma distribution, is suitable for non-negative random variables. In the former case the approximating models, known as Gram–Charlier Type A series, are of the form

$$g_\theta(x) = \sum_{i \in A} \theta_i H_i(x)\varphi(x),$$

where $\varphi(x)$ is the standard normal density and $H_i(x)$ is the ith Chebyshev–Hermite polynomial given by $H_0(x) = 1$, $H_1(x) = x$, and

$$H_i(x) = xH_{i-1}(x) - (i-1)H_{i-2}(x), \qquad i \geqslant 2.$$

In our previous notation $\psi_i(x) = H_i(x)\varphi(x)$, $w(x) = \varphi(x)^{-1}$, and $K(i) = i!$. The discrepancy which leads to orthogonality is therefore given by

$$\int_{-\infty}^{\infty} \frac{(f(x) - g_\theta(x))^2}{\varphi(x)} \, dx.$$

It can be seen that this discrepancy depends strongly on the units which are used to measure the observations. A change of units causes a different aspect of the fit to be emphasized. Consequently, the discrepancy is generally only of use for assessing fits to the standardized operating density, the density of $(x - Ex)/\mathrm{Var}^{1/2}(x)$.

It is possible to develop a criterion for selecting a model for the standardized density in the usual way. The identical criterion emerges if one treats the standardized observations, $(x_i - \bar{x})/s$, $i = 1, 2, \ldots, n$, as though they were realizations of independent random variables having the standardized operating distribution.

Suppose, in what follows, that $z_1, z_2, \ldots, z_n$ are independent observations arising from a standardized operating density $f(x)$. Then it follows immediately that

$$\theta_{0i} = \frac{1}{i!} \int_{-\infty}^{\infty} f(x)H_i(x)dx, \qquad i \in A.$$

Since $H_i(x)$ is a polynomial, θ_{0i} is a linear combination of the moments $(\mu_h, h = 1, 2, \ldots)$ of the standardized operating distribution. More specifically $\theta_{00} = 1$, $\theta_{01} = 0$, $\theta_{02} = 0$, and, for $i \geqslant 3$,

$$\theta_{0i} = \sum_{j=0}^{[i/2]} \lambda_j^{(i)} \mu_{i-2j},$$

where $\lambda_0^{(i)} = 1/i!$,

$$\lambda_j^{(i)} = \frac{(-1)^j i(i-1)\cdots(i-2j+1)}{2^j i! j!}, \qquad j \geqslant 1,$$

and $[x]$ denotes the largest integer smaller than or equal to x.

The corresponding linear combinations of the sample moments

$$m_h = \frac{1}{n} \sum_{i=1}^{n} z_i^h$$

are unbiased estimators, $\hat{\theta}_i$, of the θ_{0i}. An estimator of the variance of $\hat{\theta}_i$, which is required to evaluate the criterion, is given by

$$\widehat{\mathrm{Var}}\,\hat{\theta}_i = \frac{1}{n-1} \sum_{j,k=0}^{[i/2]} \lambda_j^{(i)} \lambda_k^{(i)} (m_{2(i-j-k)} - m_{i-2j} m_{i-2k}).$$

Clearly, for this analysis to make sense, one has to assume that the moments of the standardized operating distribution up to a sufficiently high order exist.

It should be noted that the fitted density, $g_{\hat{\theta}}(x)$, could be negative for certain arguments, particularly near the tails. A more detailed discussion of the properties of finite Gram–Charlier Type A series is given in Kendall and Stuart (1979).

A selection procedure for discrete distributions using orthogonal systems based on the Poisson distribution is given in Hall (1983a).

4.3. A TYPICAL BOOTSTRAP ALGORITHM

To illustrate the use of Bootstrap methods in univariate model selection we give the outline of an algorithm to select between the Poisson and negative binomial distributions. A concrete example of this type is Example 4.4.3.

Suppose that the observations $x_1, x_2, \ldots, x_n$ are available and that they are associated with an unknown discrete operating model described by some probability function $f(x)$ [and corresponding distribution function $F(x)$], $x = 0, 1, 2, \ldots$. The two approximating families which we consider have probability functions:

$$g_\theta(x) = \frac{\theta^x e^{-\theta}}{x!}, \qquad x = 0, 1, 2, \ldots,$$

and

$$g_{\alpha,\beta}(x) = \frac{(1 - \alpha)^\beta \alpha^x \Gamma(\beta + x)}{\Gamma(\beta)\Gamma(x + 1)}, \qquad x = 0, 1, 2, \ldots.$$

Suppose also that the Kullback–Leibler discrepancy is to be used for the purpose of selecting between these two models. Then the minimum discrepancy estimators of the parameters are simply the maximum likelihood estimators. In the Poisson case the maximum likelihood estimator, $\hat{\theta}$, is the sample mean which is easy to compute, but in the negative binomial case it is necessary to solve two simultaneous nonlinear equations to compute $\hat{\alpha}$ and $\hat{\beta}$. These equations are given in Section 4.1.1. We will assume that an algorithm to compute the estimates is available.

The expected discrepancies in this example are, respectively:

$$E_F(-(E_F \log g_\theta(x))_{\theta = \hat{\theta}})$$

and

$$E_F(-(E_F \log g_{\alpha,\beta}(x))_{\alpha = \hat{\alpha}, \beta = \hat{\beta}}).$$

Consider the Poisson case. In the inner expectation θ is fixed and x is the random variable. The result is a function of θ, but not of x. The θ is then replaced by $\hat{\theta}$ and the outside operation is the expectation of the resulting random variable.

The Bootstrap estimator of this expected discrepancy is

$$E_{F_n}(-(E_{F_n} \log g_\theta(x))_{\theta = \hat{\theta}}).$$

The inner expectation is easily calculated:

$$-E_{F_n} \log g_\theta(x) = -\frac{1}{n} \sum_{i=1}^{n} \log g_\theta(x_i)$$

$$= -\sum_{x=0}^{m} f_n(x) \log g_\theta(x)$$

$$= \Delta_n(\theta),$$

where $f_n(x)$ is the empirical probability function (i.e., the relative frequency of observations with value x) and m is the largest observation in the sample.

There remains the problem of evaluating

$$E_{F_n}\left(- \sum_{x=0}^{m} f_n(x) \log g_\theta(x) \right)_{\theta=\hat{\theta}}.$$

Analytically, this is a difficult problem. The random variable is $\hat{\theta}$, the maximum likelihood estimator of θ. Not even the distribution of $\hat{\theta}$ under F_n is accessible, not to mention that of $\log g_{\hat{\theta}}(x)$. The way out is to use Monte Carlo methods.

One generates a large number of (Bootstrap) samples under F_n. This can be achieved efficiently by randomly sampling n values, with replacement, from the original observations $x_1, x_2, \ldots, x_n$. For each resulting sample, $x_1^*, x_2^*, \ldots, x_n^*$, one computes an estimate $\hat{\theta}^*$ and the corresponding $\Delta_n(\hat{\theta}^*) = - \sum_x f_n(x) \log g_{\hat{\theta}^*}(x)$. [Note that $f_n(x)$ is the empirical probability function of the original sample, not of the Bootstrap sample.] The *average* of the computed values of $\Delta_n(\hat{\theta}^*)$ is then a Bootstrap estimator of the expected discrepancy, the Bootstrap criterion for the Poisson approximating family, $C_B(\text{Poisson})$.

The computations above must be carried out for each candidate family, in our example for the Poisson and the negative binomial. The computations are described by the following algorithm:

STEP 1.   Select a (Bootstrap) sample of size n, $x_1^*, x_2^*, \ldots, x_n^*$ (*with replacement*) from the original observations $x_1, x_2, \ldots, x_n$.

STEP 2.   Compute the maximum likelihood estimates using the Bootstrap sample: $\hat{\theta}^*$ for the Poisson case and $\hat{\alpha}^*, \hat{\beta}^*$ for the negative binomial case.

STEP 3.   Compute the observed (Bootstrap) discrepancies for this sample:

$$\Delta_n(\hat{\theta}^*) = - \sum_x f_n(x) \log g_{\hat{\theta}^*}(x)$$

and

$$\Delta_n(\hat{\alpha}^*, \hat{\beta}^*) = - \sum_x f_n(x) \log g_{\alpha^*, \beta^*}(x).$$

STEP 4.   Repeat Step 1 to Step 3 a large number of times, say M, and store the computed values of $\Delta_n(\hat{\theta}_j^*)$ and $\Delta_n(\hat{\alpha}_j^*, \hat{\beta}_j^*)$, where these represent the values of $\Delta_n(\hat{\theta}^*)$ and $\Delta_n(\hat{\alpha}^*, \hat{\beta}^*)$ at the jth iteration.

STEP 5.   Compute

$$\frac{1}{M} \sum_{j=1}^{M} \Delta_n(\hat{\theta}_j^*) = C_B(\text{Poisson})$$

and

$$\frac{1}{M} \sum_{j=1}^{M} \Delta_n(\hat{\alpha}_j^*, \hat{\beta}_j^*) = C_{\mathrm{B}} \, (\text{neg. bin.}).$$

The approximation at Step 5 can be made arbitrarily accurate by selecting M large enough. For moderate sample sizes $M = 200$ leads to sufficient accuracy for practical purposes. A reasonable alternative is to iterate say 100 times and compute the standard deviation of the $\Delta_n(\hat{\theta}^*)$ and $\Delta_n(\hat{\alpha}^*, \hat{\beta}^*)$. This will give an indication of whether the required accuracy has been achieved. If it has not then the number of iterations can be increased.

The above algorithm illustrates how easy it is to apply the Bootstrap method in this type of application. If a different method of estimation is required one simply modifies Step 2, or if a different discrepancy is used then Step 3 is modified.

Finally, an estimator of the distribution of the overall discrepancies is obtained using a histogram of the $\Delta_n(\hat{\theta}^*)$ and $\Delta_n(\hat{\alpha}^*, \hat{\beta}^*)$.

4.4. EXAMPLES OF APPLICATION

EXAMPLE 4.4.1. Consider the data in Table 1.1 together with question 1 in Section 1.1.3. We wish to estimate the distribution of flows and as mentioned it is not unreasonable to assume that the flows are independently distributed. Further evidence to support this assumption is that the estimated serial correlation coefficient (-0.046) is well within 1 standard error of its expected value (zero) under the hypothesis that the flows are independently distributed. For the purposes of estimating the flow distribution we will regard the flows as being realizations of independently and identically distributed random variables.

We consider three approximating families: normal, lognormal, and gamma. A histogram of the flows reveals that the operating model has a positive skew, and consequently the normal distribution is unlikely to yield a good fit. It is included to illustrate the point that the simpler criteria deviate from those using $\operatorname{tr} \Omega_n^{-1} \Sigma_n / n$ when the discrepancy due to approximation is large. The fact that the normal distribution yields a positive probability for the event that the flow is negative is not of practical importance unless this probability is substantial.

For the purposes of comparing the asymptotic criteria of Section 4.1 with Bootstrap methods we will first use the Kullback–Leibler discrepancy. The maximum likelihood estimates of the parameters are

Normal: $\hat{\mu} = 5.453 \times 10^2$, $\hat{\lambda}(= \hat{\sigma}^2) = 1.978 \times 10^5$,

Lognormal: $\hat{\mu} = 6.026$, $\hat{\lambda}(= \hat{\sigma}^2) = 5.577 \times 10^{-1}$,

Gamma: $\hat{\lambda} = 3.602 \times 10^{-3}$, $\hat{v} = 1.964$.

The values of the asymptotic criteria (given in Section 4.1.1) as well as the estimates of the standard deviation of $\Delta(\hat{\theta})$ [given in brackets, based on (III) of Section 2.4] are

Normal: $7.516 + 4.745/65 = 7.589$ (0.052),

Lognormal: $7.153 + 1.859/65 = 7.182$ (0.020),

Gamma: $7.191 + 2.194/65 = 7.225$ (0.024).

The lognormal distribution performs best with the gamma distribution a close second.

It is interesting to note that tr $\Omega_n^{-1}\Sigma_n$ is in the vicinity of 2 for the two distributions which perform well but is nearly 5 for the normal distribution. The simpler criterion (based on the assumption that the operating model is a member of the approximating family) would use $p = 2$ instead of tr $\Omega_n^{-1}\Sigma_n$. We have here an instance (normal distribution) in which the simpler form is inaccurate.

The remarks above also apply to the estimates of the asymptotic standard deviation of $\Delta(\hat{\theta})$. Using the analogous approximation [namely, (III)* in Section 2.4] one obtains a common estimate of $1/65 = 0.015$. This is quite close to the estimates which we obtained for the lognormal and gamma distributions, but is less than half as large as that obtained for the normal distribution.

The results of 200 Bootstrap replications yielded:

Approximating Model	Bootstrap	
	Mean	Standard Deviation
Normal	7.564	0.066
Lognormal	7.168	0.017
Gamma	7.210	0.021

The Bootstrap estimates, including the estimated standard deviations, are all quite close to those computed from asymptotic criteria. Certainly, the order of preference is preserved. The histograms of the Bootstrap overall discrepancies, which are estimates of the density of $\Delta(\hat{\theta})$, are given in Figure 4.1.

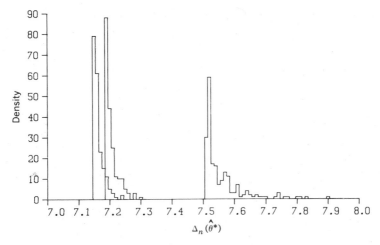

Figure 4.1. Bootstrap distribution of $\Delta_n(\hat{\theta}^*)$ for (from left to right) lognormal, gamma, and normal approximating models.

The Kolmogorov discrepancy

$$\Delta(\theta) = \sup\{|F(x) - G_\theta(x)| : x \in R\}$$

was applied to the same three families, but with maximum likelihood estimation rather than minimum discrepancy estimation in the fitting procedure. With the empirical discrepancy discussed in Example 4.4.5 (with $h = 1$), the following Bootstrap estimates of the mean and standard deviation of $\Delta(\hat{\theta})$ were obtained:

Approximating Model	Bootstrap (200 replications)	
	Mean	Standard Deviation
Normal	0.162	0.039
Lognormal	0.080	0.028
Gamma	0.098	0.030

The change in discrepancy has not altered the order of preference. The lognormal is estimated to fit a little better than the gamma, with the normal yielding the worst fit. The selected density function together with a histogram of the flows is given in Figure 4.2.

EXAMPLE 4.4.2. Linhart and van der Westhuyzen (1963) compared the lognormal and gamma models for fitting the length-biased distributions of

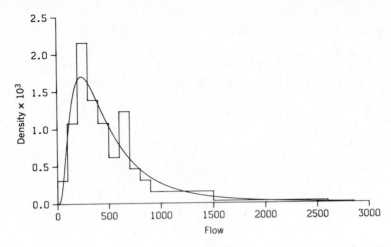

Figure 4.2. Histogram and approximating lognormal density for annual flow of Vaal River. River.

diameters of raw Merino wool. The length bias is a consequence of the method used to sample the fibers under a projection microscope. The probability that a fiber is sampled is proportional to its length. The purpose of the study was to assess which one of these two models provides a closer approximation to the operating distribution, which changes according to environmental and other factors. The selected model was to be used in further theoretical investigations in wool-textile technology.

A total of 47 different samples of about 1000 diameters were available. The samples came from various wool-producing districts and were associated with varying climatic and feeding conditions. Both approximating models were found to be satisfactory, with the lognormal being marginally preferable.

The data for two of the samples are given below. The first (case 20, $n = 1016$) is typical of the samples which led to a poor fit, the second (case 35, $n = 1007$) of those which led to a relatively good fit. The analysis is based on Pearson's chi-squared discrepancy and maximum likelihood estimation.

Diameter ($\mu m/2$)	6	7	8	9	10	11	12	13
Case 20	1	19	47	121	180	189	172	102
Case 35		7	8	37	85	147	188	155

Diameter ($\mu m/2$)	14	15	16	17	18	19	20	21
Case 20	80	30	21	23	14	4	13	
Case 35	144	102	58	34	26	8	4	4

This is an example of an application in which the criterion for selection should be based on the discrepancy due to approximation rather than on the expected discrepancy. The aim is to establish which model provides a better fit *in principle*, not just for the given sample size. However, since we will be making use of simpler criteria and both competing families have two parameters, it makes no difference whether we compare $\chi_P^2(\hat{\theta}) + p$ or $\chi_P^2(\hat{\theta}) + 2p$. It is sufficient to simply compare the values of the first term, namely, the ordinary chi-squared statistics.

The calculated values of χ_P^2 were:

Case	n	Lognormal	Gamma
20	1016	48.5	94.4
35	1007	21.6	15.8

The gamma distribution is slightly better than the lognormal distribution in case 35 and much worse in case 20.

This example of application highlights a number of general issues, which we will take the opportunity to discuss.

Although the χ_P^2 values can be compared within each case, the comparison between cases is not so straightforward. In general, criteria are constructed to estimate only those aspects of a discrepancy which are essential for comparing rival approximating models. Inessential parts, which may depend on the operating model and on the sample size, are omitted. However, the latter must be taken into account when comparing between cases.

Another factor which can spoil the comparison is specifically associated with discrepancies which are applied to grouped observations. The class intervals, which determine how the observations are to be grouped, are an integral part of the approximating family. By changing the intervals one in fact alters the approximating family, even if the distribution type (e.g., lognormal) remains unchanged. The fitted density is discretized and its rôle in the discrepancy is determined exclusively by the probabilities which it associates with the given intervals.

In this particular application circumstances are such that a rough comparison between cases is meaningful. The sample sizes are almost equal, the intervals are almost identical, and the *number* of intervals is the same in each case. We can therefore say that the discrepancy due to approximation is substantially higher in case 20 than in case 35, for both approximating families.

It is worthwhile to reflect on the classical use of chi-squared statistics in such situations. One would test the hypothesis that the operating family is lognormal and also the hypothesis that it is gamma. Both hypotheses would have to be rejected in case 20, whereas neither can be rejected in case

35. In this (and many other) application such tests do not address the pertinent question, which concerns the relative merits of the rival families and *not* whether the operating model is precisely lognormal, or gamma. It is commonplace that such hypotheses are almost always rejected whenever very large samples are available.

The approach proposed here is to admit at the outset that the operating model will be neither a lognormal nor a gamma distribution. It would be astonishing if the operating models for the 47 cases studied, and for many similar cases not studied, were all members of the same simple family. Our object here is to determine which of two rival families is more appropriate in the majority of cases. At the same time we can decide whether the selected family yields tolerable estimates in those cases where the fit is poor.

EXAMPLE 4.4.3.   Given below is the frequency distribution of the annual number of storms observed in Pretoria (Station 513/404) for the 71 years 1906 to 1976.

Number of storms	0	1	2	3	4	5	6	7	8	9	10	11	12	13	14	15	16	17
Frequency		0	2	2	3	4	10	8	10	6	5	7	5	3	0	2	2	1	1

For the purpose of deriving the distribution of large storms it is of interest to hydrologists to estimate the distribution of the annual number of storms (e.g., see Adamson and Zucchini, 1984). In this context the two approximating families of interest (for both theoretical and practical reasons) are the Poisson and negative binomial.

We used the three criteria of Section 4.1.2:

$$G^2(\hat{\theta}) + 2p, \qquad \chi_N^2(\hat{\theta}) + 2p, \qquad \chi_P^2(\hat{\theta}) + 2p,$$

where $\hat{\theta}$ is always the corresponding minimum discrepancy estimator. For the two chi-squared discrepancies the grouping $\{0, 1\}$ was used for the lower tail and $\{17, 18, \dots\}$ for the upper tail. The minimum discrepancy estimates of the parameters were:

	Poisson	Negative Binomial	
	$\hat{\theta}$	$\hat{\alpha}$	$\hat{\beta}$
Maximum likelihood	7.66	0.385	12.24
Minimum χ_N^2	7.44	0.389	12.31
Minimum χ_P^2	7.75	0.434	10.13

The corresponding criteria are given by

Approximating Model	p	$G^2 + 2p$	$\chi_N^2 + 2p$	$\chi_P^2 + 2p$
Poisson	1	23.8	14.5	31.9
Negative binomial	2	16.3	11.0	11.5

The Bootstrap estimators—for maximum likelihood estimation throughout and 100 replications—were:

Discrepancy	Approximating Model	Bootstrap Estimator of	
		$E\Delta(\hat{\theta})$	$\mathrm{Var}^{1/2}\Delta(\hat{\theta})$
Kullback–Leibler	Poisson	23.3	2.5
	Negative binomial	13.8	1.7
Neyman χ^2	Poisson	13.4	1.2
	Negative binomial	8.5	1.8
Pearson χ^2	Poisson	35.6	10.2
	Negative binomial	10.1	2.9

The negative binomial distribution is more appropriate for the given sample size. The fitted distribution, having $\hat{\alpha} = 0.377$ and $\hat{\beta} = 12.63$, is illustrated in Figure 4.3.

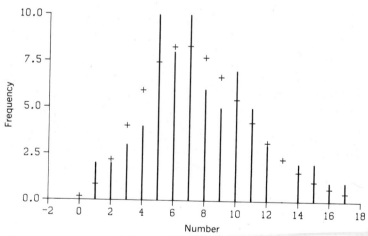

Figure 4.3. Annual number of storms in Pretoria, 1906–1976, and approximating negative binomial distribution.

EXAMPLE 4.4.4. Bliss (1953) gives numbers of adult female European red
mites observed on 150 randomly selected apple leaves:

Number of mites	0	1	2	3	4	5	6	7	8$^+$
Frequency	70	38	17	10	9	3	2	1	0

An inspection of the relative frequencies suggests that the geometric distribution should fit these data quite well. We will compare this to the more
flexible negative binomial distribution. The maximum likelihood estimates
(which will be used throughout in this example) are $\hat{\theta} = 0.466$ for the geometric and $\hat{\alpha} = 0.528$, $\hat{\beta} = 1.025$ for the negative binomial.

The three asymptotic criteria are given by

Approximating Model	$G^2 + 2p$	$\chi_N^2 + 2p$	$\chi_P^2 + 2p$
Geometric	6.2	4.6	4.9
Negative binomial	8.2	6.6	6.9

For the two chi-squared criteria the upper tail of the distribution was grouped
by combining all observations greater than or equal to 7. All three criteria
select the geometric distribution.

The Bootstrap estimators of the expectations and standard deviations
(which are given in parentheses) of $\Delta(\hat{\theta})$, obtained in 100 replications, are

Approximating Model	Kullback–Leibler	Neyman χ^2	Pearson χ^2
Geometric	5.2 (1.6)	3.8 (1.9)	4.1 (2.2)
Negative binomial	6.1 (2.3)	4.7 (2.3)	5.0 (3.0)

The Bootstrap estimators are smaller than the corresponding asymptotic
criteria but they lead to the same selection.

EXAMPLE 4.4.5. The standard statistical method of estimating design storms
(and other design events such as floods) from observed annual maximum
storms can be described as follows, Zucchini and Adamson (1984b).

Let F be the distribution function of annual maximum storms. Then
under the assumption that these storms are independently and identically
distributed, the distribution function of the largest storm in h years is F^h.
So the storm associated with a design horizon of h years and risk (i.e., probability) of occurrence r is the solution to the equation

$$F^h(x) = 1 - r.$$

The design storm, x, is estimated by fitting a model to F, say G_θ, where θ are the parameters, and then using

$$\hat{x} = G_{\hat{\theta}}^{-1}((1 - r)^{1/h}).$$

The risk, r, is given and is seldom more than 0.20 in practice, so one is dealing with the *upper* tail of the distribution function F. As h is increased, the design storm becomes associated with increasingly extreme values of this tail.

Clearly, for this application, it is important to select approximating models that fit the upper tail of the distribution function F well. The lower tail is not so important. Also, to take account of the relevant portion of the distribution, the discrepancy should be a function of design horizon. Since it is F^h which is finally of interest (rather than F), a discrepancy which satisfies the desiderata above is

$$\Delta(G_\theta, F; h) = \max \{|F(x)^h - G_\theta(x)^h|: x \in R\}.$$

In practice, the sample size may be as low as 30 and the design horizon as high as 20, or even 50, years. Such high values of h lead to criteria that have the undesirable property of being determined almost exclusively by a single observation (the largest).

A reasonable compromise is to examine values of the criteria associated with a number of smaller values of h, for example, $h = 1, 5, 10$, and then base the final selection on the results as a whole. This procedure was applied to select an approximating model for the annual maximum 1 day (24 hour) storm depths (mm) at Vryheid for the period 1951–1980.

Year	Depth	Year	Depth	Year	Depth
1951	45.2	1961	52.5	1971	84.5
1952	66.5	1962	50.0	1972	74.5
1953	142.0	1963	170.0	1973	94.0
1954	83.9	1964	62.0	1974	80.0
1955	61.1	1965	43.5	1975	74.0
1956	60.6	1966	60.0	1976	64.0
1957	84.5	1967	60.0	1977	60.0
1958	80.0	1968	53.5	1978	51.5
1959	79.0	1969	58.0	1979	58.5
1960	137.5	1970	93.0	1980	88.0

The following families were considered:

Gamma:
$$g(x) = \frac{x^{\alpha-1} e^{-x\beta} \beta^{\alpha}}{\Gamma(\alpha)}.$$

Normal:
$$g(x) = \frac{e^{-(x-\mu)^2/2\sigma^2}}{(2\pi)^{1/2}\sigma}.$$

Lognormal:
$$g(x) = \frac{e^{-(\log x - \mu)^2/2\sigma^2}}{(2\pi)^{1/2}\sigma x}.$$

Exponential: $G(x) = 1 - e^{-\lambda x}.$

Weibull: $G(x) = 1 - e^{-(x/\delta)^\rho}.$

Extreme (Type I): $G(x) = \exp(-e^{-(x-\xi)/\eta}).$

The following empirical discrepancy, based on the Weibull plotting position, was used:

$$\Delta_n(\theta) = \max\{|(i/(n+1))^h - G_\theta(x_{(i)})^h|: i = 1, 2, \ldots, n\}.$$

Note that $x_{(i)}, i = 1, \ldots, n$ are the order statistics.

The maximum likelihood estimates of the parameters were:

Gamma: $\hat{\alpha} = 8.97,$ $\hat{\beta} = 0.119.$

Normal: $\hat{\mu} = 75.7,$ $\hat{\sigma} = 28.6.$

Lognormal: $\hat{\mu} = 4.27,$ $\hat{\sigma} = 0.567.$

Exponential: $\hat{\lambda} = 0.0132.$

Weibull: $\hat{\rho} = 2.68,$ $\hat{\delta} = 85.1.$

Extreme (I): $\hat{\xi} = 64.1,$ $\hat{\eta} = 17.9.$

Table 4.1 gives the Bootstrap estimates of the expectation and standard deviation of the overall discrepancy [i.e., the mean and standard deviation of $\Delta_n(\hat{\theta}^*)$] for $h = 1, 5,$ and 10.

On the basis of this criterion the extreme (type I) distribution should be used for $h = 1$, the lognormal for $h = 5$, and the Weibull for $h = 10$. However, for $h = 1$ and $h = 10$ the lognormal distribution leads to criteria which are quite close to the minima and consequently it would not be unreasonable to use the lognormal distribution for each case.

Table 4.1.   Estimates of Expectation and Standard Deviation of Overall Discrepancy Based on 200 Bootstrap Replications

Distribution	Moment	Design Horizon		
		$h = 1$	$h = 5$	$h = 10$
Gamma	Mean	0.168	0.313	0.449
	Standard deviation	0.035	0.070	0.093
Normal	Mean	0.192	0.332	0.470
	Standard deviation	0.035	0.066	0.093
Lognormal	Mean	0.156	0.301	0.426
	Standard deviation	0.035	0.070	0.086
Exponential	Mean	0.420	0.383	0.457
	Standard deviation	0.025	0.079	0.073
Weibull	Mean	0.180	0.317	0.425
	Standard deviation	0.029	0.049	0.071
Extreme (I)	Mean	0.154	0.326	0.459
	Standard deviation	0.036	0.088	0.106

EXAMPLE 4.4.6.   Proschan (1963) gives the following time intervals (hours) between failures of the air conditioning system of an airplane:

23, 261, 87, 7, 120, 14, 62, 47, 225, 71, 246, 21, 42, 20, 5, 12, 120, 11, 3, 14, 71,

11, 14, 11, 16, 90, 1, 16, 52, 95.

We use this example to illustrate how criteria based on the Cramér–von Mises discrepancy are derived. In particular, we will compare the exponential and lognormal distributions as approximating families for the unknown operating distribution of intervals between failures.

For the exponential distribution $G_\lambda(x) = 1 - e^{-\lambda x}$ and $G'_\lambda(x) = \lambda e^{-\lambda x}$. For the lognormal distribution

$$G_\theta(x) = \phi\left(\frac{\log x - \mu}{\lambda^{1/2}}\right), \qquad \theta = (\mu, \lambda)^\mathsf{T}$$

$$\frac{\partial G_\theta(x)}{\partial \mu} = -\frac{1}{\lambda^{1/2}}\, \varphi\left(\frac{\log x - \mu}{\lambda^{1/2}}\right),$$

$$\frac{\partial G_\theta(x)}{\partial \lambda} = -\frac{(\log x - \mu)}{2\lambda^{3/2}}\, \varphi\left(\frac{\log x - \mu}{\lambda^{1/2}}\right),$$

where $\varphi(t) = e^{-t^2/2}(2\pi)^{-1/2}$ and $\phi(t) = \int_{-\infty}^{t} \varphi(u)du$ are the normal density and distribution functions.

The minimum discrepancy estimate of the exponential distribution parameter is $\hat{\lambda} = 0.02073$. The criterion is given by

$$n^2 \Delta_n(\hat{\lambda}) + n \text{ tr } \Omega_n^{-1}\Sigma_n = 4.37 + 5.78 = 10.15.$$

For the lognormal distribution $\hat{\mu} = 3.380$, $\hat{\lambda} = 2.041$, $n^2\Delta_n(\hat{\theta}) = 1.93$, n tr $\Omega_n^{-1}\Sigma_n = 7.27$, and hence the criterion is equal to 9.20. Since this is smaller than the value obtained for the exponential distribution, the lognormal is estimated to be the more appropriate model.

The bias correction term, C_C, is omitted from the criterion (see Section 4.1.3). Under the assumption that the operating model belongs to the approximating family, this term is asymptotically equal to $-n/6 = -30/6 = -5.00$ in this example. We note that, for the selected distribution, this value is very close to that obtained by direct estimation, namely, -4.99. The corresponding estimate for the exponential distribution is -5.63.

Figure 4.4 gives the histogram and the approximating lognormal density.

EXAMPLE 4.4.7. Table 3.1 in Example 3.3.1 gives the yields of barley (g) for 400 plots. We illustrate the theory discussed in Section 4.2 by selecting a model to estimate the operating distribution of the yields. In particular, we begin by standardizing the data, and then select a model for the

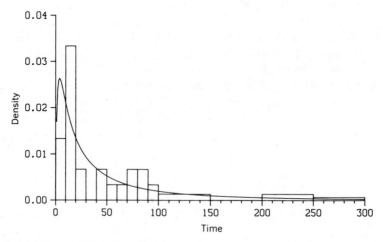

Figure 4.4. Time intervals between failures of the air conditioning system of an airplane: histogram and the approximating lognormal density.

standardized data from families of the form:

$$g_\theta(x) = \sum_{i \in A} \theta_i H_i(x) \varphi(x).$$

The estimators of θ_i are functions of the sample moments of the standardized data, m_i, which are also required for computing $\widehat{\mathrm{Var}}\ \hat{\theta}_i$ and hence the criteria $\hat{\theta}_i^2/\widehat{\mathrm{Var}}\ \hat{\theta}_i,\ i \geqslant 3$. In this application one obtains $\bar{x} = 151.92$, $s = 31.10$, and

i	m_i	$\hat{\theta}_i$	$\widehat{\mathrm{Var}}\ \hat{\theta}_i$	$\hat{\theta}_i^2/\widehat{\mathrm{Var}}\ \hat{\theta}_i$
3	-0.40145453	-0.06691	0.0010074	4.44
4	3.4907501	0.02045	0.0005132	0.81
5	-4.7246697	-0.00592	0.0001682	0.21
6	26.575814	0.00585	0.0000293	1.17
7	-65.195073			
8	317.48223			
9	-1028.3110			
10	4683.9880			
11	-17261.643			
12	75751.374			

It is only for $i = 3$ that $\hat{\theta}_i^2/\widehat{\mathrm{Var}}\ \hat{\theta}_i > 2$ and hence only θ_3 should be retained in

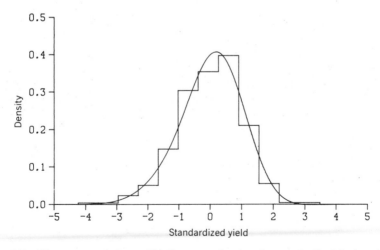

Figure 4.5. Histogram and Gram–Charlier approximation for standardized barley yields.

the series. The corresponding approximating model for the standardized data is

$$g_{\hat{\theta}}(x) = (1/\sqrt{2\pi})e^{-x^2/2}(1 - 0.0669(x^3 - 3x)),$$

and that for the original (unstandardized) data is given by $g_{\hat{\theta}}(\bar{x} + sx)$, where $\bar{x}$ and s are given above. A histogram with the approximating model is given in Figure 4.5.

CHAPTER 5

Simple Regression, One Replication Only

In this chapter we discuss the problem of model selection in regression analysis. We suppose that n pairs of observations (x_1, y_1), (x_2, y_2), ..., (x_n, y_n) are available. Essentially, two situations arise: either both x and y are random variables, or y is a random variable and x is a factor. In the former case we are concerned with properties of the joint distribution of x and y, and in the latter with properties of the distribution of y for the different levels of x.

In either case there will generally be more than one model available for describing some aspect of joint behavior of x and y. For example, we may wish to estimate $E(y|x)$ by means of some polynomial in x. It is then first necessary to select the degree of the polynomial which should be fitted. Alternatively, we may wish to select between a polynomial and some other parametric function of x.

In this chapter we consider cases where we do not have more than one observation on the response variable, y, for each observed value of the variable x. Such cases are particularly problematic from the point of view of model selection. The basic difficulty is that of specifying a suitable operating family. Entirely arbitrary operating families lead to expected discrepancies which can not be estimated. One could make assumptions about the form of the operating mean function such that it *becomes* possible to estimate the expected discrepancy. But such assumptions, which should not of course be based on the observations, can seldom be based on objective information about the subject under observation and they have a strong influence on the final selection of the approximating model.

In Section 5.1 we describe the special case in which a linear regression is assumed in the operating model. No fundamental difficulties arise and it is quite obvious how one could proceed in cases in which the operating mean function is assumed to belong to some other parametric family of functions.

Fundamental difficulties do arise if one wants to avoid assumptions on the form of the mean function. One could be of the opinion that one should assume a polynomial of "sufficiently high degree" as operating mean function. Unfortunately, as a rule, the criteria depend *heavily* on the assumed degree and there is seldom an objective reason to suppose that any particular degree is correct. The essential parameter that dominates the analysis is the *variance* in the operating model and if a polynomial of a certain *degree* is assumed for the operating mean function, implicitly a certain operating *variance* is assumed.

If *repeated* observations $y_{ij}, j = 1, 2, \ldots, J_i > 1$, are available for all—or at least some—x_i, then there is no problem; the variance can be estimated. But if this is *not* the case, then the dependence of the methods on the operating variance, which can not be estimated, constitutes a fundamental difficulty.

In Sections 5.2, 5.3, and 5.4 we will outline three methods which provide some general guidance for selecting approximating models in cases in which one wants to avoid unfounded assumptions about the form of the operating mean function. They are

(I)   *Akaike's information criterion.*

(II) *Cross-validation.*

(III) *Bootstrap methods* if both variables, x and y, can be considered random variables in the operating model.

5.1.   SHOULD THE INDEPENDENT VARIABLE BE USED AT ALL?

A simple but important question which we discuss in this section is whether the predictor variable, x, should be used at all for predicting future values of the response variable, y. We deal here specifically with the case in which an operating model having the property

$$E(y|x) = \alpha + \beta x$$

is considered to be plausible, that is, the case of a linear regression. In situations where this assumption is thought to be unrealistic, the methods described in the later sections should be used. The question which we consider is whether the simpler model of the form

$$E(y|x) = \alpha,$$

where α does not depend on x, leads to more accurate predictions, on average.

It is not possible, on the basis of data alone, to answer this question with certainty. However, one can develop methods that indicate with the usual statistical uncertainty whether this simple approximating family or the operating family will provide better predictions in the long run, that is, one can construct a criterion for selecting between the two families. Under the assumption that the random variables are *normally* distributed one can also carry out a test of the hypothesis that the operating family leads to worse predictions.

In this section we will only state the methods. They are derived in Chapter 7 where the more general multivariate case is treated.

Because we are concerned with prediction, it is reasonable to use a discrepancy that depends on the error of prediction. For the case in which x is taken to be random, we will consider the discrepancy

$$E_F(y - E_{G_\theta}(y|x))^2,$$

and in the case in which x is a factor

$$\frac{1}{n} \sum_{i=1}^{n} E_{F_i}(y_i - E_{G_\theta}(y_i))^2,$$

where F is a bivariate and the F_i are univariate operating distribution functions.

The first is simply the expected squared error of prediction if the minimum mean squared error predictor $E_{G_\theta}(y|x)$ is used, and the second the *average* expected error of prediction if the "best" predictors are used for the x values that appeared in the original sample of size n.

One gets the following criteria: Use the variable x to predict, that is, use $\hat{\alpha} + \hat{\beta}x$, where $\hat{\alpha}$ and $\hat{\beta}$ are the least squares estimators, *only if*

Two random variables: One random variable, one factor:

$$r^2 > \frac{1}{n-2}, \qquad\qquad r^2 > \frac{2}{n},$$

where $r^2 = (\sum(y_i - \bar{y})(x_i - \bar{x}))^2 / \sum(y_i - \bar{y})^2 \sum(x_i - \bar{x})^2$ is the squared correlation coefficient in the sample. If one is prepared to assume that the random variables are *normally* distributed, one can also carry out a test of the hypothesis that using x leads to an increase in the expected discrepancy. This hypothesis can be *rejected* (in which case one should use x) if the test statistic,

$$F = \frac{r^2(n - 2)}{1 - r^2}$$

which has $n_1 = 1$ and $n_2 = n - 2$ degrees of freedom, is larger than the critical value given in Tables A.4.1, A.4.2, A.4.3, A.4.4 (two random variables) and Tables A.4.5, A.4.6, A.4.7, A.4.8 (one random variable, one factor).

EXAMPLE 5.1. Verwey (1957) gave 24 apprentices a spatial vision test, x, and later recorded their post-training examination score, y. The idea was to predict the performance of future applicants. The observed correlation was $r = 0.39$ and the question arises as to whether the information in the test result improves predictions.

A convenient and plausible operating model is in this case that the (y_i, x_i) are jointly distributed with $E(y_i|x_i) = \alpha + \beta x_i$ and Var $(y_i|x_i) =$ constant. The approximating family could use $E(y_i|x_i) = \alpha^*$. The question is whether the family of the operating model should be fitted, that is, whether $\hat{\alpha} + \hat{\beta} x_i$ should be the predictor, where $\hat{\alpha}$ and $\hat{\beta}$ are the usual least-squares estimators, or just $\hat{\alpha}^* = \bar{y}$.

In this example $1/(n - 2) = 1/22 = 0.0455$ and $r^2 = 0.39^2 = 0.152$. Since r^2 is larger than $1/(n - 2)$, the predictor $\hat{\alpha} + \hat{\beta} x$ is estimated to lead to better predictions, on average.

Assuming *normality*, a test of the hypothesis that the expected discrepancy is *larger* if one uses x is possible at a significance level of 0.5. One needs the median of the distribution of r^2 for the case in which $\text{Corr}^2(x, y) = 1/(n - 2)$. (If the correlation has this value the expected discrepancy remains the same, irrespective of whether one uses x or not.) The hypothesis is rejected if r^2 is larger than the median. The median in the distribution of $r^2(n - 2)/(1 - r^2)$ for this case is 1.21 (see Table A.4.1 with $n_1 = 1$ and $n_2 = 22$); the median in the distribution of r^2 is thus 0.052. Since the observed r^2 (it has the value 0.152) is larger than this, the conclusion is again to use x.

Strong evidence that the prediction using x is preferable, however, is not available. The hypothesis that $\text{Corr}^2(x, y) < 1/(n - 2)$ (which is the hypothesis that the predictor using x leads to a larger expected discrepancy) can not be rejected at a small significance level. The critical value for r^2 for a level of 0.10 is 0.21 (see Table A.4.2) and the observed $r^2(0.152)$ is smaller than this.

5.2. A METHOD BASED ON THE KULLBACK–LEIBLER DISCREPANCY

Consider the operating model

$$y_i = \mu(x_i) + e_i = \mu_i + e_i,$$

where the e_i are uncorrelated with mean zero and variance σ_i^2. A possible approximating model is

$$y_i = h(x_i, \theta) + u_i,$$

where the u_i, $i = 1, 2, \ldots, I$, are NID$(0, \sigma^2)$. In the sequel we will sometimes write $h_i(\theta)$ instead of $h(x_i, \theta)$.

The Kullback–Leibler discrepancy is

$$-E_F \log\left[(2\pi\sigma^2)^{-I/2}\exp\left(\frac{-\sum(y_i - h_i(\theta))^2}{2\sigma^2}\right)\right]$$

$$= \frac{I}{2}\log 2\pi\sigma^2 + \frac{\sum\sigma_i^2}{2\sigma^2} + \frac{\sum(\mu_i - h_i(\theta))^2}{2\sigma^2}.$$

This discrepancy depends on σ^2, the variance under the approximating model, and on the σ_i^2, the variances in the operating model.

For the purpose of comparing approximating models it is sufficient to use the essential part, namely,

$$\Delta(\theta, \sigma^2) = I \log \sigma^2 + \frac{\sum\sigma_i^2}{\sigma^2} + \frac{\sum(\mu_i - h_i(\theta))^2}{\sigma^2}.$$

As empirical discrepancy one uses the unbiased estimator

$$\Delta_n(\theta, \sigma^2) = I \log \sigma^2 + \frac{\sum(y_i - h_i(\theta))^2}{\sigma^2}.$$

The maximum likelihood estimators of θ and σ^2 minimize Δ_n:

$$\hat{\theta} = \arg\min\left\{\sum(y_i - h_i(\theta))^2 : \theta \in \Theta\right\},$$

$$\hat{\sigma}^2 = \frac{\sum(y_i - h_i(\hat{\theta}))^2}{I},$$

and one obtains

$$\Delta_n(\hat{\theta}, \hat{\sigma}^2) = I(\log \hat{\sigma}^2 + 1).$$

The simpler criterion (AIC, see Appendix A.2.1.) is then

$$I(\log \hat{\sigma}^2 + 1) + 2(p + 1),$$

where p is the number of elements of θ. For this criterion it is not necessary to estimate the operating variances σ_i^2.

If one wants to avoid the approximation tr $\Omega^{-1}\Sigma \approx p + 1$ a criterion *can not* be found. To estimate Σ and Ω one would have to estimate the σ_i^2 and this is not possible.

Sawa (1978) proposes a criterion BIC which [to the order $O(1/n)$] is an asymptotically unbiased estimator of the expected (Kullback–Leibler) discrepancy. The difficulty is that the variance in the operating model (all σ_i^2 are assumed to be equal) must be estimated and the advantage of AIC (where this is *not* necessary) disappears. This criterion has a tendency to be more parsimonious than AIC *but it can not be used if only a single replication is available* since it is impossible to estimate the operating variance.

This state of affairs is surprising; it is hard to understand why a simple criterion can be obtained but other similar criteria can not. Some reflection shows that this is possibly so because one implicitly introduces some assumptions about the form of the operating model by using the approximation tr $\Omega^{-1}\Sigma \approx p + 1$. This approximation is only reasonable if the approximating family can almost accommodate the operating model. So by using $p + 1$ for tr $\Omega^{-1}\Sigma$ one implies that the operating model is of the form which the approximating family suggests. In many applications the approximation leads to no difficulties. Here, however, the criterion can lead to poor selections. In some cases the criterion simply continues to decrease as the number of parameters in the approximating model is increased.

5.3. A CROSS-VALIDATORY METHOD

5.3.1. A Cross-Validatory Criterion

The essentials of the method of cross-validation are described in Section 2.6. This method could also be applied in an attempt to solve the problem of this chapter.

Assume that the operating model is as in Section 5.2 and let the approximating models have expectations $h_i(\theta) \equiv h(x_i, \theta)$, where dim $(\theta) \leqslant I - 1$.

Denote by $x^{(k)}$ the vector obtained by omitting the kth element of $x = (x_1, x_2, \ldots, x_I)^\mathsf{T}$ and let

$$\hat{\theta}^{(k)} = \arg\min\left\{\sum_{i \neq k, i=1}^{I} (y_i - h_i(\theta))^2 : \theta \in \Theta\right\}$$

be the least-squares estimator of θ obtained from the $I - 1$ observations (y_i, x_i), $i = 1, 2, \ldots, k - 1, k + 1, \ldots, I$.

The (one item out) cross-validatory criterion on the basis of the Gauss discrepancy $(\Sigma_i(\mu_i - h_i(\theta))^2/I)$ is then

$$\frac{1}{I} \sum_{i=1}^{I} (y_i - h_i(\hat{\theta}^{(i)}))^2.$$

One selects the approximating family which leads to the smallest criterion.

The example to follow shows that cross-validation leads to plausible results. Again the question is why the method leads to a solution of a problem which seems to be unsolvable in principle. The impression is that the method prescribes to a certain extent the form of a desirable approximation. A certain degree of smoothness is required of the approximating model and since, in most applications, smoothness is plausible for operating models, the method leads to acceptable results.

5.3.2. Selection of Linear Approximating Models with a Cross-Validatory Criterion

We consider the special case in which the approximating model is linear in its parameters:

$$y = A\theta + e,$$

where A is $I \times p$, known and nonsingular, $Ee = 0$, $Eee^{\mathsf{T}} = \sigma^2 \mathscr{I}$.

Then $\hat{\theta}^{(k)}$ is the least squares estimator based on the $I - 1$ observations $y^{(k)} = (y_1, \ldots, y_{k-1}, y_{k+1}, \ldots, y_I)^{\mathsf{T}}$:

$$\hat{\theta}^{(k)} = C^{(k)-1} A^{(k)\mathsf{T}} y^{(k)},$$

where $C^{(k)} = A^{(k)\mathsf{T}} A^{(k)}$ and $A^{(k)}$ is obtained by eliminating the kth row from the matrix A.

The criterion is then given by

$$\frac{1}{I} \sum_{i=1}^{I} (y_i - a_i^{\mathsf{T}} \hat{\theta}^{(i)})^2,$$

where a_i^{T} are the row vectors of A. Now $C^{(k)-1} = C^{-1} + C^{-1} a_k a_k^{\mathsf{T}} C^{-1}/(1 - \alpha_k)$, where $\alpha_k = a_k^{\mathsf{T}} C^{-1} a_k$. (See, for instance, Press, 1972, p. 23.) Thus

$$y_i - a_i^\mathsf{T} \hat{\theta}^{(i)} = y_i - a_i^\mathsf{T} C^{(i)-1}(A^\mathsf{T} y - y_i a_i)$$

$$= y_i - a_i^\mathsf{T} C^{-1} A^\mathsf{T} y - \frac{\alpha_i a_i^\mathsf{T} C^{-1} A^\mathsf{T} y}{1 - \alpha_i}$$

$$+ \alpha_i y_i + \frac{\alpha_i^2 y_i}{1 - \alpha_i}$$

$$= \frac{y_i - a_i^\mathsf{T} \hat{\theta}}{1 - \alpha_i},$$

where $\hat{\theta} = C^{-1} A^\mathsf{T} y$ is the least-squares estimator of θ based on *all* observations.

The criterion thus reduces to

$$\frac{1}{I} \sum_{i=1}^{I} \frac{(y_i - a_i^\mathsf{T} \hat{\theta})^2}{(1 - a_i^\mathsf{T} C^{-1} a_i)^2}.$$

This is exactly Allen's (1971, 1974) PRESS (prediction error sum of squares) statistic.

The result is very convenient if C is the unit matrix which is the case, for example, if orthogonal polynomials are used. Unfortunately, it is not possible in this case to express the criterion as the sum of components due to individual parameters associated with particular polynomials. Nevertheless, since the elements of θ need not be re-estimated if the order of the approximating polynomial is changed, the criteria for the various possible linear combinations of orthogonal polynomials are easily obtained by replacing some of the columns in A (i.e., some of the elements in a_i) by zero.

5.4. BOOTSTRAP METHODS

In general, Bootstrap methods can only be meaningfully applied for model selection if both x and y are random variables. In this case the bivariate distribution function, F, can be estimated by the empirical distribution function, F_n, based on the n observed points in the plane, $(x_1, y_1), (x_2, y_2)$, ..., (x_n, y_n). The Bootstrap samples are generated by selecting (at random and with replacement) n of these points, that is, n *pairs* of observations.

In certain circumstances it is also possible to apply Bootstrap methods when x is a factor. Here a procedure due to Efron (1982) can be used. One begins by fitting a curve through the data points and then computes the

n residuals. Bootstrap samples are generated by superimposing random samples of the residuals on the fitted curve. To apply this procedure it is first necessary to decide on the form of the mean function which is to be fitted. In other words, one needs to specify the operating model and, moreover, this model must depend on fewer than n free parameters, otherwise there would be no residuals available for resampling. Thus this type of Bootstrapping is not applicable for model selection if one wants to avoid assumptions about the form of the operating model. (Otherwise, in the situation dealt with in Section 5.1, for example, this method could be applied.) A further difficulty here is that the residuals based on the estimated (operating) mean function do not constitute a random sample from the distribution of residuals from the actual operating model. This has the effect of biasing the selection in favor of more flexible approximating models.

We return now to the case where both x and y are random. Let $h(x, \theta)$ be an approximating conditional expectation of y given x, where θ is a vector of parameters. As before we use the notation $h_i(\theta)$ to represent $h(x_i, \theta)$. Let $\Delta(\theta, F)$ be the required discrepancy and $\hat{\theta}$ an estimator of θ. The Bootstrap estimator of the expected overall discrepancy, $E_F \Delta(\hat{\theta}, F)$ is $E_{F_n} \Delta(\hat{\theta}, F_n)$. If necessary, the latter can be computed by Monte Carlo methods to any desired accuracy. One proceeds as follows:

STEP 1.   Select a random sample of n pairs, with replacement, from (x_1, y_1), $(x_2, y_2), \ldots, (x_n, y_n)$. We denote this Bootstrap sample by $(x_1^*, y_1^*), (x_2^*, y_2^*), \ldots, (x_n^*, y_n^*)$.

STEP. 2.   Estimate the vector of parameters, θ, in the approximating family using the Bootstrap sample. We denote this estimate by $\hat{\theta}^*$.

STEP 3.   Compute the Bootstrap overall discrepancy, $\Delta(\hat{\theta}^*, F_n)$.

STEP 4.   Repeat Steps 1 to 3 a large number of times, M, and compute the mean of the discrepancies obtained in Step 3.

The mean computed in Step 4 is an estimator of the required quantity $E_{F_n}(\hat{\theta}, F_n)$ which can be made arbitrarily accurate by choosing M large enough. Typically, a few hundred iterations are sufficient, but it is possible to avoid unnecessary computation by keeping track of the sample standard deviation of the discrepancies obtained in Step 3. One can then assess when the required standard error (standard deviation/$M^{1/2}$) has been achieved.

The above procedure has to be repeated for each approximating family. The family which leads to the smallest value of $E_{F_n}(\hat{\theta}, F_n)$ is selected.

For the Gauss discrepancy,

$$\Delta(\theta, F) = E_F(y - h(x, \theta))^2,$$

an empirical discrepancy is

$$\Delta_n(\theta) = \frac{1}{n} \sum_{i=1}^{n} (y_i - h(x_i, \theta))^2 = \Delta(\theta, F_n).$$

The overall discrepancy in Step 3 is then $\Delta_n(\hat{\theta}^*)$.

5.5. EXAMPLES OF APPLICATION

In all three examples least-squares estimation was used and the cross-validatory and Bootstrap criteria are based on the Gauss discrepancy.

EXAMPLE 5.5.1.   C. E. Don (1975) determined the water content x (%) and the corresponding calorific value y (kJ/g) of 34 samples of bagasse. Both variables have to be considered as random in the operating model.

It can be seen in Figure 5.1 that, except for observations 22 and 23, the calorific values can be determined quite accurately by using the (least squares)

Table 5.1.   Water Content and Calorific
Value of Samples of Bagasse

i	x_i	y_i	i	x_i	y_i
1	6.3	18.2	18	29.9	12.9
2	6.3	18.0	19	30.4	13.4
3	7.3	18.2	20	32.3	14.1
4	11.3	17.4	21	32.3	12.7
5	12.2	16.9	22	25.0	12.4
6	12.2	16.9	23	25.6	12.3
7	13.7	16.6	24	41.0	11.7
8	15.5	16.4	25	41.7	11.5
9	15.5	16.2	26	42.3	11.4
10	15.5	15.5	27	55.0	8.6
11	15.8	16.7	28	57.2	8.5
12	16.5	16.2	29	57.6	8.5
13	19.0	15.7	30	57.8	8.5
14	20.3	15.7	31	58.7	8.0
15	23.2	15.0	32	60.3	7.8
16	23.7	15.5	33	61.0	7.5
17	27.0	14.2	34	61.7	7.4

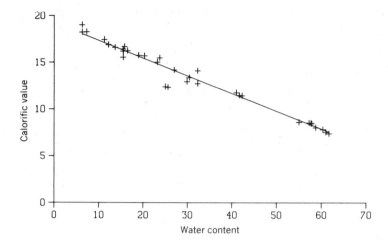

Figure 5.1.   Calorific value versus water content of bagasse and fitted straight line.

straight line

$$y = 19.2 - 0.189x.$$

Two of the three criteria discussed in this chapter indicate that this line is preferable to polynomials of higher degree. The results (cross-validation and Bootstrap based on the Gauss discrepancy) are given by

Degree of polynomial	1	2	3	4
Number of parameters	2	3	4	5
Kullback–Leibler criterion	5.57	6.68	7.32	8.65
Cross-validatory criterion	0.390	0.406	0.402	0.411
Bootstrap criterion	0.377	0.379	0.372	0.376

The Bootstrap criterion increases a little if one fits a quadratic but then decreases to achieve a minimum for a cubic. However, if the "outlying" observations, 22 and 23, are omitted then all three criteria lead to the same selection, namely, a straight line:

Degree of polynomial	1	2	3	4
Number of parameters	2	3	4	5
Kullback–Leibler criterion	− 32.2	− 30.5	− 29.0	− 27.0
Cross-validatory criterion	0.122	0.131	0.132	0.143
Bootstrap criterion	0.117	0.120	0.121	0.131

This example illustrates the fact that criteria can be quite sensitive to outliers from the approximating model and that their influence should be taken into account before a final selection is made.

EXAMPLE 5.5.2. This example relates to the relationship between the length, x (mm), and the age, y (days), of juvenile South African anchovies. The age is measured by counting the daily rings on the otoliths (ear bone) under an electron microscope. Prosch (1985) recorded the following observations:

i	x_i	y_i	i	x_i	y_i
1	50	78	9	97	268
2	38	83	10	107	332
3	52	109	11	90	256
4	55	119	12	107	314
5	52	114	13	112	333
6	65	144	14	97	286
7	61	129	15	112	355
8	57	128	16	97	313

One of the objects of taking these measurements is to fit a model which can be used to estimate the age from the length.

Although the data seem to scatter about the straight line $y = -95.6 + 3.92x$ (see the dotted line in Figure 5.2), all three criteria which were used in Example 5.5.1 indicate that a higher degree polynomial is preferable to describe the conditional expectation of y given x in this length range. The results are:

Degree of polynomial	1	2	3	4
Number of parameters	2	3	4	5
Kullback–Leibler criterion	108	107	105	107
Cross-validatory criterion	287	325	243	527
Bootstrap criterion	241	235	237	2927

The Bootstrap criterion (based on 500 replications) selects a quadratic while the other two criteria select a cubic. The fitted cubic has the equation $y = 215.4 - 8.609x + 0.1573x^2 - 0.000628x^3$; it is given by the solid line in Figure 5.2.

A number of features of the data combine to produce these apparently anomalous selections. Point 2 is somewhat high. All three criteria select models which attempt to accommodate this observation by bending the

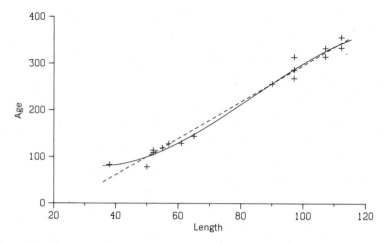

Figure 5.2. Age versus length of anchovies with approximating straight line and cubic.

fitted line a little. Since there are no observations in the middle of the x range, that is, in the interval $65 < x < 90$, this bending has little effect on the fit for the remaining observations. That observation 2 is responsible for the curvature can be seen by omitting it and recomputing the criteria. The results are:

Degree of polynomial	1	2	3	4
Number of parameters	2	3	4	5
Kullback–Leibler criterion	96	98	99	100
Cross-validatory criterion	194	232	287	324
Bootstrap criterion	167	183	220	1778

All three criteria now select a straight line and one might thus be tempted to regard observation 2 as an "outlier." In this application, however, some additional observations are available:

i	x_i	y_i	i	x_i	y_i
1	39	75	9	38	81
2	39	73	10	36	69
3	30	37	11	35	75
4	37	84	12	39	77
5	36	62	13	36	60
6	39	73	14	36	74
7	35	67	15	40	68
8	37	59	16	37	73

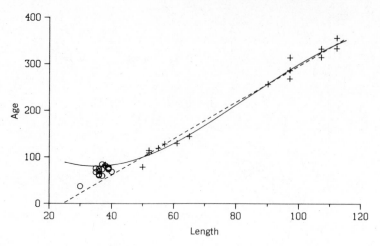

Figure 5.3. Age versus length of anchovies. Original data with fitted straight line and cubic, and additional observations.

These are results of measurements made with a light microscope (Prosch, 1985). They are illustrated—together with the original data and the originally fitted straight line and cubic—in Figure 5.3.

These observations clearly indicate that observation 2 is not an outlier and that, in the interval $35 \leqslant x \leqslant 40$, one of the originally selected models would have been more accurate than a straight line.

This example illustrates that observations at one of the extremes of the x range can have a strong influence on the selection. In the absence of additional information, there seems to be no objective method for deciding whether an influential observation should be regarded as an outlier for the purpose of model selection.

EXAMPLE 5.5.3. This example relates to the utilization of trees (Populus deltoides) for the production of matches. To estimate the matchwood volume of such trees Tingle (1965) gives a graph of expected utilization, y (% of volume), as a function of diameter at breast height, x (inches). The graph was based on the observations summarized in Table 5.2 which gives the average utilization of trees belonging to a number of diameter classes. A total of 1790 trees were measured.

As shown in Figure 5.4 the observations follow a smooth curve, except for observation 15 which is a little out of line. The shape of the curve suggests an exponential model of the form $\alpha + \beta \rho^x$. The method of least squares gives $\hat{\alpha} = 93.7$, $\hat{\beta} = -86.5$, and $\hat{\rho} = 0.666$. As can be seen in Figure 5.4 this curve fits the observations quite accurately.

Table 5.2. Average Matchwood Utilization of Populus Deltoides in the Given Diameter Classes

i	x_i	y_i	i	x_i	y_i
1	8	9	9	16	92
2	9	33	10	17	92
3	10	55	11	18	93
4	11	69	12	19	94
5	12	77	13	20	92
6	13	82	14	21	94
7	14	87	15	22	89
8	15	90	16	23	93

We compare the fit of this exponential model with that obtained using polynomials. In particular, the results based on the Kullback–Leibler discrepancy (Section 5.2) are compared with those based on cross-validation (Section 5.3):

Degree of polynomial	2	3	4	5	6	7	Exponential
Number of parameters	3	4	5	6	7	8	3
Kullback–Leibler criterion	81	49	37	36	33	34	40
Cross-validatory criterion	76	9	8	15	14	93	8

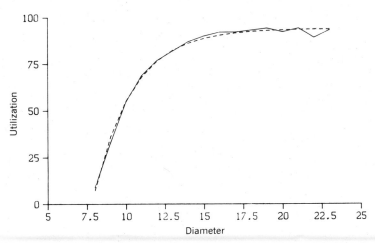

Figure 5.4. Average matchwood utilization versus diameter of Populus deltoides: data (—) and fitted exponential function (---).

Whereas the cross-validatory criterion achieves a minimum for a polynomial of degree 4 (which is rated as marginally better than the exponential model), the Kullback–Leibler criterion in fact only achieves a local minimum for a polynomial of degree 6, increases for degree 7, and then more or less continues to decrease.

CHAPTER 6

Simple Regression, Replicated Observations

The difficulty that we encountered in the last chapter disappears if several observations of y are available for each value of x. The operating model is then given by

$$y_{ij} = \mu(x_i) + e_{ij}, \qquad i = 1, 2, \ldots, I,$$
$$= \mu_i + e_{ij} \qquad j = 1, 2, \ldots, J_i, J_i > 1,$$

e_{ij} independently and for each i identically distributed, $E e_{ij} = 0$, $\operatorname{Var} e_{ij} = \sigma_i^2$. Since at least $2I$ observations are available the parameters μ_i and σ_i^2 can be estimated.

In this chapter we discuss selection methods based on the Gauss discrepancy. For *linear* approximating models convenient finite sample methods can be derived. These are discussed in Section 6.1. For approximating models with mean functions which do not depend linearly on their parameters, asymptotic methods discussed in Section 6.3 are available.

6.1. LINEAR APPROXIMATING MODELS

The assumed operating model is

$$y = A\mu + e,$$

where

$$y = (y_{11}, y_{12}, \ldots, y_{1J_1}, y_{21}, \ldots, y_{2J_2}, \ldots, y_{IJ_I})^\mathsf{T},$$

$A = \{a_{ijk}: i, j = 1, 1; 1, 2; \ldots; I, J_I; k = 1, 2, \ldots, I\}$ is a $J. \times I$ matrix with $a_{ijk} = \delta_{ik}$ and the e_{ij} are independently and for each i identically distributed with $Ee_{ij} = 0$ and $\text{Var } e_{ij} = \sigma_i^2$.

The approximating families which we use have mean

$$Ey = AB\theta,$$

where B is $I \times p$ and of full rank, $p \leq I$.

The Gauss discrepancy is

$$\Delta(\theta) = (A\mu - AB\theta)^{\mathsf{T}}(A\mu - AB\theta)$$

and an empirical discrepancy is

$$\Delta_n(\theta) = (A\hat{\mu} - AB\theta)^{\mathsf{T}}(A\hat{\mu} - AB\theta),$$

where

$$\hat{\mu} = (A^{\mathsf{T}}A)^{-1}A^{\mathsf{T}}y = \text{Diag } \{1/J_i\}y. = \bar{y}.,$$

$$y. = (y_{1.}, \ldots, y_{I.})^{\mathsf{T}}, \qquad \bar{y}. = (\bar{y}_{1.}, \ldots, \bar{y}_{I.})^{\mathsf{T}}.$$

The ordinary least-squares estimator

$$\hat{\theta} = (B^{\mathsf{T}}A^{\mathsf{T}}AB)^{-1}B^{\mathsf{T}}A^{\mathsf{T}}y = (B^{\mathsf{T}} \text{ Diag } \{J_i\}B)^{-1}B^{\mathsf{T}}y.$$

is also the minimum discrepancy estimator.

The expected discrepancy is

$$\mu^{\mathsf{T}}A^{\mathsf{T}}A\mu - \mu^{\mathsf{T}}A^{\mathsf{T}}AB(B^{\mathsf{T}}A^{\mathsf{T}}AB)^{-1}B^{\mathsf{T}}A^{\mathsf{T}}A\mu + \text{tr } AB(B^{\mathsf{T}}A^{\mathsf{T}}AB)^{-1}B^{\mathsf{T}}A\Sigma,$$

where $\Sigma = Eee^{\mathsf{T}}$ is a $J. \times J.$ diagonal matrix with the σ_i^2 in the diagonal. An unbiased estimator of this expectation, that is, a criterion, is

$$y^{\mathsf{T}}A(A^{\mathsf{T}}A)^{-1}A^{\mathsf{T}}y - \text{tr}(A^{\mathsf{T}}A)^{-1}A^{\mathsf{T}}\Sigma A$$

$$- y^{\mathsf{T}}AB(B^{\mathsf{T}}A^{\mathsf{T}}AB)^{-1}B^{\mathsf{T}}A^{\mathsf{T}}y + 2 \text{ tr}(B^{\mathsf{T}}A^{\mathsf{T}}AB)^{-1}B^{\mathsf{T}}A^{\mathsf{T}}\hat{\Sigma}AB,$$

where $\hat{\Sigma}$ is obtained from Σ by replacing σ_i^2 by

$$\hat{\sigma}_i^2 = \frac{\sum_j(y_{ij} - \bar{y}_{i.})^2}{J_i - 1}.$$

The criterion can also be written as

$$\sum_i \frac{y_i^2}{J_i} - \sum_i \hat{\sigma}_i^2 - y^{\mathsf{T}} B (B^{\mathsf{T}} \text{ Diag } \{J_i\} B)^{-1} B^{\mathsf{T}} y.$$

$$+ 2 \text{ tr}(B^{\mathsf{T}} \text{ Diag } \{J_i\} B)^{-1} B^{\mathsf{T}} \text{ Diag } \{J_i \hat{\sigma}_i^2\} B.$$

Let b_{ij} be the elements of B, then the elements (j, k) of B^{T} Diag $\{J_i\} B$ and of B^{T} Diag $\{J_i \hat{\sigma}_i^2\} B$ are $\sum_i J_i b_{ij} b_{ik}$ and $\sum_i J_i \hat{\sigma}_i^2 b_{ij} b_{ik}$, respectively.
For the operating family $B = \mathscr{I}$ and the criterion reduces to $\sum_i \hat{\sigma}_i^2$.

EXAMPLE 6.1.   Oman and Wax (1984) considered the problem of estimating gestation age, x (weeks), by ultrasound measurements of fetal bone lengths [the femur length, FL (mm), and the biparietal diameter, BPD (mm)]. The data are summarized in Table 6.1. Although this paper is mainly concerned with multivariate calibration, one of the steps in the analysis is to fit a model to the separate regressions of FL on x and BPD on x. Quadratic approximating models were selected in both cases.

Suppose that we wish to select a *smooth* approximating mean function, say a polynomial of degree four or less, with the methods of this section. The matrix B has then, for an approximating polynomial of degree $p - 1$, the elements $b_{ij} = x_i^{j-1}$, $i = 1, 2, \ldots, 28$, $j = 1, 2, \ldots, p$, where $x_i = 13 + i$ gives the gestation age. The elements (j, k) of B^{T} Diag $\{J_i\} B$ and B^{T} Diag $\{J_i \hat{\sigma}_i^2\} B$ are in this case $\sum_i J_i x_i^{j+k-2}$ and $\sum_i J_i \hat{\sigma}_i^2 x_i^{j+k-2}$.

One obtains for FL: $\sum_i y_{i.}^2/J_i = 3214098.51$, $\sum_i \hat{\sigma}_i^2 = 201.23$ and for BPD: $\sum_i y_{i.}^2/J_i = 5682670.34$, $\sum_i \hat{\sigma}_i^2 = 322.59$. The criterion for the saturated model is simply $\sum_i \hat{\sigma}_i^2$. For the other models one has to add $\sum_i y_{i.}^2/J_i - \sum_i \hat{\sigma}_i^2$ to the difference between the second and the first column of the following table which gives $y^{\mathsf{T}} B (B^{\mathsf{T}} \text{ Diag } \{J_i\} B)^{-1} B^{\mathsf{T}} y$ (column 1) and $2 \text{ tr}(B^{\mathsf{T}} \text{ Diag } \{J_i\} B)^{-1} B^{\mathsf{T}} \text{ Diag } \{J_i \hat{\sigma}_i^2\} B$ (column 2).

	Model	1	2	Criterion
FL	$p = 2$	3209986.20	27.32	3938
	$p = 3$	3213993.48	42.22	−54
	$p = 4$	3214004.35	56.24	−51
	$p = 5$	3214008.18	72.70	−38
	saturated			201
BPD	$p = 2$	5676048.98	44.26	6343
	$p = 3$	5682050.64	66.94	364
	$p = 4$	5682622.89	88.12	−187
	$p = 5$	5682625.87	107.36	−171
	saturated			323

Table 6.1. Femur Length and Biparietal Diameter for Different Gestation Ages

	Femur Length			Biparietal Diameter		
Week	J_i	Mean	Standard Deviation	J_i	Mean	Standard Deviation
14	12	14.42	2.50	12	27.33	2.19
15	23	16.09	2.02	23	29.87	2.62
16	35	19.14	2.51	35	33.97	2.04
17	38	20.92	2.16	38	36.53	2.61
18	25	24.12	1.99	25	39.36	2.38
19	31	28.65	1.62	31	43.36	2.20
20	25	32.04	2.32	25	47.08	2.58
21	28	34.61	2.31	28	50.18	2.86
22	30	37.37	2.79	30	52.80	3.25
23	42	39.12	2.36	41	54.07	3.07
24	39	42.13	2.96	39	58.08	3.58
25	50	45.30	2.77	49	62.04	3.27
26	33	47.45	2.92	33	65.09	4.25
27	53	49.36	3.31	52	66.62	4.57
28	47	51.57	3.19	47	70.91	3.71
29	48	54.02	2.54	48	73.25	3.40
30	59	56.32	2.97	59	75.85	3.90
31	62	57.82	2.95	62	78.18	4.03
32	67	61.21	2.65	66	81.08	3.64
33	62	62.24	2.41	61	82.00	4.16
34	70	64.27	2.63	68	84.57	3.63
35	55	65.45	2.20	54	85.70	3.65
36	61	67.67	2.96	61	87.61	3.37
37	43	69.19	3.42	43	89.26	3.68
38	35	70.57	2.62	35	89.89	4.10
39	34	71.74	2.56	33	92.06	3.65
40	10	72.50	3.17	10	91.00	3.89
41	6	74.50	3.27	6	91.17	2.86

On the basis of this criterion a quadratic mean function is selected for the regression of FL on x and a cubic mean function for the regression of BDP on x. The selected models are illustrated in Figure 6.1.

6.2. LINEAR APPROXIMATING MODELS USING ORTHOGONAL FUNCTIONS: ANALYSIS OF DISCREPANCY

In applications based on the results of Section 6.1 the criterion has to be calculated for all envisaged approximating mean functions. Since this can

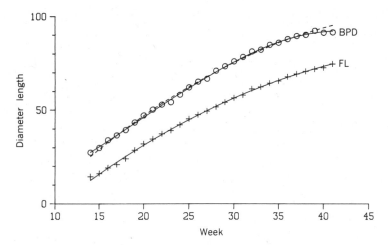

Figure 6.1. Observed means and approximating mean functions for femur length (FL, quadratic) and biparietal diameter [BPD, cubic (—) and quadratic (---)].

be tedious, the use of approximating mean functions which are linear combinations of *orthogonal functions* is recommended.

We assume now that J observations are available for each value of x and that the operating model is

$$y_{ij} = \mu_i + e_{ij}, \qquad i = 1, 2, \ldots, I,$$

$$j = 1, 2, \ldots, J,$$

$$Ee_{ij} = 0 \quad \text{and} \quad E\bar{e}_{i.}\bar{e}_{j.} = \sigma_{ij}.$$

An important special case occurs if the e_{ij} are uncorrelated. In such a case $\sigma_{ij} = \delta_{ij}\sigma_i^2/J$, where σ_i^2 is the variance of e_{ij}.

The Gauss discrepancy is the square of a *distance* in R_I and there are systems of functions which are orthogonal under the corresponding inner product. A typical example are polynomials of degree 0, 1, 2, ..., $I - 1$, denoted by $P_1, P_2, \ldots, P_I$, which are constructed so that

$$\sum_{r=1}^{I} P_i(x_r)P_j(x_r) = \delta_{ij},$$

where the x_r are (usually equidistant) values of a variable x. For every subset of $P_1, P_2, \ldots, P_I$ an approximating model can be constructed with mean function

$$h(x, \theta^S) = \sum_{i \in S} \theta_i P_i(x).$$

Here S denotes a subset of $\{1, 2, \ldots, I\}$ with $1 \leqslant p \leqslant I$ elements and θ^S has elements θ_i for $i \in S$ and zero for $i \notin S$. For $p = I$ we shall use θ for θ^S.

We show how the most appropriate approximating family (i.e., the most appropriate set S) can be found by determining the contribution to the expected discrepancy of the individual parameters *separately*. To find the set S it is not necessary to calculate the criterion for each of the $2^I - 1$ different sets; it is sufficient to calculate the I contributions to the criterion of the parameters θ_i.

The orthogonal functions generate orthogonal base vectors

$$P_r = (P_r(x_1), \ldots, P_r(x_I))^\mathsf{T}, \qquad r = 1, 2, \ldots, I,$$

which define a new coordinate system in R_I. It is convenient to work in this new system. If ζ denotes the coordinates of a point in the new system and z the coordinates of the point in the old system, one has

$$z_i = \sum_r \zeta_r P_r(x_i) \quad \text{and} \quad \zeta_i = \sum_r z_r P_i(x_r).$$

The Discrepancy Due to Approximation

Each mean vector

$$h(\theta^S) = (h(x_1, \theta^S), \ldots, h(x_I, \theta^S))^\mathsf{T}$$

defines a point in R_I and since

$$h(x_i, \theta^S) = \sum_{r \in S} \theta_r P_r(x_i)$$

this point has the coordinates θ^S in the new system. It follows immediately that this point is in the p-dimensional subspace spanned by the p vectors $\{P_i : i \in S\}$.

We denote the coordinates of the operating mean in the new system by $\theta_0 = (\theta_{01}, \theta_{02}, \ldots, \theta_{0I})^\mathsf{T}$:

$$\mu_i = h(x_i, \theta_0) = \sum_r \theta_{0r} P_r(x_i),$$
$$\theta_{0i} = \sum_r \mu_r P_i(x_r).$$

Since the discrepancy used here is the squared Euclidean distance, the "best" θ^S, denoted by θ_0^S, is the projection of θ_0 into the subspace spanned by $\{P_i : i \in S\}$, that is, the elements of θ_0^S are θ_{0i} for $i \in S$ and zero for $i \notin S$. The squared distance between θ_0 and θ_0^S is then the discrepancy due to approximation:

$$\Delta(\theta_0^S) = \sum_{i \notin S} \theta_{0i}^2.$$

If an element r is *added* to S the discrepancy due to approximation *decreases* by θ_{0r}^2. The contribution of each parameter to the discrepancy due to approximation can thus be individually assessed. One obtains the following analysis of the discrepancy of approximation:

Discrepancy due to approximation if no parameter is used		$\sum_i \theta_{0i}^2$
Contribution to discrepancy due to approximation of	θ_1	$-\theta_{01}^2$
	θ_2	$-\theta_{02}^2$
	$\vdots$	$\vdots$
	θ_I	$-\theta_{0I}^2$
Discrepancy due to approximation if all I parameters are used		0

The Empirical Discrepancy and the Discrepancy due to estimation

The discrepancy used here is the squared distance between the operating and approximating mean points in R_I:

$$\sum_i (\mu_i - h(x_i, \theta^S))^2 = \sum_{i \in S} (\theta_{0i} - \theta_i)^2 + \sum_{i \notin S} \theta_{0i}^2.$$

The vector $\bar{y} = (\bar{y}_{1.}, \ldots, \bar{y}_{I.})^\mathsf{T}$ estimates μ. An empirical discrepancy is the squared distance between $\bar{y}$ and $h(\theta^S)$. If the coordinates of $\bar{y}$ in the new system are denoted by $\hat{\theta}$, with elements

$$\hat{\theta}_i = \sum_r \bar{y}_r . P_i(x_r),$$

the empirical discrepancy can be expressed as

$$\sum_{i \in S} (\hat{\theta}_i - \theta_i)^2 + \sum_{i \notin S} \hat{\theta}_i^2.$$

It can immediately be seen that this is minimized by $\theta_i = \hat{\theta}_i$ for $i \in S$. The notation was justified: the minimum discrepancy estimator $\hat{\theta}^S$ has elements $\hat{\theta}_i$ for $i \in S$ and zero for $i \notin S$. In other words, $\hat{\theta}^S$ is the projection of $\hat{\theta}$ into the space spanned by $\{P_i : i \in S\}$.

The discrepancy due to estimation is the squared distance between θ_0^S and $\hat{\theta}_0^S$ and is given by

$$\sum_{i \in S} (\hat{\theta}_i - \theta_{0i})^2.$$

The Expected Discrepancy

The overall discrepancy is the squared distance between θ_0 and $\hat{\theta}^S$:

$$\Delta(\hat{\theta}^S) = \sum_{i \in S} (\hat{\theta}_i - \theta_{0i})^2 + \sum_{i \notin S} \theta_{0i}^2$$

and is the sum of the discrepancies due to estimation and due to approximation. The expected discrepancy is

$$E\Delta(\hat{\theta}^S) = \sum_{i \in S} \text{Var}(\hat{\theta}_i) + \sum_{i \notin S} \theta_{0i}^2.$$

By adding an element r to S the expected discrepancy changes by

$$E(\theta_{0r} - \hat{\theta}_r)^2 - \theta_{0r}^2 = \text{Var}\,\hat{\theta}_r - \theta_{0r}^2$$

$$= \sum_{ij} \sigma_{ij} P_r(x_i) P_r(x_j) - \theta_{0r}^2.$$

This contribution can be positive or negative. The best set S contains all elements r corresponding to parameters which lead to *negative* contributions.
One has the following analysis of expected discrepancy:

Expected discrepancy if no parameter is used		$\sum_i \theta_{0i}^2$
Contribution to expected discrepancy of	θ_1	$\text{Var}\,\hat{\theta}_1 - \theta_{01}^2$
	$\vdots$	$\vdots$
	θ_I	$\text{Var}\,\hat{\theta}_I - \theta_{0I}^2$
Expected discrepancy if all I parameters are used		$\sum_i \text{Var}\,\hat{\theta}_i$

The Criterion

The contribution to the expected discrepancy of a parameter θ_r, that is $\text{Var}\,\hat{\theta}_r - \theta_{0r}^2$, is estimated without bias by

$$2\,\widehat{\text{Var}}\,\hat{\theta}_r - \hat{\theta}_r^2,$$

where

$$\widehat{\text{Var}}\ \hat{\theta}_r = \sum_{ij} \hat{\sigma}_{ij} P_r(x_i) P_r(x_j),$$

and $\hat{\sigma}_{ij}$ is an unbiased estimator of σ_{ij}. The analysis of the criterion becomes:

Criterion if no parameter is used	$\sum_i \hat{\theta}_i^2 - \sum_i \widehat{\text{Var}}\ \hat{\theta}_i$
Contribution to criterion of $\qquad\theta_1$	$2\ \widehat{\text{Var}}\ \hat{\theta}_1 - \hat{\theta}_1^2$
$\qquad\qquad\qquad\vdots$	
$\qquad\qquad\qquad\theta_I$	$2\ \widehat{\text{Var}}\ \hat{\theta}_I - \hat{\theta}_I^2$
Criterion if all I parameters are used	$\sum_i \widehat{\text{Var}}\ \hat{\theta}_i$

Only those parameters whose estimated contributions are *negative* should be fitted.

This method is summarized by the following simple rule: Use a parameter θ_r only if

$$F = \frac{\hat{\theta}_r^2}{\widehat{\text{Var}}\ (\hat{\theta}_r)} > 2$$

or, to put it another way, only if

$$\hat{\theta}_r > \sqrt{2} \times (\text{standard error of } \hat{\theta}_r).$$

The results so far are general. When the e_{ij} are uncorrelated

$$\sigma_{ij} = \frac{\delta_{ij}\sigma_i^2}{J},$$

$$\text{Var}\ \hat{\theta}_r = \sum_i \left(\frac{\sigma_i^2}{J}\right) P_r^2(x_i),$$

$$\widehat{\text{Var}}\ \hat{\theta}_r = \sum_i \left(\frac{\hat{\sigma}_i^2}{J}\right) P_r^2(x_i),$$

where $\hat{\sigma}_i^2$ is estimated by the sample variance of the observations y_{ij},

$j = 1, 2, \ldots, J$:

$$\hat{\sigma}_i^2 = \frac{\sum_j (y_{ij} - \bar{y}_{i.})^2}{J - 1}.$$

If one is also prepared to assume that $\sigma_i^2 = \sigma^2$, then Var $\hat{\theta}_r = \sigma^2/J$ and the decision rule reduces to: use the parameter θ_r only if

$$F = \frac{J\hat{\theta}_r^2}{\hat{\sigma}^2} = \frac{J\hat{\theta}_r^2}{\text{MSE}} > 2.$$

Here MSE is the residual mean square in the analysis of variance:

$$\text{MSE} = \frac{\sum_{ij} (y_{ij} - \bar{y}_{i.})^2}{I(J - 1).}$$

Furthermore, if one is willing to assume normality in the operating model one may test the hypothesis that the contribution of a certain parameter to the expected discrepancy is positive and use the parameter only if this hypothesis has to be rejected. In this case Var $\hat{\theta}_r = \sum_i P_r^2(x_i)\sigma^2/J = \sigma^2/J$ for all r and the obvious test statistic to test the hypothesis that $\sigma^2/J - \theta_{0r}^2 > 0$, that is, $J\theta_{0r}^2/\sigma^2 < 1$, is

$$F = \frac{J\hat{\theta}_r^2}{\text{MSE}}.$$

Under the hypothesis that $J\theta_{0r}^2/\sigma^2 = 1$, the statistic F has a noncentral F distribution with $n_1 = 1$ and $n_2 = I(J - 1)$ degrees of freedom and noncentrality parameter $\lambda = 1$. The mentioned hypothesis has to be rejected if F is larger than the corresponding critical value, given in Tables A.4.5 to A.4.8 in the Appendix.

If a significance level of 0.5 is used, this method corresponds roughly to the $F > 2$ decision procedure which was outlined above. For a *small* significance level (0.10; 0.05; 0.01) the selection method is *stricter*: one is asking for *strong* evidence that the parameter in question does not increase the expected discrepancy. This results in models with fewer parameters than in the decision procedure that was described initially.

In analysis of variance $J\hat{\theta}_r^2$ is known as the sum of squares for θ_r and

$$F = \frac{J\hat{\theta}_r^2}{\text{MSE}} = \frac{\text{SS}\theta_r}{\text{MSE}}$$

is the statistic used to test the hypothesis that $\theta_r = 0$. Here we test a different hypothesis with the same statistic and we therefore require other critical values.

EXAMPLE 6.2.1. Table A.3.1 in the Appendix gives the monthly gross evaporation at Matatiele for the seasons 1937/38 to 1966/67. (South African Department of Water Affairs, 1967.) The season begins in October (month 1) and ends in September (month 12). We represent these data using the operating model

$$y_{ij} = \mu_i + e_{ij}, \qquad i = 1, 2, \ldots, 12,$$
$$j = 1, 2, \ldots, 30,$$

where y_{ij} is the gross evaporation (mm) in month i of year j (where $j = 1$ represents 1937/38), and μ_i is the mean gross evaporation in month i. We assume that the e_{ij} are independently distributed with $Ee_{ij} = 0$ and Var $e_{ij} = \sigma_i^2$.

One can of course estimate the mean monthly gross evaporation μ_i by $\bar{y}_{i.}$. However, with this type of data one can often improve the estimates, particularly for short data records, by making use of the knowledge that the μ_i follow an approximately sinusoidal pattern. It is well known that in such situations truncated Fourier series often lead to good approximating models. We begin by representing the μ_i in terms of their Fourier series

$$\mu_i = \sum_{r=1}^{12} \theta_r P_r(i),$$

where $\theta_1, \theta_2, \ldots, \theta_{12}$ are the Fourier coefficients and

$$P_1(i) = (\tfrac{1}{12})^{1/2},$$
$$P_{2r}(i) = (\tfrac{2}{12})^{1/2} \cos \omega_r i,$$
$$P_{2r+1}(i) = (\tfrac{2}{12})^{1/2} \sin \omega_r i, \qquad r = 1, 2, \ldots, 5,$$
$$P_{12}(i) = (\tfrac{1}{12})^{1/2} (-1)^i,$$
$$\omega_r = \frac{2\pi r}{12}.$$

It is convenient here to use a different notation for the parameters $\theta_1, \theta_2, \ldots, \theta_{12}$, namely, $\alpha_0 = \theta_1$, $\alpha_r = \theta_{2r}$, and $\beta_r = \theta_{2r+1}$, for $r = 1, 2, \ldots, 5$, and $\alpha_6 = \theta_{12}$.

The point of reparameterizing the operating model in this fashion is that one can obtain quite accurate approximating models by truncating the Fourier series representation of μ_i. In this way the number of parameters which have to be estimated, and hence the discrepancy due to estimation, can be reduced. Least-squares estimators of the parameters in the operating model are given by

$$\hat{\alpha}_0 = (\tfrac{1}{12})^{1/2} \sum_i \bar{y}_{i.},$$

$$\hat{\alpha}_r = (\tfrac{2}{12})^{1/2} \sum_i \bar{y}_{i.} \cos \omega_r i,$$

$$\hat{\beta}_r = (\tfrac{2}{12})^{1/2} \sum_i \bar{y}_{i.} \sin \omega_r i, \qquad r = 1, 2, \ldots, 5$$

$$\hat{\alpha}_6 = (\tfrac{1}{12})^{1/2} \sum_i (-1)^i \bar{y}_{i.}.$$

For the purposes of selection the pairs of parameters $(\alpha_r, \beta_r), r = 1, 2, \ldots, 5$, are considered jointly because they are coefficients belonging to the same frequency. By pairing the parameters in this way we ensure that the selections do not depend on the starting month, that, is, we obtain the same selection if $i = 1$ corresponds to, say, January instead of October.

It follows from the results given in this section that the pair (α_r, β_r) should be retained in the model if

$$2 \widehat{\text{Var}} \, \hat{\alpha}_r - \hat{\alpha}_r^2 + 2 \widehat{\text{Var}} \, \hat{\beta}_r - \hat{\beta}_r^2$$

$$= \frac{2 \sum_i P_{2r}^2(i) \hat{\sigma}_i^2}{J} - \hat{\alpha}_r^2 + \frac{2 \sum_i P_{2r+1}^2(i) \hat{\sigma}_i^2}{J} - \hat{\beta}_r^2 < 0,$$

that is, if

$$\frac{J(\hat{\alpha}_r^2 + \hat{\beta}_r^2)}{\sum_i [P_{2r}^2(i) + P_{2r+1}^2(i)] \hat{\sigma}_i^2} > 2,$$

where J is the number of observations in each month; here $J = 30$.

Since $P_{2r}^2(i) + P_{2r+1}^2(i) = \tfrac{2}{12}$ for all i, it follows that the denominator is simply twice the residual mean square

$$\text{MSE} = \frac{1}{12} \sum_i \hat{\sigma}_i^2 = \frac{1}{12(J - 1)} \sum_{ij} (y_{ij} - \bar{y}_{i.})^2.$$

The pair (α_r, β_r) should thus be retained in the approximating model if

$$F = \frac{J(\hat{\alpha}_r^2 + \hat{\beta}_r^2)/2}{\text{MSE}} = \frac{\text{MS} \, \alpha_r, \beta_r}{\text{MSE}} > 2.$$

The parameters α_0 and α_6 have to be dealt with separately. There is usually no question of omitting α_0 from the approximating model. On the other hand, α_6 should be retained if

$$F = \frac{J\hat{\alpha}_6^2}{\text{MSE}} = \frac{\text{MS } \alpha_6}{\text{MSE}} > 2.$$

The estimates of the parameters in the operating model are given by

$$\hat{\alpha}_0 = 454.384,$$

$$\hat{\alpha}_1 = 3.194, \quad \hat{\beta}_1 = 135.539,$$

$$\hat{\alpha}_2 = 10.962, \quad \hat{\beta}_2 = 1.379, \; \cdot$$

$$\hat{\alpha}_3 = 4.410, \quad \hat{\beta}_3 = -7.824,$$

$$\hat{\alpha}_4 = 4.648, \quad \hat{\beta}_4 = 0.932,$$

$$\hat{\alpha}_5 = -9.440, \quad \hat{\beta}_5 = -1.333,$$

$$\hat{\alpha}_6 = 3.608,$$

and the basic analysis of variance is

Variation	d.f.	SS	MS
Months	12	6755255	
Residuals	348	131451	377.73
Total	360	6886706	

The analysis of the criterion is

Parameter	d.f.	SS	F
α_0	1	6193954	16397.8
α_1, β_1	2	551428	729.9
α_2, β_2	2	3662	4.9
α_3, β_3	2	2420	3.2
α_4, β_4	2	674	0.9
α_5, β_5	2	2727	3.6
α_6	1	391	1.0

According to this criterion only (α_4, β_4) and α_6 should be omitted from the approximating model.

The selected model is disappointingly complex. In particular, the inclusion of the high-frequency component corresponding to (α_5, β_5) is, at first glance, unexpected. However, since the data consist of monthly totals and the number of days in each month varies, it is probable that this high-frequency oscillation is, in fact, present in the operating model. To correct for the effect of different numbers of days in each month one can divide each observation by the number of days in the corresponding month and select a model for the mean gross evaporation *per day* in each month. If this is done one gets the estimators

$$\hat{\alpha}_0 = 14.958,$$

$$\hat{\alpha}_1 = -0.021, \qquad \hat{\beta}_1 = 4.491,$$

$$\hat{\alpha}_2 = 0.474, \qquad \hat{\beta}_2 = -0.033,$$

$$\hat{\alpha}_3 = 0.126, \qquad \hat{\beta}_3 = -0.055,$$

$$\hat{\alpha}_4 = 0.074, \qquad \hat{\beta}_4 = -0.181,$$

$$\hat{\alpha}_5 = 0.017, \qquad \hat{\beta}_5 = -0.057,$$

$$\hat{\alpha}_6 = 0.010.$$

The analysis of the criterion, with MSE = 0.404, is given by

Parameter	d.f.	SS	F
α_0	1	6712.467	167592.8
α_1, β_1	2	605.179	748.0
α_2, β_2	2	6.776	8.4
α_3, β_3	2	0.569	0.7
α_4, β_4	2	1.142	1.4
α_5, β_5	2	0.105	0.1
α_6	1	0.003	0.01

The selected model has no high-frequency component and contains the five parameters α_0, α_1, β_1, α_2, and β_2. The observed and fitted means are given below and are illustrated in Figure 6.2.

	Observed	Fitted
Oct	5.19	5.31
Nov	5.82	5.79
Dec	5.98	5.96
Jan	5.82	5.83
Feb	5.37	5.35
Mar	4.49	4.52
Apr	3.45	3.49
May	2.71	2.63
Jun	2.32	2.29
Jul	2.50	2.64
Aug	3.58	3.50
Sep	4.59	4.50

In some applications there is sufficient variation in the data to mask the type of effect (due to different numbers of days in each month) which was encountered in the above example. Table A.3.2 (see Appendix) gives the monthly precipitation for Matatiele for the same period as in the last example.

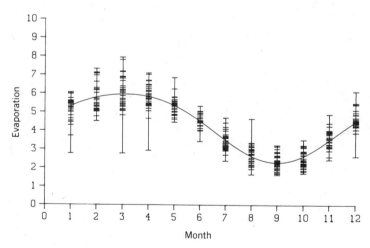

Figure 6.2. Monthly average gross evaporation per day at Matatiele, October (month 1) 1937 to September 1967, and approximating mean function.

The analysis of discrepancy, with MSE $= 1293.6$, is given by

Parameter	d.f.	SS	F
α_0	1	1329332.0	1027.7
α_1, β_1	2	64734.6	250.2
α_2, β_2	2	25087.9	9.7
α_3, β_3	2	532.1	0.2
α_4, β_4	2	121.5	0.1
α_5, β_5	2	807.8	0.3
α_6	1	504.1	0.4

Even without corrections for the number of days the criterion selects a five-parameter approximating model with

$$\hat{\alpha}_0 = 210.5,$$
$$\hat{\alpha}_1 = -71.2, \qquad \hat{\beta}_1 = 128.5,$$
$$\hat{\alpha}_2 = -5.8, \qquad \hat{\beta}_2 = -28.3.$$

The approximating mean function and the observations are illustrated in Figure 6.3.

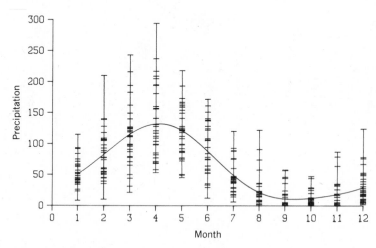

Figure 6.3. Monthly precipitation at Matatiele, October (month 1) 1937 to September 1967, and approximating mean function.

EXAMPLE 6.2.2. Problems arise when one is selecting from a large number of approximating models. This occurs, for example, if the operating model contains a large number of parameters, and approximating models are constructed by eliminating some of them. For each parameter one makes a decision as to whether the parameter should be included in the model. Now as the number of decisions increases so the expected number of incorrect decisions also increases. This problem is analogous to that encountered in testing of multiple hypotheses.

We consider the problem of estimating the mean daily rainfall at Roma, Lesotho. Available are daily observations for the years 1966 to 1979, that is, $J = 14$ years, as given in Table A.3.3 in the Appendix. The operating model is taken to be of the form

$$y_{ij} = \mu_i + e_{ij}, \qquad i = 1, 2, \ldots, 365,$$
$$j = 1, 2, \ldots, 14,$$

where y_{ij} is the rainfall on day i of year j and the e_{ij} will be taken to be independently distributed with $Ee_{ij} = 0$ and Var $e_{ij} = \sigma_i^2$.

As in Example 6.2.1 the approximating models will be based on the Fourier representation of the 365 means. The estimators and criteria are obtained by making the obvious modifications to those given in Example 6.2.1.

The initial analysis of variance is

Variation	d.f.	SS	MS
Days	365	49821.03	
Residuals	4745	192619.97	40.59
Total	5110	242441.00	

and the analysis of criterion is given in Table 6.2. This is an example where one would not apply the decision rule ($F > 2$) strictly. Here one would normally require strong evidence for the inclusion of each parameter, but since the data are not clearly normally distributed the test outlined above is not applicable.

Of course, one can arbitrarily raise the critical value of F from 2 to some higher value. Alternatively, one can restrict the class of approximating models considered in some way. In this example it is reasonable to assume that the μ_i series does not have any high-frequency component. One policy here would be to begin with the low-frequency components and select all

Table 6.2. Analysis of Criterion for Rainfall in Roma

J	SS	F	J	SS	F	J	SS	F
0	25633.66	631.5	61	5.27	0.1	122	136.89	1.7
1	10427.33	128.4	62	90.54	1.1	123	37.87	0.5
2	457.59	5.6	63	102.08	1.3	124	65.33	0.8
3	371.92	4.6	64	150.92	1.9	125	184.97	2.3
4	235.01	2.9	65	64.88	0.8	126	90.54	1.1
5	58.33	0.7	66	17.50	0.2	127	140.54	1.7
6	156.55	1.9	67	53.57	0.7	128	22.65	0.3
7	165.32	2.0	68	84.64	1.0	129	4.15	0.1
8	95.57	1.2	69	22.56	0.3	130	42.16	0.5
9	240.33	3.0	70	180.16	2.2	131	43.36	0.5
10	164.50	2.0	71	216.97	2.7	132	48.88	0.6
11	8.62	0.1	72	45.43	0.6	133	241.38	3.0
12	166.38	2.0	73	105.24	1.3	134	250.89	3.1
13	51.61	0.6	74	39.96	0.5	135	9.59	0.1
14	13.25	0.2	75	101.13	1.2	136	56.32	0.7
15	18.91	0.2	76	3.77	0.0	137	77.16	1.0
16	29.68	0.4	77	24.33	0.3	138	58.12	0.7
17	29.85	0.4	78	26.34	0.3	139	44.82	0.6
18	10.59	0.1	79	1.21	0.0	140	92.95	1.1
19	26.40	0.3	80	32.54	0.4	141	18.14	0.2
20	36.23	0.4	81	2.36	0.0	142	25.35	0.3
21	29.10	0.4	82	41.69	0.5	143	8.65	0.1
22	109.67	1.4	83	44.09	0.5	144	18.31	0.2
23	72.38	0.9	84	68.22	0.8	145	8.23	0.1
24	28.92	0.4	85	68.58	0.8	146	87.00	1.1
25	34.64	0.4	86	143.43	1.8	147	174.67	2.2
26	22.62	0.3	87	2.34	0.0	148	102.27	1.3
27	84.75	1.0	88	6.82	0.1	149	114.66	1.4
28	31.53	0.4	89	61.57	0.8	150	87.58	1.1
29	21.25	0.3	90	127.06	1.6	151	73.46	0.9
30	82.54	1.0	91	19.57	0.2	152	78.88	1.0
31	10.82	0.1	92	31.14	0.4	153	2.62	0.0
32	188.32	2.3	93	145.51	1.8	154	74.86	0.9
33	172.32	2.1	94	378.95	4.7	155	87.78	1.1
34	98.63	1.2	95	204.52	2.5	156	85.30	1.1
35	75.69	0.9	96	48.01	0.6	157	97.21	1.2
36	72.64	0.9	97	60.51	0.7	158	163.59	2.0
37	203.68	2.5	98	181.56	2.2	159	46.81	0.6
38	224.83	2.8	99	40.84	0.5	160	53.50	0.7
39	386.16	4.8	100	65.83	0.8	161	24.57	0.3
40	69.40	0.9	101	88.21	1.1	162	1.89	0.0
41	62.60	0.8	102	41.35	0.5	163	56.81	0.7
42	226.79	2.8	103	75.51	0.9	164	37.97	0.5
43	49.88	0.6	104	8.62	0.1	165	29.63	0.4
44	28.99	0.4	105	21.74	0.3	166	17.10	0.2
45	107.85	1.3	106	27.58	0.3	167	9.09	0.1
46	56.44	0.7	107	73.01	0.9	168	21.25	0.3
47	65.81	0.8	108	6.90	0.1	169	86.32	1.1
48	154.15	1.9	109	95.84	1.2	170	46.80	0.6
49	152.07	1.9	110	44.66	0.6	171	13.40	0.2
50	33.19	0.4	111	169.27	2.1	172	23.06	0.3
51	179.04	2.2	112	46.48	0.6	173	84.75	1.0
52	30.30	0.4	113	35.17	0.4	174	82.67	1.0
53	17.13	0.2	114	29.13	0.4	175	22.92	0.3
54	33.50	0.4	115	8.91	0.1	176	35.05	0.4
55	164.66	2.0	116	21.74	0.3	177	70.58	0.9
56	58.36	0.7	117	22.97	0.3	178	7.88	0.1
57	10.16	0.1	118	0.68	0.0	179	5.33	0.1
58	17.89	0.2	119	4.25	0.1	180	20.25	0.2
59	7.66	0.1	120	27.86	0.3	181	27.19	0.3
60	31.11	0.4	121	150.26	1.9	182	22.28	0.3

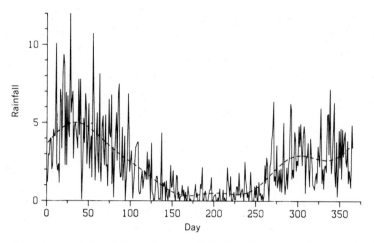

Figure 6.4. Average daily rainfall at Roma, 1966–1979, and approximating mean function.

components up to (but not including) the first component which is excluded by the decision rule. This would lead to the approximating model with parameters α_0, (α_r, β_r), $r = 1, 2, 3, 4$. The average daily rainfall and the corresponding approximating mean are illustrated in Figure 6.4.

6.3. NONLINEAR APPROXIMATING MODELS

In this section we will again consider the operating model

$$y_{ij} = \mu_i + e_{ij}, \qquad j = 1, 2, \ldots, J_i,$$

where the e_{ij} have zero mean, are independently and for each i identically distributed, and Var $e_{ij} = \sigma_i^2$.

If the approximating mean function $h(x, \theta)$ is not linear in its parameters, finite sample methods are not available and one has to use either Bootstrap or asymptotic methods. No new aspects arise with Bootstrap methods. We discuss here asymptotic methods based on the Gauss discrepancy. The discrepancy is

$$\Delta(\theta) = \sum_{ij} (\mu_i - h(x_i, \theta))^2$$

and an empirical discrepancy is

$$\Delta_n(\theta) = \sum_{ij} (\bar{y}_{i.} - h(x_i, \theta))^2.$$

The minimum discrepancy estimator is the least squares estimator $\hat{\theta}$. Numerical methods are applied for the minimization. In Appendix A.2.5 it is shown that under some regularity conditions a criterion is

$$\Delta_n(\hat{\theta}) + \text{tr } \Omega_n^{-1}\Sigma_n,$$

where

$$\Omega_n = 2 \sum_i J_i[h''(x_i, \hat{\theta})(h(x_i, \hat{\theta}) - \bar{y}_{i.}) + h'(x_i, \hat{\theta})h'(x_i, \hat{\theta})^\top]$$

and

$$\Sigma_n = 4 \sum_i J_i \hat{\sigma}_i^2 h'(x_i, \hat{\theta})h'(x_i, \hat{\theta})^\top.$$

Here $\hat{\sigma}_i^2$ is again the sample variance of the J_i observations y_{ij}.

If the operating model belongs to the approximating family *and* $\sigma_i^2 = \sigma^2$ for all i then tr $\Omega^{-1}\Sigma = 2p\sigma^2$ and in this case a suitable criterion is

$$\sum_{ij} (\bar{y}_{i.} - h(x_i, \hat{\theta}))^2 + 2p\hat{\sigma}^2,$$

where

$$\hat{\sigma}^2 = \text{MSE} = \frac{\sum_{ij} (y_{ij} - \bar{y}_{i.})^2}{(J_. - I)}.$$

For the operating family the simpler criterion becomes $2I$ MSE.

EXAMPLE 6.3. To demonstrate the methods we generated artificial data. Following Zwanzig (1980) we used

$$y_{ij} = 1 + e^{2x_i}(1 + 3x_i) + e_{ij}, \qquad i = 1, \ldots, 21,$$

$$j = 1, 2, 3,$$

where $x_i = (i - 1)/10 - 1$ and the e_{ij} were generated as independent $N(0, 2)$-distributed variables (see Table 6.3).

Families which could be used are

$$M_1 : h(x, \theta) = \theta_1 e^{\theta_2 x} + \theta_3,$$

$$M_2 : h(x, \theta) = \theta_1 x^2 + \theta_2 x + \theta_3,$$

Table 6.3. Data Generated
According to Operating Model
of Example 6.3

i	y_{i1}	y_{i2}	y_{i3}
1	2.21	0.18	0.98
2	1.47	0.84	0.91
3	1.74	0.52	0.57
4	−1.65	3.06	0.48
5	1.61	2.47	−1.46
6	−0.52	1.59	1.84
7	−2.77	1.89	2.66
8	−1.12	1.62	0.97
9	−0.70	−0.40	0.66
10	0.55	−1.50	0.30
11	1.64	1.95	2.71
12	1.73	4.25	1.46
13	4.41	3.46	4.54
14	6.10	4.78	3.74
15	8.34	5.42	5.55
16	7.97	5.94	5.83
17	10.86	10.21	11.54
18	12.92	15.17	13.39
19	19.32	19.15	17.83
20	25.46	24.15	23.68
21	29.26	31.58	27.92

and

$$M_3: h(x, \theta) = e^{\theta_1 x}(\theta_2 + \theta_3 x) + \theta_4.$$

The derivatives $h'(x, \theta)$ and $h''(x, \theta)$ are

M_1		M_2	
$h'(x, \theta)$	$h''(x, \theta)$	$h'(x, \theta)$	$h''(x, \theta)$
$\begin{bmatrix} e^{\theta_2 x} \\ \theta_1 x e^{\theta_2 x} \\ 1 \end{bmatrix}$	$\begin{bmatrix} 0 & x e^{\theta_2 x} & 0 \\ x e^{\theta_2 x} & \theta_1 x^2 e^{\theta_2 x} & 0 \\ 0 & 0 & 0 \end{bmatrix}$	$\begin{bmatrix} x^2 \\ x \\ 1 \end{bmatrix}$	$\begin{bmatrix} 0 & 0 & 0 \\ 0 & 0 & 0 \\ 0 & 0 & 0 \end{bmatrix}$

$$M_3$$

$h'(x, \theta)$	$h''(x, \theta)$

$$\begin{bmatrix} x(\theta_2 + \theta_3 x)e^{\theta_1 x} \\ e^{\theta_1 x} \\ xe^{\theta_1 x} \\ 1 \end{bmatrix} \qquad \begin{bmatrix} x^2(\theta_2 + \theta_3 x)e^{\theta_1 x} & xe^{\theta_1 x} & x^2 e^{\theta_1 x} & 0 \\ xe^{\theta_1 x} & 0 & 0 & 0 \\ x^2 e^{\theta_1 x} & 0 & 0 & 0 \\ 0 & 0 & 0 & 0 \end{bmatrix}$$

In practice, one would not know the operating model and would assume the saturated model: $y_{ij} = \mu_i + e_{ij}$. There is no evidence of heteroskedasticity and one would assume $\sigma_i^2 = \sigma^2$.

For the simpler criterion one needs the residual mean square in the one-way analysis of variance of y_{ij}. Here one gets MSE = 1.9525.

For model family M_1 the estimators of the parameters were

$$\hat{\theta}_1 = 2.051, \qquad \hat{\theta}_2 = 2.702, \qquad \hat{\theta}_3 = 0.0795,$$

and the corresponding $\Delta_n(\hat{\theta}) = 40.91$. The trace of $\Omega_n^{-1}\Sigma_n$ was 11.34 and the criterion $40.91 + 11.34 = 52.25$. If the trace term is replaced by $2p\hat{\sigma}^2 = 6 \times 1.9525 = 11.72$ the criterion becomes 52.63.

For M_2 the estimators were

$$\hat{\theta}_1 = 13.89, \qquad \hat{\theta}_2 = 11.24, \qquad \hat{\theta}_3 = 1.119;$$

also $\Delta_n(\hat{\theta}) = 180.88$ and tr $\Omega_n^{-1}\Sigma_n = 11.71$, so the criterion is $180.88 + 11.71 = 192.59$.

For M_3 we obtained

$$\hat{\theta}_1 = 1.302, \qquad \hat{\theta}_2 = -3.552, \qquad \hat{\theta}_3 = 10.34, \qquad \hat{\theta}_4 = 5.052,$$

$\Delta_n(\hat{\theta}) = 24.27$, and tr $\Omega_n^{-1}\Sigma_n = 15.60$. The criterion has the value $24.27 + 15.60 = 39.87$. This model is the best of the three and also more appropriate than the saturated model which has a simple criterion of $42 \times 1.9525 = 82.00$.

It is of interest to compare tr $\Omega_n^{-1}\Sigma_n$ with the simpler estimator $2p\hat{\sigma}^2$:

Model	p	tr $\Omega_n^{-1}\Sigma_n$	$2p\sigma^2$
M_1	3	11.34	11.72
M_2	3	11.71	11.72
M_3	4	15.60	15.62

CHAPTER 7

Multiple Regression

Multiple regression is one area of statistics in which the search for model selection criteria has a long history and continues to receive a good deal of attention. A summary of research on the topic is given in Thompson (1978). (See also Hocking, 1972, 1976, and Draper and Smith, 1981.) In this area the problem of model selection usually manifests itself as a problem of *variable selection*. Typically, a *response variable* is represented as a linear combination of several "independent" or *predictor variables* and one wishes to determine which subset of the predictor variables should be included in the model to best meet some specified objective. The simplest case, in which only one predictor variable is present, was dealt with in Section 5.1.

The use of an excessive number of predictor variables in applications of regression analysis has been of concern to statisticians for several decades. An early example was reported by Hotelling (1940). Practicing statisticians have been aware for some time that, in most applications, prediction accuracy is not improved by simply increasing the number of predictors. More often the opposite effect is achieved. For an extreme example see Guttman (1954, p. 307). These problems were often seen in connection with the overestimation of the multiple correlation coefficient in the joint distribution of a few variables when these variables were selected from a large set of possible predictor variables.

Various methods to select variables were proposed. Whether explicitly stated or implicitly assumed, the underlying objective in all these methods is to minimize some expected discrepancy based on the error of prediction. The variable selection methods originally used were usually of the naive type: one tested whether certain regression coefficients are zero. The (implicit) corresponding discrepancy is the residual variance which can never increase if further variables are added.

To the best of our knowledge the first author to suggest a non-naive selection procedure was H. L. Jones. His (1946) pioneering paper seems to have been subsequently forgotten. [It appears in the list of references given by Hocking (1976) but is otherwise not mentioned in his review.] Jones' (1946) paper arose in connection with practical problems in time series analysis. He deals with the case of fixed predictor variables, admits misspecification, uses the discrepancy of Section 7.2, and points out its equivalence with the Gauss discrepancy (see the remark at the end of Section 7.2). He gives the $F > 2$ rule for the selection of variables (see Section 6.2 and Chapter 8). In principle, his paper contains both the C_p method of Mallows (1973) and (because of the indicated equivalence) the methods of Linhart and Zucchini (1982b).

The many subsequent papers on variable selection are covered by the reviews of Hocking (1976) and Thompson (1978). Discrepancies involving predictive density functions are discussed in Geisser and Eddy (1979). The situation in which several response variables are observed is considered in van der Merwe et al. (1983). A recent summary of these and other methods can be found in Sparks (1984); see also Sparks et al. (1985).

We will only consider two variable selection methods, the S_p and C_p methods.

7.1. * SELECTION OF RANDOM PREDICTOR VARIABLES: THE S_p METHOD

The S_p method is derived under the assumption that the response variable and the k predictor variables are jointly normally distributed. In other words, it is assumed that the operating model is some $(k + 1)$-dimensional normal distribution. We denote the corresponding distribution function by F.

Predictions are made by taking a regression sample of values y_i, $x_i = (x_{1i}, \ldots, x_{ki})^T$, $i = 1, \ldots, n$, and using it to estimate $E(y_{n+1}|x_{1\,n+1}, \ldots, x_{k\,n+1})$, the minimum mean squared error predictor of y_{n+1} for an independently observed *predictor set* $x_{n+1} = (x_{1\,n+1}, \ldots, x_{k\,n+1})^T$.

The families of approximating models use $(p + 1)$-dimensional normal distributions for the response variable and a subset of the predictor variables, that is, $(y_i, x_{j_1 i}, x_{j_2 i}, \ldots, x_{j_p i})$, $i = 1, 2, \ldots, n, n + 1, 1 \leqslant j_1 < j_2 \cdots < j_p \leqslant k$, are independently and identically distributed with some $(p + 1)$-dimensional normal distribution function G_θ. For $p = k$ one obtains the operating family. To simplify the notation we will assume that $j_1 = 1, j_2 = 2, \ldots, j_p = p$, unless otherwise indicated. This amounts to a renaming of the variables.

The conditional expectation, given $x_{1\,n+1}, \ldots, x_{p\,n+1}$,

$$E_F((y_{n+1} - E_{G_\theta}(y_{n+1}|x_{1\,n+1}, \ldots, x_{p\,n+1}))^2|x_{1\,n+1}, \ldots, x_{p\,n+1}),$$

is the mean squared error in predicting y_{n+1} from a given x_{n+1} if one acts as if G_θ were the true distribution function. To eliminate the dependence on $x_{1n+1}, \ldots, x_{pn+1}$ one takes the expectation over these random variables. The resulting discrepancy is

$$\Delta(G_\theta, F) = E_F(y_{n+1} - E_{G_\theta}(y_{n+1}|x_{1n+1}, \ldots, x_{pn+1}))^2.$$

It is sufficient, here, to take θ as the vector of regression coefficients. The elements of this $(p + 1)$-dimensional vector are functions of the means, variances, and covariances (e.g., see Cramér, 1946, p. 303).

The vector θ_0 minimizing $\Delta(G_\theta, F)$ is the vector of regression coefficients in the (marginal) joint distribution of the response variable and the p predictor variables under the operating model F. The minimum discrepancy estimator of θ_0 is the least-squares estimator $\hat\theta$ of the regression coefficients calculated from the regression sample using the p predictor variables only. The overall discrepancy is

$$\Delta(\hat\theta) = \Delta(G_{\hat\theta}, F) = E_F(y_{n+1} - E_{G_{\hat\theta}}(y_{n+1}|x_{1n+1}, \ldots, x_{pn+1}))^2.$$

This is a random variable since the expectation E_F, as before, is taken only over the variables y_{n+1} and x_{n+1}.

The expected overall discrepancy is then the expectation of $\Delta(G_{\hat\theta}, F)$ over the variables in the regression sample:

$$E\Delta(\hat\theta) = E_{F*}(y_{n+1} - E_{G_{\hat\theta}}(y_{n+1}|x_{1n+1}, \ldots, x_{pn+1}))^2,$$

where E_{F*} is the expectation over *all* variables y_i, x_i, $i = 1, 2, \ldots, n + 1$.

One can show (e.g., see Thompson, 1978, p. 131) that this expected discrepancy is equal to

$$\frac{\sigma_p^2(n - 2)(n + 1)}{n(n - p - 2)},$$

where σ_p^2 is the residual variance in the multiple regression of the response variable on the p predictor variables.

For the purpose of comparing models the factor $(n - 2)(n + 1)/n$ is inessential and can be omitted because it does not depend on the approximating family. The well-known S_p criterion is simply an estimator of the essential part of the expected discrepancy:

$$S_p = \frac{\text{MSE}_p}{n - p - 2}, \qquad p = 1, 2, \ldots, k,$$

where

$$\text{MSE}_p = \frac{\text{SSE}_p}{n-p-1} = \frac{1}{n-p-1} \sum_{i=1}^{n} (y_i - \hat{\theta}_0 - \hat{\theta}_1 x_{1i} - \cdots - \hat{\theta}_p x_{pi})^2$$

is the residual mean square in the regression analysis of the y_i on $x_{1i}, x_{2i}, \ldots,$ x_{pi}. By a simple modification to the above argument one can show that this criterion is also applicable for the case $p = 0$.

If two sets K_1 and K_2 (of size q and p, respectively, $0 \leqslant q \leqslant p \leqslant k$) of predictor variables are under consideration then one would choose that set which leads to the smaller criterion. *For this it is immaterial whether one of these sets is contained in the other or not.*

The rule given in Section 5.1, relating to the question as to whether a single predictor variable should be used at all, is simply the special case with $p = 1$ and $q = 0$. Since $\text{SSE}_1/\text{SSE}_0 = 1 - r^2$, where r is the correlation coefficient, it follows that the inclusion of the predictor variable leads to a smaller criterion, that is, $S_1 < S_0$, if and only if $r^2 > 2/(n-1)$.

If the q variables in K_1 form a *subset* of the p variables in K_2, a test of the hypothesis that the expected overall discrepancy *increases* if one uses all p variables is available. One would use this *test* (and not the criterion) if *strong* evidence is required that the inclusion of further variables improves predictions.

The hypothesis that the expected discrepancy is larger if the p variables in K_2 are used than if only those in K_1 are used can be expressed as

$$\frac{\sigma_q^2}{\sigma_p^2} < \frac{n-q-2}{n-p-2},$$

where σ_q^2 and σ_p^2 are the residual variances in the regression of y on the variables in K_1 and K_2.

Let

$$F = \frac{\text{SS Regr}_p - \text{SS Regr}_q}{(p-q)\text{MSE}_p}.$$

One can show (Linhart, 1958) that

$$R = \left[\frac{(p-q)F}{(p-q)F + n - p - 1} \right]^{1/2}$$

has the same distribution as the estimated multiple correlation coefficient from a $(p - q + 1)$-dimensional normal distribution using $n - q$ observations, and where the population multiple correlation coefficient is given by

$$P = \left(\frac{\sigma_q^2 - \sigma_p^2}{\sigma_q^2}\right)^{1/2}.$$

The hypothesis

$$\frac{\sigma_q^2}{\sigma_p^2} < \frac{n - q - 2}{n - p - 2}$$

is then equivalent to the hypothesis that

$$P < \left(\frac{p - q}{n - q - 2}\right)^{1/2}$$

The statistic R can thus be used to test the hypothesis that the q variables in K_1 predict better than the p variables in K_2. The critical values are the percentage points in the distribution of R if $(\sigma_q^2 - \sigma_p^2)/\sigma_q^2 = (p - q)/(n - q - 2)$, that is, if

$$P = \left(\frac{n_1}{n_1 + n_2 - 1}\right)^{1/2},$$

where $n_1 = p - q$ and $n_2 = n - p - 1$.

There is no difficulty in transforming the critical values for R into those for F. This was done for the significance levels $\alpha = 0.5, 0.1. 0.05, 0.01$ and the results are given in Tables A.4.1 to A.4.4 in the Appendix. The critical values given are for $n_1 = p - q$ and $n_2 = n - p - 1$ degrees of freedom. This test for S_p was also described by Browne (1969).

7.2. SELECTION OF FIXED PREDICTOR VARIABLES: THE C_p METHOD

For the C_p method (Mallows, 1973, and other unpublished lectures and reports since 1964) it is assumed that the predictor variables are *fixed* and not random. The operating model F is a linear regression model: the $e_i = y_i - E_F y_i$, $i = 1, \ldots, n$, are independently and identically distributed and $E_F y_i = \theta_0 x_{0i} + \theta_1 x_{1i} + \theta_2 x_{2i} + \cdots + \theta_k x_{ki}$. The $n \times (k + 1)$ matrix with elements $x_{0i}, x_{1i}, \ldots, x_{ki}$, $i = 1, \ldots, n$, is of full rank $(k + 1) < n$, $x_{0i} = 1$ for all i, and the vector $(\theta_0, \theta_1, \theta_2, \ldots, \theta_k)$ is fixed but unknown. The approximating models G_θ are: the $y_i - E_{G_\theta} y_i$, $i = 1, \ldots, n$, are independently and identically distributed and $E_{G_\theta} y_i = \theta_0 x_{0i} + \theta_1 x_{j_1 i} + \theta_2 x_{j_2 i} + \cdots + \theta_p x_{j_p i}$.

As in the last section we set $j_1 = 1, j_2 = 2, \ldots, j_p = p$. We denote now the $(p + 1)$-dimensional vector $(x_{0i}, x_{1i}, x_{2i}, \ldots, x_{pi})^\mathsf{T}$ by x_i and the parameters $(\theta_0, \theta_1, \theta_2, \ldots, \theta_p)^\mathsf{T}$ by θ, $1 \leqslant p \leqslant k$.

The discrepancy used here is the *average* of the mean squared errors of prediction taken over the n predictor sets, $x_{n+i} = x_i$, $i = 1, \ldots, n$, appearing in the regression sample:

$$\Delta(G_\theta, F) = \sum_{i=1}^{n} \frac{\Delta(G_\theta, F; x_{n+i})}{n},$$

where

$$\Delta(G_\theta, F; x_{n+i}) = E_F(y_{n+i} - \theta^\mathsf{T} x_{n+i})^2 = E_F(y_i - \theta^\mathsf{T} x_i)^2.$$

It is easy to see that the vector $\theta_0 = (\theta_{00}, \theta_{01}, \ldots, \theta_{0p})^\mathsf{T}$ which minimizes $\Delta(G_\theta, F) = \sum_i E_F(y_i - \theta^\mathsf{T} x_i)^2/n$ is obtained by finding the hyperplane in a $(p + 1)$-dimensional space giving the best fit, in the least-squares sense, to the n points $(E_F y_i, x_{1i}, \ldots, x_{pi})$. An estimator of θ_0 is the least-squares estimator $\hat{\theta}$ of the regression coefficients in the regression analysis of y_i on $x_{1i}, x_{2i}, \ldots, x_{pi}$, $i = 1, 2, \ldots, n$.

The overall discrepancy is thus

$$\Delta(G_{\hat{\theta}}, F) = \sum_{i=1}^{n} \frac{E_F(y_{n+i} - \hat{\theta}^\mathsf{T} x_i)^2}{n},$$

where the expectations are only taken over the y_{n+i}, $i = 1, 2, \ldots, n$, and the regression sample (and therefore $\hat{\theta}$) is fixed. The expected overall discrepancy is then

$$\sum_{i=1}^{n} \frac{E_{F*}(y_{n+i} - \hat{\theta}^\mathsf{T} x_i)^2}{n},$$

where the expectations are now over y_{n+i} *and* the regression sample, that is, $y_1, y_2, \ldots, y_n$.

It can be shown (e.g., see Thompson, 1978, p. 133) that this expectation is equal to

$$\frac{\sigma^2(n + p + 1) + \sum_{i=1}^{n} (E_F y_i - \theta_0^\mathsf{T} x_i)^2}{n}.$$

To obtain a criterion this expected overall discrepancy has to be estimated. The residual mean square in the regression analysis *using all k predictor variables*, MSE_k, provides an estimator of σ^2, and the residual mean square in the analysis using p variables only, $MSE_p = SSE_p/(n - p - 1) = \sum_i (y_i - \hat{\theta}^T x_i)^2/(n - p - 1)$, has expectation $\sigma^2 + \sum_i (E_F y_i - \theta_0^T x_i)^2/(n - p - 1)$. The criterion, an unbiased estimator of the expected overall discrepancy, is thus

$$\frac{SSE_p + 2(p + 1)MSE_k}{n}.$$

The C_p criterion is actually slightly different. It is an estimator of (n/σ^2) times the expected overall discrepancy, minus n. Since both n and σ^2 remain unchanged for all subsets of variables the resulting procedure is equivalent.

If two sets of variables, K_1 and K_2, are under consideration one chooses the set with the smaller criterion. It is immaterial for this procedure whether one of these two sets is contained in the other or not. If, however, $K_1 \subset K_2$, and if one is prepared to assume *normally* distributed $e_i = y_i - E_F y_i$ in the operating model, a test of the hypothesis that the expected discrepancy for the variables in K_2 is larger than for the variables in K_1 is available.

We assume again that K_1 contains q predictor variables and K_2 contains p predictor variables, $1 \leq q < p \leq k$, and denote by the subscripts q, p, and k the sums of squares and mean squares belonging to the variables in K_1, K_2 and to all k variables, respectively. Since the criterion is an unbiased estimator of the expected discrepancy, the hypothesis is

$$\frac{E(SSE_p + 2(p + 1)MSE_k)}{n} > \frac{E(SSE_q + 2(q + 1)MSE_k)}{n},$$

which is equivalent to

$$\frac{E(SS\ Regr_p - SS\ Regr_q)}{(p - q)E(MSE_k)} < 2.$$

The test statistic to be used is

$$F = \frac{SS\ Regr_p - SS\ Regr_q}{(p - q)MSE_k}.$$

The numerator of F is distributed as $\sigma^2 \chi^2_{p-q; \lambda}$ and—independently of it—the denominator as $\sigma^2(p - q)\chi^2_{n-k-1}/(n - k - 1)$. Therefore, the statistic F has a noncentral F distribution with $p - q$ and $n - k - 1$ degrees of freedom and noncentrality parameter given by

$$\lambda = \frac{E(\text{SS Regr}_p - \text{SS Regr}_q)}{\sigma^2} - (p - q).$$

The hypothesis to be tested is that

$$\lambda < p - q.$$

Let $n_1 = p - q$ and $n_2 = n - k - 1$. The percentage points of $F_{n_1, n_2; \lambda}$ for $\lambda = n_1$ which are given in Tables A.4.5 to A.4.8 in the Appendix provide the critical values for this test.

This test for C_p was given by Browne (1969); it is also essentially covered by the work of Toro-Vizcarrondo and Wallace (1968).

Remark.   The criterion which was obtained here is based on a discrepancy involving errors of *prediction*. It is easy to see, however, that this discrepancy is equivalent to the Gauss discrepancy which characterizes the *internal* fit of a regression hyperplane. The Gauss discrepancy is

$$\sum_i (E_F y_i - E_{G\theta} y_i)^2,$$

whereas the discrepancy based on the error of prediction is

$$\frac{1}{n} \sum_{i=1}^{n} E_F(y_i - \theta^\top x_i)^2 = \frac{1}{n} \sum_{i=1}^{n} E_F(E_F y_i + e_i - \theta^\top x_i)^2$$

$$= \frac{1}{n} \sum_{i=1}^{n} (E_F y_i - E_{G\theta} y_i)^2 + \sigma^2.$$

Since n and σ^2 do not depend on the approximating model the two discrepancies are equivalent.   ∎

7.3.   COMPUTATIONS FOR TESTS FOR VARIABLE SELECTION

In this section we give a unified computing method for the test in the random and the fixed case (S_p and C_p). Here $q < p \leqslant k$, $q = 0, 1, 2, \ldots, p - 1$, and the q variables are assumed to form a *subset* of the p variables. One computes the following analysis of variance:

Source of Variation	d.f.	SS	MS
(1) Regression (q variables)	q	SS(1)	MS(1)
(2) Difference [(3) − (1)]	$p - q$	SS(2)	MS(2)
(3) Regression (p variables)	p	SS(3)	MS(3)
(4) Difference [(5) − (3)]	$k - p$	SS(4)	MS(4)
(5) Regression (k variables)	k	SS(5)	MS(5)
(6) Residual (k variables)	$n - k - 1$	SS(6)	MS(6)
(7) Total	$n - 1$		

If $q = 0$, row (1) must be set equal to zero and rows (2) and (3) are identical. If $p = k$, row (4) must be set equal to zero and rows (3) and (5) are identical. Computing formulas for the required sums of squares can be found in Draper and Smith (1981) and other texts on multiple regression analysis.

The test statistic for the hypothesis that the q variables predict better than the p variables is as follows:

Random Case

$F = $ MS(2)/MS(6 + 4), with $n_1 = p - q$, $n_2 = n - p - 1$, and where

$$MS(6 + 4) = \frac{SS(6) + SS(4)}{d.f.(6) + d.f.(4)}.$$

Fixed Case

$F = $ MS(2)/MS(6), with $n_1 = p - q$ and $n_2 = n - k - 1$.

If $q = 0$, $p = k = 1$, the test statistic is, in *both* cases,

$$F = r^2(n - 2)/(1 - r^2)$$

with $n_1 = 1$ and $n_2 = n - 2$ degrees of freedom, where r is the usual estimator of the correlation between y and x.

EXAMPLE 7.3. Knüppel *et al.* (1958) applied multiple regression analysis to a problem in steel production. At the end of a melting process, a certain blowing time (y) is necessary to achieve a given desirable phosphorus content in the end-product. The necessary blowing time varies with the conditions prevailing at the melting and these conditions can be characterized by a number of variables (such as Ph, C, Si, Mg content of the mixing iron, weight of the mixing iron, $CaCO_3$ addition, temperature, etc.).

Results for 40 meltings were available. The question is which of the variables should in future be measured and used for the determination of the necessary blowing time and, of course, *how* the blowing time should be determined from these variables. On the basis of previous experience the authors assumed a *linear* regression model. Using their own method to choose the best variables they eliminated 5 of the original 12 variables in a first step, and a further variable in a second step, thus retaining 6 variables.

We apply the significance test to decide whether the elimination of the sixth variable in the second step is justified. As the predictor variables can be taken as random, the S_p criterion is used.

The analysis of variance table, which is constructed from the data summarized in the original paper, is given below:

Source of Variation	d.f.	SS	MS
Regression (on variables 1, 6, 8, 9, 11, 12)	6	255454.6050	
Difference	1	17119.6535	17119.6535
Regression (on variables 1, 6, 8, 9, 10, 11, 12)	7	272574.2585	
Difference	5	20062.9769	
Regression (on variables 1, 2, ..., 12)	12	292637.2354	
Residual	27	351002.7396	13000.1015
Total	39	643639.9750	

The test on the basis of S_p gives

$$F = \frac{17119.6535}{(351002.7396 + 20062.9769)/32} = 1.48.$$

The percentage points for 1 and 32 d.f. are 1.18 (50%) and 5.74 (90%). According to this criterion one would therefore normally include the seventh variable (variable 10), unless one decided on a policy of including it only if there is *strong* evidence that such an inclusion would not lead to inferior predictions. Such evidence is not available for variable 10 and so under this policy one would *exclude* the variable.

7.4. SELECTION OF PRINCIPAL COMPONENTS

The analysis of discrepancy discussed in Section 6.2 is available if one uses the Gauss discrepancy, and the design matrix X is such that $X^T X$ is diagonal. This requirement is often *not* met, but if one is prepared to work with *principal components* rather than with the variables themselves, then by first orthogonalizing the model one can ensure that the requirement *is* met.

The following analysis is based on the discrepancy of Section 7.2, the average of the mean squared errors of prediction belonging to the n predictor sets in the regression sample. By the remark at the end of the mentioned section the results are also valid for the Gauss discrepancy.

Let the operating model be

$$y = X\theta_0 + e,$$

where $y = (y_1, y_2, \ldots, y_n)^T$, $e = (e_1, e_2, \ldots, e_n)^T$; the $n \times k$ matrix X with elements $x_{1i}, x_{2i}, \ldots, x_{ki}$ has rank k, $E_F e = 0$, and $E_F e e^T = \sigma^2 \mathscr{I}$.

Since $X^T X$ is symmetric there exists an orthogonal matrix P such that $P^T X^T X P = \text{Diag}\{\lambda_i\}$, where the λ_i are the (positive) latent roots of $X^T X$. The operating model can then also be written

$$y = X P \eta_0 + e,$$

$$\eta_0 = P^T \theta_0.$$

Families of approximating models are generated by setting some of the elements of η to zero. Let S be some subset of the set $\{1, 2, \ldots, k\}$. We denote by η^S the vector with elements η_i for $i \in S$ and zero for $i \notin S$. The approximating model associated with the index set S approximates the expectation of y by $X P \eta^S$.

The least-squares estimator of η_0^S, which we denote by $\hat{\eta}^S$, has as ith element $\hat{\eta}_i$ for $i \in S$, and zero for $i \notin S$. Here

$$\hat{\eta} = \text{Diag}\{1/\lambda_i\} P^T X^T y$$

and

$$\hat{\eta}_i = (1/\lambda_i) z_i^T y,$$

where z_i is the ith column of XP, that is, the vector of the ith principal component. It follows from standard regression theory that

$$\text{Var}\, \hat{\eta}_i = \frac{\sigma^2}{\lambda_i}.$$

The expected discrepancy is (compare Section 7.2)

$$E\Delta(\hat{\eta}^S) = \frac{\sigma^2(n + p) + (XP\eta_0 - XP\eta_0^S)^\top(XP\eta_0 - XP\eta_0^S)}{n}$$

$$= \frac{\sigma^2(n + p) + \sum_{i \notin S} \lambda_i \eta_{0i}^2}{n}.$$

This leads to the following analysis of expected discrepancy:

Expected discrepancy if no parameter is used	$\dfrac{1}{n}\left(n\sigma^2 + \displaystyle\sum_{i=1}^{k} \lambda_i \eta_{0i}^2\right)$

Contribution (to expected discrepancy) of		
	η_1	$(\sigma^2 - \lambda_1 \eta_{01}^2)/n$
	η_2	$(\sigma^2 - \lambda_2 \eta_{02}^2)/n$
	$\vdots$	
	η_k	$(\sigma^2 - \lambda_k \eta_{0k}^2)/n$

Expected discrepancy if all parameters are used	$(n + k)\sigma^2/n$

The contribution of η_i to the expected discrepancy is given by

$$\left(\frac{\lambda_i}{n}\right)(\text{Var } \hat{\eta}_i - \eta_{0i}^2)$$

and can be estimated by

$$\left(\frac{\lambda_i}{n}\right)(2\,\widehat{\text{Var }} \hat{\eta}_i - \hat{\eta}_i^2),$$

where

$$\widehat{\text{Var }} \hat{\eta}_i = \frac{\hat{\sigma}^2}{\lambda_i} = \frac{\text{MSE}_k}{\lambda_i}.$$

The analysis of the criterion is thus

Criterion if no parameter is used		$\dfrac{1}{n}\left(2n\hat{\sigma}^2 + \sum_{i=1}^{k}\lambda_i\hat{\eta}_i^2\right)$
Contribution (to criterion) of	η_1	$(2\hat{\sigma}^2 - \lambda_1\hat{\eta}_1^2)/n$
	η_2	$(2\hat{\sigma}^2 - \lambda_2\hat{\eta}_2^2)/n$
	$\vdots$	
	η_k	$(2\hat{\sigma}^2 - \lambda_k\hat{\eta}_k^2)/n$
Criterion if all k parameters are used		$2(n + k)\hat{\sigma}^2/n$

A parameter η_i should be used if its contribution to the criterion is *negative*, that is, if

$$F = \frac{\lambda_i\hat{\eta}_i^2}{\mathrm{MSE}_k} > 2.$$

If $\sigma^2 - \lambda_i\eta_{0i}^2 = 0$ and the errors are normally distributed, this statistic has a noncentral F distribution with 1 and $n - k$ degrees of freedom, and noncentrality parameter $\lambda = 1$. Thus one can test the hypothesis that the contribution of a particular parameter to the expected discrepancy is positive. Critical values for the test statistic, F, are given in Tables A.4.5 to A.4.8.

A disadvantage of the approximating families proposed in this section is that the selected models contain all k predictor variables. This is a consequence of selecting principal components rather than variables. The orthogonality property of principal components, however, does lead to a particularly convenient selection procedure because each parameter can be treated separately. The approximating families of Section 7.2 which are based on *variable* selection lead to a more cumbersome selection procedure.

EXAMPLE 7.4. Rypstra (1982) investigated weathering properties of treated wood. He studied the dependence of moisture content (y) on a number of meteorological variables: x_1 [rain (mm)], x_2 [evaporation (mm)], x_3 [solar radiation (h)], x_4 [wind velocity (km/D)], x_5 [maximum relative humidity (%)], x_6 [minimum relative humidity (%)], x_7 [average temperature (°C)], x_8 [maximum temp. (°C)], and x_9 [minimum temp. (°C)]. For this purpose he took 24 cubes of Pinus pinaster (100 mm × 50 mm × 50 mm), treated 12 of them with water-repellent solution, and left the remaining 12 untreated (control). All variables (y and $x_1, \ldots, x_9$) were measured in 52 consecutive weeks; the data are given in Table A.3.4 in the Appendix.

The purpose here is not prediction, but as we pointed out in the remark of Section 7.2, the discrepancy of that section (average mean squared error of prediction) is equivalent to the Gauss discrepancy. This measures in a reasonable way how well (in ten-dimensional space) a nine-dimensional approximating surface differs from the nine-dimensional operating mean surface $E(y|x_1, \ldots, x_9)$.

In practice, ten free parameters are usually too many for 52 observations; the expected discrepancy is too large because of the considerable discrepancy due to estimation. One can reduce the number of parameters in two ways: by omitting some of the x variables ("variable selection") or by omitting some of the principal components. The difference between these two procedures is that one is in effect selecting from two different classes of approximating families.

There are two advantages in selecting principal components rather than variables in this example. First, the selection procedure is computationally much easier. (Scanning each possible approximating model in the variable selection case is expensive.) Second, the two selected models—for the treated case and control case—will contain estimates of the coefficients for each of the predictor variables. The two selected models are easier to compare if they contain the same set of predictor variables.

Table 7.1 gives the analysis of variance for the two cases, control and treated. The dependent variable is the average moisture content of the 12 cubes. The discussion of the results is simpler if all independent variables are standardized by subtracting their mean and dividing by their standard deviation. This was done here.

The sums of squares for the individual principal components are $\lambda_i \hat{\eta}_i^2$; they are the numerator of F. The denominator, MSE_k, the estimated residual variance in the operating model, can also be found in the analysis of variance table. For completeness the latent roots λ_i and the estimated coefficients $\hat{\eta}_i$ are listed below:

i	λ_i	$\hat{\eta}_i$ (Control)	$\hat{\eta}_i$ (Treated)
1	224.721	3.260	0.8650
2	83.399	-3.623	-0.4883
3	62.885	-0.277	-0.1460
4	40.725	0.860	0.7627
5	22.963	2.292	0.9126
6	13.636	0.633	-0.1840

i	λ_i	$\hat{\eta}_i$ (Control)	$\hat{\eta}_i$ (Treated)
7	6.396	2.697	1.4198
8	4.273	-12.212	-2.7460
9	0.001	113.39	25.5557

Table 7.1. Analysis of Variance for Moisture Content of Cubes of Pinus Pinaster

	d.f.	SS	MS	F
Control:				
Principal component 1	1	2388.65		41.0
2	1	1094.68		18.8
8	1	637.29		10.9
5	1	120.62		2.1
7	1	46.49		<1
4	1	30.13		
9	1	18.90		
6	1	5.47		
3	1	4.84		
Regression	9	4347.07		
Residual	42	2449.55	58.323	
Total	51	6796.62		
Treated:				
Principal component 1	1	168.14		23.8
8	1	32.23		4.6
4	1	23.69		3.4
2	1	19.88		2.8
5	1	19.13		2.7
7	1	12.89		1.8
3	1	1.34		<1
9	1	0.96		
6	1	0.46		
Regression	9	278.72		
Residual	42	296.89	7.069	
Total	51	575.61		

The selected model has the components numbered 1, 2, 8, and 5 in the control case, and the components 1, 8, 4, 2, and 5 in the treated case. For these models the estimated regression coefficients of the (standardized) original nine variables can be obtained from the $\hat{\eta}_i$ and the matrix P of latent vectors of $X^T X$. Since $\hat{\theta} = P\hat{\eta}^S$, one obtains for controls:

$$\hat{E} y_1 = 15.84 + 2.22x_1 - 1.95x_2 + 0.34x_3 - 1.54x_4 - 0.32x_5$$
$$- 1.31x_6 - 2.15x_7 - 9.19x_8 + 8.72x_9,$$

and for the treated case:

$$\hat{E} y_2 = 10.71 - 0.13x_1 - 0.27x_2 + 0.28x_3 - 0.21x_4 + 0.19x_5$$
$$+ 0.05x_6 - 0.75x_7 - 2.39x_8 + 1.85x_9.$$

The most important fact which this analysis reveals is the drop in the average moisture content from 16 to 11 and the marked reduction of the residual variance from 58 to 7 which the treatment induced. So the influence of those factors which affect the moisture content but were *not* measured (and which contribute to error) is strongly reduced.

The measured climatic variables also lose their importance after treatment. Every regression coefficient in the selected model for the treated case is smaller in absolute value than the corresponding coefficient in the control case.

7.5. LINEAR OPERATING AND APPROXIMATING MODELS

Not all model selection problems in multiple regression are variable selection problems. This becomes abundantly clear in Chapters 8 and 9 which are on the special cases of analysis of variance and covariance. Example 7.5 to come also involves model selection but not variable selection.

The approximating models in Section 7.2 are obtained from the operating model

$$y = A\theta + e$$

by imposing restrictions of the type

$$Q\theta = 0,$$

where A is a $n \times k$ matrix of rank $k < n$, $Ee = 0$, $Eee^T = \Sigma$, and Q is a $b \times k$ matrix of rank $b < k$.

If one partitions $\theta^\mathsf{T} = (\theta_1^\mathsf{T}, \theta_2^\mathsf{T})$ such that θ_2 has b rows, and $A = (A_1, A_2)$ and $Q = (Q_1, Q_2)$ correspondingly, and arranges the elements of θ and Q such that Q_2 is nonsingular, the resulting approximating model has $p = k - b$ parameters and is given by

$$y = B\theta_1 + e,$$

where

$$B = A_1 - A_2 Q_2^{-1} Q_1.$$

The discrepancy considered in Section 7.2 was the average of the mean squared errors of prediction. For convenience we will work with the Gauss discrepancy. As was pointed out before, these two discrepancies lead to equivalent criteria.

The overall discrepancy for the approximating model is then

$$(A\theta - B\hat{\theta}_1)^\mathsf{T}(A\theta - B\hat{\theta}_1) = \theta^\mathsf{T} C\theta - 2\hat{\theta}_1^\mathsf{T} B^\mathsf{T} A\theta + \hat{\theta}_1^\mathsf{T} D\hat{\theta}_1,$$

where

$$C = A^\mathsf{T} A, \qquad D = B^\mathsf{T} B, \qquad \text{and} \qquad \hat{\theta}_1 = D^{-1} B^\mathsf{T} y.$$

The expected discrepancy is

$$\theta^\mathsf{T} C\theta - \theta_{10}^\mathsf{T} D\theta_{10} + \operatorname{tr} BD^{-1}B^\mathsf{T}\Sigma,$$

where

$$\theta_{10} = D^{-1} B^\mathsf{T} A\theta.$$

An estimator of the expected discrepancy, a criterion, is

$$\hat{\theta}^\mathsf{T} C\hat{\theta} - \hat{\theta}_1 D\hat{\theta}_1 - \operatorname{tr} AC^{-1}A^\mathsf{T}\hat{\Sigma} + 2\operatorname{tr} BD^{-1}B^\mathsf{T}\hat{\Sigma},$$

where

$$\hat{\theta} = C^{-1} A^\mathsf{T} y$$

and $\hat{\Sigma}$ is an estimator of Σ.

Now $\hat{\theta}^\mathsf{T} C\hat{\theta}$ is the sum of squares for regression under the operating model and $\hat{\theta}_1^\mathsf{T} D\hat{\theta}_1$ the sum of squares for regression under the approximating model.

These two quantities can be obtained by standard multiple regression calculations. The criterion for the restricted model can thus also be written

SS Regr(op. mod.) − SS Regr(appr. mod.)

$$- \mathrm{tr}\, AC^{-1}A^\mathsf{T}\hat{\Sigma} + 2\mathrm{tr}\, BD^{-1}B^\mathsf{T}\hat{\Sigma}.$$

The trace terms are very simple if $\Sigma = \sigma^2 \mathscr{I}$, where $\mathscr{I}$ is a $n \times n$ unit matrix. Then $\mathrm{tr}\, AC^{-1}A^\mathsf{T}\Sigma = \sigma^2\, \mathrm{tr}\, AC^{-1}A^\mathsf{T} = \sigma^2\, \mathrm{tr}\, A^\mathsf{T}AC^{-1} = \sigma^2 k$. Since in this case the residual mean square under the operating model is an unbiased estimator of σ^2, the criterion becomes

SS Regr(op. mod.) − SS Regr(appr. mod.) + $(2p - k)$MSE(op. mod.).

It is not necessary to assume that the errors in the operating model are uncorrelated and have equal variances as long as estimators for the trace terms in the criterion can be found. Examples of this type appear in the later chapters and in Example 7.5 to follow.

It is often convenient to take the *difference* between the criterion for the approximating family and the criterion for the operating family. The former family is selected if this difference is negative. This leads to the rule: select the approximating family if

$$\frac{\text{SS Regr(op. mod.)} - \text{SS Regr(appr. mod.)}}{\mathrm{tr}\, AC^{-1}A^\mathsf{T}\hat{\Sigma} - \mathrm{tr}\, BD^{-1}B^\mathsf{T}\hat{\Sigma}} < 2.$$

For $\Sigma = \sigma^2 \mathscr{I}$ this inequality reduces to

$$F = \frac{\text{SS Regr(op. mod.)} - \text{SS Regr(appr. mod.)}}{(k - p)\text{MSE(op. mod.)}} < 2,$$

where F is the usual test statistic in the test of the hypothesis $Q\theta = 0$.

For operating models with normally distributed e, and $Eee^\mathsf{T} = \sigma^2 \mathscr{I}$, a test of the hypothesis that the operating family leads to a larger expected discrepancy than the approximating family is also available. If the two expected discrepancies are equal, F has a noncentral F distribution with $n_1 = n$ and $n_2 = n - k$ degrees of freedom and noncentrality parameter $\lambda = n_1$. Thus the mentioned hypothesis has to be rejected if the observed F is larger than the critical value given in Tables A.4.5 to A.4.8.

For the variable selection problem discussed in Section 7.2 the restriction which leads to the approximating model was always that some of the elements of θ are zero. (Q_1 is a zero matrix and Q_2 a unit matrix.) As was pointed out this is a very special case.

Here one must also point out that it is not necessary that the approximating family is obtained by restricting the operating family. Any linear approximating model could be used. The above analysis and the criteria (but not the outlined test of hypothesis) remain valid if the operating model is

$$y = A\theta + e$$

and the approximating family

$$y = B\theta_1 + u,$$

where B is not obtained from A by restricting θ (as was the case above).

EXAMPLE 7.5. Many hydrological records have gaps and hydrologists often have to estimate missing data values. The monthly rainfall totals (mm) for Mount Edgecombe and for Nshongweni Dam are given in Tables 7.2 and 7.3. Suppose that we wish to estimate the missing values in the Mount Edgecombe record (target record) using values from the Nshongweni Dam record (control record) by means of linear regression.

Table 7.2. Mount Edgecombe: Monthly Rainfall Totals (mm)

	Oct	Nov	Dec	Jan	Feb	Mar	Apr	May	Jun	Jul	Aug	Sep
1936/37	74.4	276.4	38.4	52.3	157.0	55.1	89.4	10.4	37.3	28.2	65.5	12.2
1937/38	51.8	70.1	232.9	90.7	224.3	10.2	103.4	17.5	86.4	82.0	21.3	8.9
1938/39	122.9	110.7	86.1	86.1	203.2	130.6	34.0	107.7	10.9	64.3	23.6	108.0
1939/40	77.7	140.7	104.9	33.8	57.2	41.7	47.5	232.2	89.2	0.5	5.3	43.7
1940/41	84.1	212.1	102.6	57.4	36.6	87.1	125.2	1.5	9.1	13.2	13.5	75.2
1941/42	53.8	107.7	43.2	179.6	76.2	128.8	34.8	68.6	31.0	20.1	61.2	56.9
1942/43	86.9	179.6	232.9	96.8	148.8	154.9	170.4	84.8	22.4	96.3	88.9	23.4
1943/44	188.7	146.3	72.9	32.8	111.0	170.9	41.4	22.9	39.4	9.7	15.7	136.9
1944/45	87.9	70.4	31.2	76.7	146.8	237.0	42.9	49.3	4.1	0.3	4.3	21.1
1945/46	47.8	21.8	97.3	123.4	58.9	134.1	114.3	12.7	3.3	4.8	0.0	44.5
1946/47	56.1	96.8	101.6	104.9	199.4	89.9	68.8	3.3	80.8	11.9	28.2	-
1947/48	-	-	-	-	-	-	-	-	-	-	-	-
1948/49	-	-	-	-	-	-	-	-	-	-	-	-
1949/50	-	-	-	-	-	-	-	-	-	-	-	-
1950/51	-	-	-	-	-	-	-	-	-	-	-	-
1951/52	-	-	-	-	-	-	-	-	-	-	-	-
1952/53	-	-	-	373.1	190.2	41.1	27.4	10.2	1.0	5.6	65.5	87.1
1953/54	55.1	68.8	166.6	88.9	110.2	90.7	24.4	54.9	18.5	11.2	42.7	150.4
1954/55	285.0	100.3	29.7	194.3	36.6	164.8	85.9	16.8	15.5	15.5	21.8	103.9
1955/56	73.9	107.2	56.6	4.8	200.9	170.4	63.2	19.3	14.2	2.8	77.5	68.3
1956/57	38.9	108.2	302.8	97.0	137.4	103.9	242.8	7.6	1.5	15.0	16.3	129.0
1957/58	114.8	115.1	113.8	238.8	259.1	105.9	247.9	20.1	22.9	4.8	2.8	117.1
1958/59	45.2	130.8	92.2	68.3	84.3	17.3	22.9	111.3	0.5	21.3	73.7	38.9
1959/60	111.3	81.3	108.5	78.0	78.7	93.5	55.6	20.8	6.4	11.4	9.9	39.9
1960/61	68.8	126.0	194.8	117.3	56.4	102.6	256.5	24.4	155.4	11.4	19.3	82.8
1961/62	74.9	103.1	54.1	114.3	63.2	113.8	47.8	21.8	0.5	4.1	98.0	10.4
1962/63	65.8	122.4	69.3	206.8	58.9	136.4	66.3	1.8	47.2	103.1	-	27.4
1963/64	61.0	83.3	96.0	263.1	97.8	51.6	87.4	8.6	33.5	23.6	7.9	36.6
1964/65	148.8	72.9	82.8	75.7	37.1	8.1	20.6	92.5	114.8	35.3	76.7	79.0
1965/66	82.8	107.2	103.9	116.1	52.3	22.1	53.1	108.0	26.7	7.6	37.6	27.4
1966/67	59.7	105.4	100.6	156.7	75.4	142.7	96.5	4.1	8.6	22.9	0.3	16.3
1967/68	123.4	152.1	71.4	248.7	76.5	118.4	-	-	-	-	112.3	59.4

Table 7.3.   Nshongweni Dam: Monthly Rainfall Totals (mm)

	Oct	Nov	Dec	Jan	Feb	Mar	Apr	May	Jun	Jul	Aug	Sep
1936/37	50.8	232.9	46.5	65.5	105.2	21.3	6.4	5.3	12.7	6.9	11.4	25.9
1937/38	57.4	56.4	144.3	88.9	241.8	9.7	131.1	2.8	6.1	41.1	21.8	5.6
1938/39	85.6	71.1	156.7	90.9	155.4	62.5	22.6	62.2	0.8	19.3	16.8	68.1
1939/40	75.2	123.7	90.9	25.1	50.3	56.9	29.7	209.6	66.0	10.2	5.8	53.8
1940/41	31.5	212.9	88.9	61.5	43.4	102.1	92.5	0.0	4.1	11.4	7.9	48.5
1941/42	33.3	87.9	59.2	144.5	95.3	177.8	61.5	51.8	6.9	12.2	52.1	34.3
1942/43	127.3	152.7	186.7	71.6	72.9	107.2	206.2	66.8	16.3	90.7	87.9	10.9
1943/44	193.3	147.1	94.0	39.6	79.0	184.2	11.7	12.7	23.9	8.6	7.4	112.3
1944/45	105.7	75.4	36.3	47.5	91.2	195.8	13.7	30.5	1.5	4.6	1.5	16.0
1945/46	100.8	22.6	69.1	119.9	98.6	142.2	70.1	9.9	6.4	1.5	0.8	19.8
1946/47	50.0	95.3	79.2	68.8	124.0	87.1	73.4	6.6	72.6	23.6	20.6	17.3
1947/48	43.4	140.0	126.2	123.4	126.0	90.9	48.8	8.1	0.0	0.0	1.8	22.1
1948/49	135.6	88.6	77.7	72.1	85.9	57.2	56.6	9.1	0.5	8.6	4.8	52.1
1949/50	74.9	177.5	85.6	131.8	68.1	46.5	15.2	25.4	2.3	35.6	39.1	6.9
1950/51	17.8	39.4	108.0	113.5	33.8	109.7	29.5	4.6	5.1	0.0	99.1	31.0
1951/52	46.0	36.3	74.4	129.0	48.8	64.5	50.3	25.9	2.5	28.7	14.2	21.3
1952/53	36.6	79.5	101.9	167.9	143.3	30.2	13.0	12.2	18.0	1.3	72.6	42.2
1953/54	73.9	68.1	130.0	53.1	97.5	66.3	40.6	41.9	13.5	13.2	10.2	64.8
1954/55	177.0	76.5	27.9	211.3	91.9	87.6	51.3	3.0	9.9	2.3	14.0	42.7
1955/56	40.1	79.2	81.5	16.8	167.1	164.6	52.1	15.5	11.4	4.3	28.4	29.0
1956/57	40.1	84.6	200.7	81.3	60.2	69.1	164.1	14.0	0.0	5.6	24.9	113.3
1957/58	85.1	92.7	50.8	68.6	142.0	33.5	97.0	1.0	11.2	2.5	6.9	77.0
1958/59	21.1	115.6	62.7	98.6	67.3	19.3	23.1	212.9	3.8	18.0	35.1	10.4
1959/60	52.6	81.0	62.2	48.0	65.8	61.2	45.0	9.1	4.1	1.8	5.6	31.5
1960/61	53.3	88.9	91.2	119.4	60.2	83.1	232.2	24.4	18.3	6.9	23.1	53.1
1961/62	29.5	52.8	52.3	141.2	77.0	127.0	47.0	13.0	0.5	0.0	37.6	1.8
1962/63	46.2	118.9	60.5	146.1	41.9	158.0	23.9	0.8	9.9	119.6	0.8	15.7
1963/64	51.1	39.9	67.1	191.3	55.6	67.3	76.5	6.1	30.7	17.0	6.4	-
1964/65	142.0	60.5	196.3	50.0	29.5	16.8	7.6	186.7	91.2	26.9	60.7	74.4
1965/66	73.9	96.5	64.0	129.0	54.9	12.4	32.0	34.5	14.0	3.8	23.9	28.7
1966/67	36.6	93.7	58.7	134.6	207.8	122.2	81.8	8.4	8.6	9.1	2.5	17.3
1967/68	77.2	81.5	58.9	119.1	48.3	60.7	12.2	5.1	11.4	0.5	63.5	47.5

In general, we can not assume that the regression coefficients (and the corresponding residual variances) will be the same for each month. However, rather than estimate a separate regression line for each of the 12 months, it may be preferable to group the months and use a single regression line for all the months within a group. In this way the number of parameters that need to be estimated is reduced. This may lead to more accurate estimation of the missing values, Zucchini and Sparks (1984).

Denote the target data by y_{ij} and the control data by x_{ij}, where i denotes the month ($i = 1, 2, \ldots, 12$) and j the year. For a particular month i, the index j ranges over all years in which both target and control records are available, $j = 1, 2, \ldots, n(i)$. For example, $n(1) = 26$, $n(4) = 27$, $n(12) = 25$. The operating model is of the form

$$y_{ij} = \theta_i x_{ij} + e_{ij}, \qquad i = 1, 2, \ldots, 12$$
$$j = 1, 2, \ldots, n(i),$$

where the e_{ij} are uncorrelated, $E e_{ij} = 0$, and Var $e_{ij} = \sigma_i^2$. To write it in matrix form, $y = A\theta + e$, one has to use

$$y = (y_{11}, \ldots, y_{1n(1)}, y_{21}, \ldots, y_{12n(12)})^{\mathsf{T}},$$
$$e = (e_{11}, \ldots, e_{12n(12)})^{\mathsf{T}},$$
$$\theta = (\theta_1, \theta_2, \ldots, \theta_{12})^{\mathsf{T}}.$$

$$A = \begin{bmatrix} x_1 & 0 & \cdots & 0 \\ 0 & x_2 & \cdots & 0 \\ \vdots & & & \\ 0 & 0 & \cdots & x_{12} \end{bmatrix},$$

where $x_i = (x_{i1}, \ldots, x_{in(i)})^{\mathsf{T}}$.

The dimension of A is $n \times 12$, where $n = \sum_i n(i)$. The covariance matrix of e has the form

$$\Sigma = \begin{bmatrix} \sigma_1^2 \mathscr{I}_{n(1)} & 0 & \cdots & 0 \\ 0 & \sigma_2^2 \mathscr{I}_{n(2)} & \cdots & 0 \\ \vdots & & & \\ 0 & 0 & & \sigma_{12}^2 \mathscr{I}_{n(12)} \end{bmatrix}$$

where $\mathscr{I}_i$ is an $i \times i$ unit matrix.

We will consider the following possible groupings of months: twelve groups of one month each, six groups of two months, four groups of three months, three groups of four months, two groups of six months, one group of twelve months, and use the letter p to denote the number of groups. We also deal with one special case, for which we will use $p = 0$. This represents the model with zero intercept and slope equal to 1, where the control record data point is transferred to the target record untransformed. This degenerate case is considered for the purpose of comparison to gauge how much, if at all, is to be gained by performing a regression analysis.

We now derive the criterion which can be used to decide which number of groups should be used in the analysis.

The approximating model with p parameters ($p = 12, 6, 4, 3, 2, 1$) is given by

$$y = B_{(p)}\theta_{(p)} + u_{(p)},$$

where

$$B_{(p)} = \begin{bmatrix} x_{(p),1} & 0 & \cdots & 0 \\ 0 & x_{(p),2} & \cdots & 0 \\ \vdots & & & \\ 0 & 0 & \cdots & x_{(p),p} \end{bmatrix}$$

and

$$x_{(p),j} = (x_{12(j-1)/p+1}^{\mathsf{T}}, \ldots, x_{12j/p}^{\mathsf{T}})^{\mathsf{T}}, \qquad j = 1, 2, \ldots, p.$$

These approximating families are in fact obtained by restricting the operating family. (The conditions $Q\theta = 0$ stipulate the equality of some of the 12 original parameters θ_i.)

Using the results of the analysis of this section, the criterion for the case of p parameters, denoted by $C_{(p)}$, is

$$C_{(p)} = \hat{\theta}^{\mathsf{T}} C \hat{\theta} - \hat{\theta}_{(p)}^{\mathsf{T}} D_{(p)} \hat{\theta}_{(p)}$$
$$- \operatorname{tr} A C^{-1} A^{\mathsf{T}} \hat{\Sigma} + 2 \operatorname{tr} B_{(p)} D_{(p)}^{-1} B_p^{\mathsf{T}} \hat{\Sigma},$$

where

$$A^{\mathsf{T}} A = C, \ B_p^{\mathsf{T}} B_{(p)} = D_{(p)},$$
$$\hat{\theta} = C^{-1} A^{\mathsf{T}} y,$$
$$\hat{\theta}_{(p)} = D_{(p)}^{-1} B_{(p)}^{\mathsf{T}} y$$

and $\hat{\Sigma}$ is obtained from Σ by replacing σ_i^2 by $\hat{\sigma}_i^2$, the residual mean square in the regression analysis of y on x (straight line through the origin) for the ith month.

It is not recommended that the criterion be computed using the above matrix expressions because simpler forms are available. In particular, it is straightforward (though quite tedious) to demonstrate that

$$\hat{\theta}^{\mathsf{T}} C \hat{\theta} = \sum_{i=1}^{12} \frac{S_{xy}^2(i)}{S_{xx}(i)},$$

$$\operatorname{tr} A C^{-1} A^{\mathsf{T}} \hat{\Sigma} = \sum_{i=1}^{12} \frac{C_{xy}(i)}{S_{xx}(i)},$$

$$\hat{\theta}_{(p)}^{\mathsf{T}} D_p \hat{\theta}_{(p)} = \sum_{i=1}^{p} \frac{S_{xy}^2(i, p)}{S_{xx}(i, p)},$$

$$\operatorname{tr} B_p D_p^{-1} B_p^{\mathsf{T}} \hat{\Sigma} = \sum_{i=1}^{p} \frac{C_{xy}(i, p)}{S_{xx}(i, p)},$$

$$\hat{\sigma}_i^2 = \frac{C_{xy}(i)}{S_{xx}(i)},$$

$$\hat{\theta}_{(p)i} = \frac{S_{xy}(i, p)}{S_{xx}(i, p)},$$

where

$$S_{xx}(i) = \sum_{j=1}^{n(i)} x_{ij}^2,$$

$$S_{xy}(i) = \sum_{j=1}^{n(i)} x_{ij}y_{ij},$$

$$S_{yy}(i) = \sum_{j=1}^{n(i)} y_{ij}^2,$$

$$C_{xy}(i) = \frac{S_{xx}(i)S_{yy}(i) - S_{xy}^2(i)}{n(i) - 1},$$

$$S_{xx}(i, p) = \sum_{j=12(i-1)/p+1}^{12i/p} S_{xx}(j),$$

$$S_{xy}(i, p) = \sum_{j=12(i-1)/p+1}^{12i/p} S_{xy}(j),$$

$$C_{xy}(i, p) = \sum_{j=12(i-1)/p+1}^{12i/p} C_{xy}(j).$$

Collecting all these terms the criterion can be written as

$$C(p) = \sum_{i=1}^{12} \frac{S_{xy}(i)^2 - C_{xy}(i)}{S_{xx}(i)}$$

$$- \sum_{i=1}^{p} \frac{S_{xy}(i, p)^2 - 2C_{xy}(i, p)}{S_{xx}(i, p)}.$$

For the case where $p = 12$, that is, the operating model, it reduces to

$$C(12) = \sum_{i=1}^{12} \frac{C_{xy}(i)}{S_{xx}(i)}.$$

The case $p = 0$ has to be derived separately. It can be shown that

$$C(0) = \sum_{i=1}^{12} \frac{S_{xy}(i)^2 - C_{xy}(i)}{S_{xx}(i)} + \sum_{i=1}^{12} \{S_{xx}(i) - 2S_{xy}(i)\}.$$

For the data in Tables 7.2 and 7.3 one obtains the sums of squares and products given below:

i	Month	$n(i)$	$S_{xy}(i)$	$S_{xx}(i)$	$S_{yy}(i)$
1	Oct	26	218621	190204	280674
2	Nov	26	346124	298319	412844
3	Dec	26	306762	267010	411362
4	Jan	27	406941	316668	604123
5	Feb	27	347940	315004	451100
6	Mar	27	304550	282966	358389
7	Apr	26	235092	200486	318780
8	May	26	115381	140895	119397
9	Jun	26	30473	21624	69404
10	Jul	26	29014	27399	36163
11	Aug	26	41083	30497	66565
12	Sep	25	92264	65636	142035

The whole procedure has to be repeated 12 times, once for each possible "water year." In other words, one begins with January as the first month and December as the last and computes the values of $C(p)$; next one begins with February and ends with January and computes the $C(p)$, and so on until finally one begins with December and ends with November and computes the $C(p)$. It is only in this way that we can ensure that the best grouping of months is found. The particular grouping which leads to the smallest $C(p)$ is estimated to be the best.

The resulting values of the criteria (given as multiples of 10^{-6}) are given below:

Starting Month	$C(0)$	$C(1)$	$C(2)$	$C(3)$	$C(4)$	$C(6)$	$C(12)$
Oct	616	183	203	102	188	148	158
Nov	616	183	155	165	164	173	158
Dec	616	183	122	185	110	148	158
Jan	616	183	201	163	188	173	158
Feb	616	183	184	102	164	148	158
Mar	616	183	196	165	110	173	158
Apr	616	183	203	185	188	148	158

Starting Month	$C(0)$	$C(1)$	$C(2)$	$C(3)$	$C(4)$	$C(6)$	$C(12)$
May	616	183	155	163	164	173	158
Jun	616	183	122	102	110	148	158
Jul	616	183	201	165	188	173	158
Aug	616	183	184	185	164	148	158
Sep	616	183	196	163	110	173	158

It can be seen that the model with $p = 0$ is by far the worst, that it is a poor policy simply to use untransformed rainfall totals from Nshongweni dam for estimating the missing totals at Mount Edgecombe. The lowest value of the criterion is 102×10^{-6} and occurs with $p = 3$ and October as starting month. The best policy is therefore to use three groups of four months, namely, {October, November, December, January}, {February, March, April, May}, and {June, July, August, September}. This amounts to estimating three regression coefficients, one for each group.

CHAPTER 8

Analysis of Variance

Analysis of variance models are linear models and are therefore covered by the results given in Section 7.5. However, analysis of variance models have special features which can lead to simple and convenient selection methods. This is so in the *balanced* case, that is, where there are equal numbers of observations for each treatment combination. In this chapter we deal only with balanced analysis of variance models (both fixed and mixed) and with methods based on the Gauss discrepancy.

The main advantage here is that an analysis of the criterion can be carried out, that is, the criterion can be represented as a sum of the contributions due to the individual components (effects and interactions) of the model. This makes it particularly easy to select those components which are estimated to reduce the expected discrepancy and which are therefore to be included in the approximating model.

Neither homoskedasticity nor normality are required to apply the criterion, that is, to estimate the expected discrepancy. However, as before, *both* of the above assumptions are necessary if one wishes to test hypotheses relating to the expected discrepancy under different approximating models.

Section 8.1 deals with the case of a fixed analysis of variance. For simplicity, and since no new features arise if the dimensionality of the model is increased, we consider a two-way model with K replications. Section 8.2 deals with mixed models. For simplicity we again consider a two-way model with one fixed effect and one random effect.

Section 8.3 deals with a special type of mixed models, namely, one in which the levels of the fixed effect are associated with the values of a given quantitative variable, that is, $\mu_i = \mu(x_i)$. This model can also be viewed as a regression model whose residuals have a particular correlation structure. This structure arises frequently in practice.

8.1. FIXED MODELS

We consider a two-way analysis of variance operating model:

$$y_{ijk} = \mu_{ij} + e_{ijk}$$
$$= \mu + \alpha_i + \beta_j + (\alpha\beta)_{ij} + e_{ijk},$$

where the e_{ijk} are independently distributed, $Ee_{ijk} = 0$, $\mathrm{Var}\ e_{ijk} = \sigma^2$, and

$$\mu = \bar{\mu}_{..}, \alpha_i = \bar{\mu}_{i.} - \bar{\mu}_{..}, \beta_j = \bar{\mu}_{.j} - \bar{\mu}_{..},$$
$$(\alpha\beta)_{ij} = \mu_{ij} - \bar{\mu}_{i.} - \bar{\mu}_{.j} + \bar{\mu}_{..}.$$

Unless otherwise indicated all sums in this chapter are over i, j, and k.

The analysis of the discrepancy due to approximation is based on the decomposition

$$\sum \mu_{ij}^2 = \sum \bar{\mu}_{..}^2 + \sum (\bar{\mu}_{i.} - \bar{\mu}_{..})^2 + \sum (\bar{\mu}_{.j} - \bar{\mu}_{..})^2 + \sum (\mu_{ij} - \bar{\mu}_{i.} - \bar{\mu}_{.j} + \bar{\mu}_{..})^2$$
$$= \sum \mu^2 + \sum \alpha_i^2 + \sum \beta_j^2 + \sum (\alpha\beta)_{ij}^2,$$

and is summarized below:

Discrepancy due to approximation if no parameter is used	$\sum \mu_{ij}^2$
Contribution of μ	$- \sum \mu^2$
α	$- \sum \alpha_i^2$
β	$- \sum \beta_j^2$
$(\alpha\beta)$	$- \sum (\alpha\beta)_{ij}^2$
Discrepancy due to approximation if all IJ parameters are used	0

The minimum discrepancy estimators are the least-squares estimators:

$$\hat{\mu} = \bar{y}_{...}$$
$$\hat{\alpha}_i = \bar{y}_{i..} - \bar{y}_{...}$$
$$\hat{\beta}_j = \bar{y}_{.j.} - \bar{y}_{...}$$
$$\widehat{(\alpha\beta)}_{ij} = \bar{y}_{ij.} - \bar{y}_{i..} - \bar{y}_{.j.} + \bar{y}_{...}.$$

In a standard analysis of variance the decision as to whether a given effect, say α, should be included in the approximating model is based on a test of the hypothesis that $\sum \alpha_i^2 = 0$. In other words, one is testing the hypothesis that the discrepancy due to approximation is not increased if the effect is omitted from the operating model. This is done by calculating the appropriate F statistic from the analysis of variance table:

Variation	d.f.	SS
μ	1	$\sum \hat{\mu}^2$
α	$I - 1$	$\sum \hat{\alpha}_i^2$
β	$J - 1$	$\sum \hat{\beta}_j^2$
$(\alpha\beta)$	$(I - 1)(J - 1)$	$\sum \widehat{(\alpha\beta)}_{ij}^2$
E	$IJ(K - 1)$	$\sum (y_{ijk} - \hat{\mu} - \hat{\alpha}_i - \hat{\beta}_j - \widehat{(\alpha\beta)}_{ij})^2$
Total	IJK	$\sum y_{ijk}^2$

Since

$$E \sum (\mu_{ij} - \hat{\mu} - \hat{\alpha}_i - \hat{\beta}_j - \widehat{(\alpha\beta)}_{ij})^2$$
$$= E \sum (\mu_{ij} - \mu - \alpha_i - \beta_j - (\alpha\beta)_{ij} - (\hat{\mu} - \mu) - (\hat{\alpha}_i - \alpha_i) - (\hat{\beta}_j - \beta_j)$$
$$- (\widehat{(\alpha\beta)}_{ij} - (\alpha\beta)_{ij}))^2$$
$$= \sum (\mu_{ij}^2 - \mu^2 - \alpha_i^2 - \beta_j^2 - (\alpha\beta)_{ij}^2 + \text{Var } \hat{\mu} + \text{Var } \hat{\alpha}_i + \text{Var } \hat{\beta}_j$$
$$+ \text{Var } \widehat{(\alpha\beta)}_{ij}),$$

the analysis of the expected discrepancy is given by

Expected discrepancy if no parameter is used	$\sum \mu_{ij}^2$
Contribution of μ	$\sum \text{Var } \hat{\mu} - \sum \mu^2$
α	$\sum \text{Var } \hat{\alpha}_i - \sum \alpha_i^2$
β	$\sum \text{Var } \hat{\beta}_j - \sum \beta_j^2$
$(\alpha\beta)$	$\sum \text{Var } \widehat{(\alpha\beta)}_{ij} - \sum (\alpha\beta)_{ij}^2$
Expected discrepancy if all parameters are used	$\sum (\text{Var } \hat{\mu} + \text{Var } \hat{\alpha}_i + \text{Var } \hat{\beta}_j + \text{Var } \widehat{(\alpha\beta)}_{ij})$

In general, if $\hat{\gamma}$ is an unbiased estimator of some parameter γ and $\widehat{\mathrm{Var}\,\hat{\gamma}}$ is an unbiased estimator of $\mathrm{Var}\,\hat{\gamma}$, then $2\,\widehat{\mathrm{Var}\,\hat{\gamma}} - \hat{\gamma}^2$ is an unbiased estimator of $\mathrm{Var}\,\hat{\gamma} - \gamma^2$. The estimators $\hat{\mu}$, $\hat{\alpha}_i$, $\hat{\beta}_j$, and $\widehat{(\alpha\beta)}_{ij}$ are unbiased. Thus by making use of unbiased estimators for the variances in the above analysis of expected discrepancy, we obtain the following analysis of the criterion:

Criterion if no parameter is used	$\sum \bar{y}_{ij.}^2 - (IJ)\mathrm{MSE}$
Contribution of μ	$2\sum \widehat{\mathrm{Var}\,\hat{\mu}} - \sum \hat{\mu}^2$
α	$2\sum \widehat{\mathrm{Var}\,\hat{\alpha}_i} - \sum \hat{\alpha}_i^2$
β	$2\sum \widehat{\mathrm{Var}\,\hat{\beta}_j} - \sum \hat{\beta}_j^2$
$(\alpha\beta)$	$2\sum \widehat{\mathrm{Var}\,\widehat{(\alpha\beta)}_{ij}} - \sum \widehat{(\alpha\beta)}_{ij}^2$
Criterion if all parameters are used	$\sum (\widehat{\mathrm{Var}\,\hat{\mu}} + \widehat{\mathrm{Var}\,\hat{\alpha}_i} + \widehat{\mathrm{Var}\,\hat{\beta}_j} + \widehat{\mathrm{Var}\,\widehat{(\alpha\beta)}_{ij}})$.

The sums $\sum \hat{\mu}^2$, $\sum \hat{\alpha}_i^2$, and so on are simply the sums of squares of the effects in the usual analysis of variance. Unbiased estimates of the corresponding sums of variances are given, for example, by $\sum \widehat{\mathrm{Var}\,\hat{\mu}} = \mathrm{MSE}$, $\sum \widehat{\mathrm{Var}\,\hat{\alpha}_i} = (I - 1)\mathrm{MSE}$, and so on, where $\mathrm{MSE} = \mathrm{SSE}/IJ(K - 1)$ is the error mean square. The analysis of the criterion can thus be rewritten in the form:

Criterion if no parameter is used	$\mathrm{SS(treatments)} - (IJ)\mathrm{MSE}$
Contribution of μ	$2\,\mathrm{MSE} - \mathrm{SS}\mu$
α	$2(I - 1)\mathrm{MSE} - \mathrm{SS}\alpha$
β	$2(J - 1)\mathrm{MSE} - \mathrm{SS}\beta$
$(\alpha\beta)$	$2(I - 1)(J - 1)\mathrm{MSE} - \mathrm{SS}(\alpha\beta)$
Criterion if all parameters are used	$(IJ)\mathrm{MSE}$

The selected model thus contains those effects whose contribution to the criterion is *negative*, that is, those for which

$$F = \frac{\mathrm{MS\ (effect)}}{\mathrm{MSE}} > 2.$$

Under the additional assumption that the e_{ijk} in the operating model are *normally* distributed, one can also test the hypothesis that the inclusion of a given effect *increases* the expected discrepancy (and thus leads to a poorer fit). This test should be applied if one requires strong evidence that the inclusion of the effect is advantageous.

The statistic

$$F = \frac{MS\ (\text{effect})}{MSE}$$

has a noncentral F distribution with degrees of freedom $n_1 = $ d.f. (effect) and $n_2 = $ d.f. (E), and a noncentrality parameter, λ, which is related to the expected mean squares:

$$\frac{E\ MS\ (\text{effect})}{E\ MSE} = \frac{\sigma^2(1 + \lambda/n_1)}{\sigma^2} = 1 + \frac{\lambda}{n_1}.$$

The hypothesis that the expected discrepancy *increases* if a certain effect is included in the approximating model is the hypothesis that

$$\frac{E\ MS\ (\text{effect})}{E\ MSE} < 2,$$

that is, the hypothesis that

$$\lambda < n_1.$$

The distribution of F, with $\lambda = n_1$, can therefore be used to implement the test. The upper percentage points of this distribution are given in Tables A.4.5 to A.4.8 in the Appendix.

The Heteroskedastic Case

The decision rule $(F = MS\ (\text{effect})/MSE > 2)$ for the selection of main effects and interactions remains valid if the assumption of equal error variances in the operating model is relaxed, that is, if $\text{Var}\ e_{ijk} = \sigma_{ij}^2$ is admitted. Furthermore, the analysis of discrepancy is based on the identical computing formulas as in the homoskedastic case.

To establish this we first note that the estimates $\hat{\mu}$, $\hat{\alpha}_i$, and so on remain unbiased in this more general case. Thus the analysis of expected discrepancy given above is valid whatever the variances of the e_{ijk}, also the analysis of the criterion. One must, however, calculate $\sum \text{Var } \hat{\mu}$, $\sum \text{Var } \hat{\alpha}_i$, and so on, and find unbiased estimators of these sums of variances.

Now

$$\sum \text{Var } \hat{\mu} = \frac{\sum \sigma_{ij}^2}{IJK},$$

$$\sum \text{Var } \hat{\alpha}_i = \frac{(I - 1) \sum \sigma_{ij}^2}{IJK},$$

$$\sum \text{Var } \hat{\beta}_j = \frac{(J - 1) \sum \sigma_{ij}^2}{IJK},$$

$$\sum \text{Var } \widehat{(\alpha\beta)}_{ij} = \frac{(I - 1)(J - 1) \sum \sigma_{ij}^2}{IJK},$$

and

$$\text{EMSE} = \frac{E \sum (y_{ijk} - \bar{y}_{ij.})^2}{IJ(K - 1)}$$

$$= \frac{E \sum (e_{ijk} - \bar{e}_{ij.})^2}{IJ(K - 1)} = \frac{\sum \sigma_{ij}^2}{IJK}.$$

Multiples of MSE are therefore unbiased estimators of the sums of variances, as in the homoskedastic case. The analysis of criterion table in its second form—and hence also the decision rule—remain valid in the heteroskedastic case.

Note, however, that this simple rule ($F = \text{MS (effect)}/\text{MSE} > 2$) applies only to the selection of main effects and interactions in factorial analysis of variance models. An example of a case in which one must proceed differently is the split up of a treatment sum of squares into components belonging to different orthogonal contrasts, for instance, the linear, quadratic, ... contrasts in regression analysis with replicated observations. The general results of Sections 6.2 and 7.5 must be used.

A test of the hypothesis that the expected discrepancy increases if a certain effect is used is *not* available in the heteroskedastic case.

EXAMPLE 8.1.   Frauen (1979) investigated the yield of ten different varieties of beans. The beans were grown on 60 plots of size 9.3 m^2 in a completely randomized design. Each variety was grown on six different plots, three times with a density of 20 seeds and three times with a density of 40 seeds per plot.

The observed yields, y_{ijk}, of the ith variety planted with density j are in Table 8.1 (Heise, 1981). One may think in terms of random samples of size K from IJ very large populations. One must then treat the K random variables y_{ijk}, for fixed i, j, as independently and identically distributed with mean μ_{ij} and a variance σ_{ij}^2. The result is the operating model

$$\begin{aligned} y_{ijk} &= \mu_{ij} + e_{ijk} \\ &= \mu + \tau_i + \delta_j + (\tau\delta)_{ij} + e_{ijk}, \\ i &= 1, 2, \ldots, I, \quad j = 1, 2, \ldots, J, \quad k = 1, 2, \ldots, K, \end{aligned}$$

where the e_{ijk} are independently distributed and for each i, j identically distributed, have zero mean, and variance σ_{ij}^2.

The representation $\mu_{ij} = \mu + \tau_i + \delta_j + (\tau\delta)_{ij}$ with $\sum \tau_i = 0$, $\sum \delta_j = 0$, $\sum_j (\tau\delta)_{ij} = 0$ for all i, $\sum_i (\tau\delta)_{ij} = 0$ for all j, is used in order to be able to consider simplifications of the original operating model.

In this analysis the *expectations* of the y_{ijk}, rather than the σ_{ij}^2 are of interest and thus the discrepancy considered in this chapter is appropriate. The expected overall discrepancy is, in fact, the sum of the mean squared errors of the estimators of the μ_{ij}.

Table 8.1. Yields of Varieties of Beans, Planted with Different Densities

	Density					
Variety	20 Seeds/Plot			40 Seeds/Plot		
Dacre	352	380	404	426	404	403
M. Bead	350	348	377	386	349	384
Blaze	408	442	430	436	422	402
Hera	352	336	422	480	406	434
Kristall	368	384	408	422	406	424
Russian	200	329	290	374	276	317
Felix	330	368	340	438	336	286
Minica	458	400	440	482	368	390
Wierboon	495	540	434	560	475	472
Rowena	346	372	410	420	345	359

The analysis of variance is given by

Variation	d.f.	SS	MS	F
Varieties	9	148093.8	16454.9	10.3
Density	1	5396.0	5396.0	3.4
VD	9	9857.5	1095.3	0.7
Error	40	63604.0	1590.1	
Total	59	226951.3		

The value of F for $(\tau\delta)$ is smaller than 2, all other F values are larger than 2. The selected model is thus purely additive; the means μ_{ij} are estimated by

$$\bar{y}_{i..} + \bar{y}_{.j.} - \bar{y}_{...}.$$

We note that this procedure is *not* based on the assumption that $\sigma_{ij}^2 = \sigma^2$ for all i, j. However, if one makes this assumption and one also assumes that the e_{ijk} are normally distributed, then one could test the hypothesis that the expected discrepancy *increases* if certain effects are included in the model.

In this example we may require *strong* evidence that it is advantageous to include the effect of *density*. The critical value at the significance level 0.10 for 1 and 40 degrees of freedom is 5.49. The observed F value is 3.40. At this level of significance the effect for density would also have to be omitted. The approximating model would simply be

$$y_{ijk} = \mu + \tau_i + e_{ijk}$$

and the estimator of Ey_{ijk} would be $\hat{\mu} + \hat{\tau}_i = \bar{y}_{i..}$.

8.2. MIXED MODELS

The methods outlined here use a discrepancy which is based on *expectations*. It follows immediately that they can not be used for decisions relating to *random* effects. However, with a minor modification, they can be used for the selection of *fixed* effects in a mixed model.

For the standard analysis of a mixed model the expected mean squares are needed to determine the *proper denominator* in the F statistic used to test the hypothesis that a certain effect is zero. Numerator and denominator of F must have the same expectation under the hypothesis.

For definiteness let the model be

$$y_{ijk} = \mu + \alpha_i + b_j + (ab)_{ij} + e_{ijk},$$

where the random variables b_j, $(ab)_{ij}$, and e_{ijk} are independently distributed with zero mean, Var $b_j = \sigma_b^2$, Var $(ab)_{ij} = \sigma_{ab}^2$, and Var $e_{ijk} = \sigma_e^2$. Then the expected mean square for α is $\sum \alpha_i^2/(I - 1) + K\sigma_{ab}^2 + \sigma_e^2$ and, if $\alpha_i = 0$ for all i, it is $K\sigma_{ab}^2 + \sigma_e^2$. This is also the expectation of MSab, which is therefore the *proper denominator* for F in the test that $\alpha_i = 0$ for all i.

It is not difficult to show that for *mixed* models a valid decision rule is obtained if in the statistic F the denominator MSE is replaced by MS (proper denominator):

$$F = \frac{\text{MS (effect)}}{\text{MS (proper denominator)}}.$$

We establish this for the model above which can be written as

$$y_{ijk} = \mu + \alpha_i + u_{ijk},$$

where $u_{ijk} = b_j + (ab)_{ij} + e_{ijk}$.

In Section 8.1 uncorrelated errors were assumed. The u_{ijk} are correlated, but the least-squares estimators $\hat{\mu}$, $\hat{\alpha}_i$, and so on, remain unbiased in this more general case. Also the analyses of the expected discrepancy and the criterion remain valid. So even in mixed models the effect α should remain in the model if

$$2 \sum \widehat{\text{Var}} \, \hat{\alpha}_i - \sum \hat{\alpha}_i^2 < 0,$$

where $\widehat{\text{Var}} \, \hat{\alpha}_i$ is an unbiased estimator of Var $\hat{\alpha}_i$.

Now

$$\sum \text{Var} \, \hat{\alpha}_i = \sum \text{Var} \, (\bar{u}_{i..} - \bar{u}_{...}) = (I - 1)(K\sigma_{ab}^2 + \sigma_e^2)$$

$$E \, \text{SS}ab = E \sum (\bar{u}_{ij.} - \bar{u}_{i..} - \bar{u}_{.j.} + \bar{u}_{...})^2$$

$$= (I - 1)(J - 1)(K\sigma_{ab}^2 + \sigma_e^2).$$

Thus $(I - 1)$MSab is an unbiased estimator of $\sum \text{Var} \, \hat{\alpha}_i$ and $2 \sum \widehat{\text{Var}} \, \hat{\alpha}_i - \sum \hat{\alpha}_i^2 < 0$ if and only if

$$F = \frac{\text{MS}\alpha}{\text{MS}ab} = \frac{\text{MS}\alpha}{\text{MS (proper denominator)}} > 2.$$

Also the *test* described in Section 8.1 remains valid if all random variables in the operating model are *normally* distributed. Of course n_2 is now the degrees of freedom of the proper denominator.

The Heteroskedastic Case

The decision rule ($F = MS\alpha/MS$ (proper denominator) > 2) remains the same in the *heteroskedastic* case as long as one decides on the inclusion of main effects and interactions *in factorial models*. (One could, for example, have an operating model with Var $b_j = \sigma_{b,j}^2$, Var $(ab)_{ij} = \sigma_{ab,ij}^2$, Var $e_{ijk} = \sigma_{e,ij}^2$.) This is an immediate consequence of the fact that the expected mean squares of main effects and interactions in heteroskedastic models are obtained from the expected mean squares of the corresponding homoskedastic model by replacing all *variances* (σ_b^2, σ_{ab}^2, σ_e^2) by *average variances* ($\sum_j \sigma_{b,j}^2/J$, $\sum_{ij} \sigma_{ab,ij}^2/IJ$, $\sum_{ij} \sigma_{e,ij}^2/IJ$).

Again, as in the fixed heteroskedastic case, there are situations in which the general results of Sections 6.2 and 7.5 must be used. Examples 8.3.1 and 8.3.2 illustrate this.

EXAMPLE 8.2. The Institute for Animal Husbandry at the University of Göttingen studied the triglyceride and cholesterol levels (mg/dL) in the blood of minipigs. We analyze a part of their data relating to a total of 24 pigs. For each of two races, blood samples were taken at ages 2, 4, 6, 8, and 10 weeks. (For each race 6 male and 12 female parent animals were chosen randomly. Each male had offspring with two of the females. From each of the 12 litters a male and a female offspring were selected for the experiment.) The data are in Table 8.2.

Let y_{ijklm} denote the cholesterol level at the mth age ($m = 1, 2, 3, 4, 5$) of an offspring of the jth father and the kth mother of race i, where the sex of the offspring is characterized by l ($l = 1$ (male), $l = 2$ (female)). Then the following is a plausible operating model:

$$y_{ijklm} = \mu + \alpha_i + b_{ij} + c_{ijk} + \delta_l + (\alpha\delta)_{il} + (b\delta)_{ijl} + (c\delta)_{ijkl} + \tau_m + (\alpha\tau)_{im}$$

$$+ (b\tau)_{ijm} + (c\tau)_{ijkm} + (\delta\tau)_{lm} + (\alpha\delta\tau)_{ilm} + (b\delta\tau)_{ijlm} + e_{ijklm}.$$

$$i = 1, 2; \quad j = 1, \ldots, 6; \quad k = 1, 2; \quad l = 1, 2; \quad m = 1, \ldots, 5;$$

(α = race, b = father in race, c = mother in father in race, δ = sex, τ = age. All interactions with random effects are taken as random.)

The analysis of variance for these data is given in Table 8.3 and the expected mean squares in Table 8.4.

Table 8.2. Cholesterol Levels of Minipigs

Race		i		1					2		
Mother		k		1			2		1		2
Sex		l	1	2	1	2	1	2	1	2
Father	j									
Age	m									
1	1		98	33	19	15	21	11	98	8
	2		382	307	429	401	23	20	11	99
	3		411	410	24	29	2	92	94	96
	4		24	4	408	421	82	1	89	100
	5		8	41	88	82	79	81	88	81
2	1		420	425	96	98	426	437	10	5
	2		409	498	18	100	406	401	100	96
	3		80	4	80	93	77	85	404	411
	4		79	68	92	88	78	69	418	420
	5		70	100	91	4	83	83	83	91
3	1		93	78	13	96	14	405	321	321
	2		412	494	400	498	95	91	10	8
	3		77	95	97	92	100	93	13	25
	4		80	82	41	79	3	81	100	89
	5		2	100	92	87	3	83	97	91
4	1		1	91	13	97	8	98	1	20
	2		3	73	96	97	404	494	3	97
	3		90	92	97	100	87	71	420	413
	4		83	65	92	93	93	79	1	82
	5		94	79	98	89	91	86	91	2
5	1		1	8	81	81	88	2	40	20
	2		86	77	15	4	93	13	14	13
	3		95	1	20	8	1	14	13	21
	4		95	100	26	23	89	7	29	13
	5		75	93	19	19	83	99	23	52
6	1		70	65	75	89	40	3	9	93
	2		64	67	73	92	6	5	79	98
	3		68	67	100	74	97	92	94	91
	4		67	60	10	28	95	92	47	98
	5		73	80	13	83	8	85	3	12

The proper denominators in the tests for fixed effects are as follows:

Effect	A	D	AD	T	AT	DT	ADT
Proper denominator	B in A	BD	BD	BT	BT	BDT	BDT

Table 8.3. Analysis of Variance for Cholesterol Levels of Minipigs

Variation	d.f.	SS	MS	F
A	1	8472.82	8472.82	0.12
B in A	10	712423.39	71242.34	
C in B in A	12	124985.20	10415.43	
D	1	6531.27	6531.27	2.47
AD	1	303.75	303.75	0.11
BD	10	26492.98	2649.30	
CD	12	19564.80	1630.40	
T	4	240125.30	60031.33	2.39
AT	4	123179.72	30794.93	1.23
BT	40	1003537.28	25088.43	
CT	48	1322855.30	27559.49	
DT	4	6471.27	1617.82	0.91
ADT	4	2119.71	529.93	0.30
BDT	40	70989.52	1774.74	
E	48	62862.71	1309.64	
Total	239	3730915		

Table 8.4. Expected Mean Squares in Operating Model for Cholesterol Levels of Minipigs

Variation	Expected Mean Squares[a]
A	$\sum \alpha_i^2/(I-1) + KLM\sigma_b^2 + LM\sigma_c^2 + KM\sigma_{b\delta}^2 + M\sigma_{c\delta}^2 + KL\sigma_{b\tau}^2 + L\sigma_{c\tau}^2 + K\sigma_{b\delta\tau}^2 + \sigma_e^2$
B in A	$KLM\sigma_b^2 + LM\sigma_c^2 + KM\sigma_{b\delta}^2 + M\sigma_{c\delta}^2 + KL\sigma_{b\tau}^2 + L\sigma_{c\tau}^2 + K\sigma_{b\delta\tau}^2 + \sigma_e^2$
C in B in A	$LM\sigma_c^2 + M\sigma_{c\delta}^2 + L\sigma_{c\tau}^2 + \sigma_e^2$
D	$\sum \delta_l^2/(L-1) + KM\sigma_{b\delta}^2 + M\sigma_{c\delta}^2 + K\sigma_{b\delta\tau}^2 + \sigma_e^2$
AD	$\sum (\alpha\delta)_{il}^2/(I-1)(L-1) + KM\sigma_{b\delta}^2 + M\sigma_{c\delta}^2 + K\sigma_{b\delta\tau}^2 + \sigma_e^2$
BD	$KM\sigma_{b\delta}^2 + M\sigma_{c\delta}^2 + K\sigma_{b\delta\tau}^2 + \sigma_e^2$
CD	$M\sigma_{c\delta}^2 + \sigma_e^2$
T	$\sum \tau_m^2/(M-1) + KL\sigma_{b\tau}^2 + L\sigma_{c\tau}^2 + K\sigma_{b\delta\tau}^2 + \sigma_e^2$
AT	$\sum (\alpha\tau)_{im}^2/(I-1)(M-1) + KL\sigma_{b\tau}^2 + L\sigma_{c\tau}^2 + K\sigma_{b\delta\tau}^2 + \sigma_e^2$
BT	$KL\sigma_{b\tau}^2 + L\sigma_{c\tau}^2 + K\sigma_{b\delta\tau}^2 + \sigma_e^2$
CT	$L\sigma_{c\tau}^2 + \sigma_e^2$
DT	$\sum (\delta\tau)_{lm}^2/(L-1)(M-1) + K\sigma_{b\delta\tau}^2 + \sigma_e^2$
ADT	$\sum (\alpha\delta\tau)_{ilm}^2/(I-1)(L-1)(M-1) + K\sigma_{b\delta\tau}^2 + \sigma_e^2$
BDT	$K\sigma_{b\delta\tau}^2 + \sigma_e^2$
E	σ_e^2

[a]All sums are over $i, j, k, l,$ and m.

Only D (sex) and T (age), in the analysis of variance table, have F values which are larger than 2. The selected model is thus of the form

$$y_{ijklm} = \mu + \delta_l + \tau_m + u_{ijklm},$$

where u summarizes all random effects in the operating model. The minimum discrepancy estimators of $\mu + \delta_l + \tau_m$ are $\bar{y}_{...l.} + \bar{y}_{....m} - \bar{y}_{.....}$:

l $\quad$ m	1	2	3	4	5	
1	91.8	156.7	103.3	87.8	60.6	100.0
2	102.2	167.2	113.8	98.2	71.0	110.5
	97.0	162.0	108.6	93.0	65.8	105.3

The estimators of $\mu + \delta_l$ and $\mu + \tau_m$ are in the margins.

8.3. A SPECIAL CASE: A MIXED MODEL IN REGRESSION ANALYSIS

In the examples that follow the operating model is a one-way analysis of variance model with a nested error term. This is a mixed model in the sense of this chapter but the treatment effects are determined by the values of a quantitative variable and one is really dealing with a case of regression analysis.

Assume an operating model

$$y_{ij} = \mu_i + b_j + e_{ij}$$
$$= \mu + \alpha_i + b_j + e_{ij}, \quad i = 1, 2, \ldots, I,$$
$$j = 1, 2, \ldots, J,$$
$$Eb_j = Ee_{ij} = 0, \quad \text{Var } b_j = \sigma_b^2, \quad \text{Var } e_{ij} = \sigma_i^2,$$

where $\mu_i = \mu(x_i)$ and x is an exogenous variable. We consider this model because it applies in the examples to follow; more complicated models, like $y_{ijk} = \mu_i + b_j + c_{ij} + e_{ijk}$, are treated similarly.

Suppose that we wish to approximate the mean function by some model which has fewer than I parameters. We could, for example, use orthogonal polynomials in which case the analysis amounts to a subdivision of the treatment sum of squares into components for the linear, quadratic, ... contrasts and to a decision as to which polynomial should be used.

In the homoskedastic case, that is, if $\operatorname{Var} e_{ij} = \sigma^2$, the $F > 2$ rule described in Section 8.2 can be used. One needs the expected mean squares to determine the proper denominator:

Variation	d.f.	EMS
A	$I - 1$	$J \sum_i \alpha_i^2/(I - 1) + \sigma_e^2$
B	$J - 1$	$I\sigma_b^2 + \sigma_e^2$
AB	$(I - 1)(J - 1)$	σ_e^2
Total	$IJ - 1$	

For this model the proper denominator for SSA and for its components is thus MSAB.

In the more general heteroskedastic case ($\operatorname{Var} e_{ij} = \sigma_i^2$) the rule of Section 8.2 ($F = $ SS (contrast)/MS$AB > 2$) does *not* apply if one wants to decide on the use of a particular orthogonal polynomial. One has to return to the general results of Section 6.2.

Consider the approximating model with mean function

$$h_i(\theta) = h(x_i, \theta) = \sum_{r \in S} \theta_r P_r(x_i),$$

where the $P_r(x_i)$ are orthogonal polynomials. For convenience we define $P_1(x_i) = 1/\sqrt{I}$. This, together with the orthogonality property, implies that $\sum_i P_r(x_i) = 0$ for $r = 2, 3, \ldots, I$.

Remark.   Orthogonal polynomials are tabulated to a certain extent in Fisher and Yates (1963). They can also be calculated according to formulas given in Bliss (1970, Section 14.1). See also Emerson (1968).   ∎

The decision rule of Section 6.2 states: include the parameter θ_r in the approximating model if

$$2J \widehat{\operatorname{Var}} \hat{\theta}_r - J\hat{\theta}_r^2 < 0,$$

where

$$\hat{\theta}_r = \sum_i \bar{y}_{i.} P_r(x_i)$$

is an unbiased estimator of

$$\theta_{0r} = \sum_i \mu_i P_r(x_i).$$

Using $\hat{\sigma}_{ij}$ to denote an unbiased estimator of $\sigma_{ij} = \text{Cov}\,(\bar{y}_{i.}, \bar{y}_{j.})$, it follows that the parameter θ_r should be included if

$$F = \frac{J\hat{\theta}_r^2}{J\,\widehat{\text{Var}}\,\hat{\theta}_r} = \frac{J\hat{\theta}_r^2}{J\sum_{ij}\hat{\sigma}_{ij}P_r(x_i)P_r(x_j)} > 2.$$

For the model considered here

$$\sigma_{ij} = \frac{\sigma_b^2}{J}, \qquad \text{for } i \neq j,$$

$$= \frac{\sigma_b^2 + \sigma_i^2}{J}, \qquad \text{for } i = j,$$

and the denominator of F can be simplified. For $r = 2, 3, \ldots, I$ one has that

$$J\sum_{ij}\sigma_{ij}P_r(x_i)P_r(x_j) = \sum_i \sigma_i^2 P_r^2(x_i).$$

Thus θ_r should be included in the approximating model if

$$F = \frac{J\hat{\theta}_r^2}{\sum_i \hat{\sigma}_i^2 P_r^2(x_i)} > 2, \qquad r = 2, 3, \ldots, I,$$

where $\hat{\sigma}_i^2$ is an unbiased estimator of σ_i^2.

To find an estimator $\hat{\sigma}_i^2$ we note at first that

$$\hat{\sigma}_{ii} = \frac{\sum_j (y_{ij} - \bar{y}_{i.})^2}{J(J-1)}$$

is an unbiased estimator of σ_{ii}. From the analysis of variance table,

Variation	d.f.	$E\,MS$
A	$(I-1)$	$J\sum_i \alpha_i^2/(I-1) + \sum_i \sigma_i^2/I$
B	$(J-1)$	$I\sigma_b^2 + \sum_i \sigma_i^2/I$
AB	$(I-1)(J-1)$	$\sum_i \sigma_i^2/I$

it follows that

$$\hat{\sigma}_b^2 = \frac{MSB - MSAB}{I}$$

is an unbiased estimator of σ_b^2. Thus

$$\hat{\sigma}_i^2 = J\hat{\sigma}_{ii} - \hat{\sigma}_b^2$$

is an unbiased estimator of σ_i^2.

EXAMPLE 8.3.1. Consider Martin's (1960) data which were also analyzed by Bliss (1970, Vol. 2, p. 10). Table 8.5 gives the histaminolytic index y_{ij}, taken for $J = 13$ women in the 2nd, 3rd, $\ldots$, 9th month of pregnancy.
An operating model would be

$$y_{ij} = \mu_i + b_j + e_{ij},$$

where the random variables b_j and e_{ij} are independently distributed with zero mean.
The variances for the third and fourth month are larger than for the other months; an operating model with $\sigma_i^2 = \sigma^2$ is not plausible.
We consider the problem of estimating the eight expectations, $\mu_1, \mu_2, \ldots,$ μ_8, by the values of some polynomial. To select the degree of the polynomial

Table 8.5. Histaminolytic Indices of 13 Pregnant Women

Case				Index in $(i + 1)$th Month					
j	i 1	2	3	4	5	6	7	8	Totals
1	12	40	50	61	71	77	85	96	492
2	9	27	41	61	74	79	90	99	480
3	10	29	30	51	63	70	89	98	440
4	11	37	43	57	69	79	87	96	479
5	9	38	46	57	69	74	93	102	488
6	6	40	51	60	73	80	92	99	501
7	11	39	52	64	75	79	88	99	507
8	5	37	45	56	68	76	87	99	473
9	14	41	48	58	67	72	80	99	479
10	13	37	47	57	66	72	81	91	464
11	12	38	49	59	70	78	90	98	494
12	10	22	30	55	68	73	91	101	450
13	7	37	43	62	73	79	88	106	495
$y_{i.}$	129	462	575	758	906	988	1141	1283	6242
$\bar{y}_{i.}$	9.92	35.54	44.23	58.31	69.69	76.00	87.77	98.69	60.02
$J\hat{\sigma}_{ii}$	7.24	33.44	50.53	11.56	11.90	11.50	15.19	12.06	

on the basis of the criterion one first constructs the standard analysis of variance table:

Variation	d.f.	SS	MS
Month (A)	7	77684.88	11097.84
Case (B)	12	575.71	47.98
AB	84	1265.37	15.06
Total	103	79525.96	

The usual methods are then employed to calculate the estimated coefficients $\hat{\theta}_r = \sum_i \bar{y}_{i.} P_r(x_i)$, $r = 2, 3, \ldots, I$, where the P_r are orthogonal polynomials. For the denominators of F one needs $J\hat{\sigma}_{ii}$ (the sample variance in the ith column; see bottom of Table 8.5) and

$$\hat{\sigma}_b^2 = \frac{MSB - MSAB}{I} = \frac{47.98 - 15.06}{8} = 4.12.$$

The table below summarizes the analysis of the criterion:

Parameter	$\hat{\theta}_i$	$SS(\theta_i) = J\hat{\theta}_i^2$	$J \widehat{\text{Var}}\, \hat{\theta}_i$	F
θ_2	76.32	75723.26	12.91	5865.5
θ_3	−9.03	1060.66	9.32	113.8
θ_4	6.38	529.46	17.08	31.0
θ_5	−2.07	55.72	15.45	3.6
θ_6	2.80	102.03	19.42	5.3
θ_7	−4.05	213.03	22.51	9.5
θ_8	−0.14	0.25	13.63	0.02

The sums of squares for $\theta_2, \theta_3, \ldots, \theta_8$ sum to SSA. A parameter θ_r should be used if $F = J\hat{\theta}_r^2/J \widehat{\text{Var}}\, \hat{\theta}_r > 2$. With this criterion only θ_8 should be excluded in the approximating model, that is, the following mean function is selected:

$$\mu_i = \sum_{r=1}^{7} \theta_r P_r(x_i).$$

This method corresponds roughly to a test at a level of 0.5 of the hypothesis that a certain parameter leads to an increase in the expected discrepancy.

Exact methods to test this hypothesis at a *smaller* level, which one uses if *stronger* evidence is required that a certain parameter does not result in an increase of the expected discrepancy, are not available because $\sigma_i^2 = \sigma^2$ is not plausible in the operating model.

Nevertheless, the critical values of the mentioned test (derived under the assumption of normality and homoskedasticity) give some impression of the magnitudes involved. At the significance level 0.01, the critical value is 11.7. This value suggests that the hypothesis that the inclusion of θ_5, θ_6, and θ_7 leads to an increase in the expected discrepancy can not be rejected.

In conclusion, the criterion selects a model having the seven parameters, $\theta_1, \theta_2, \ldots, \theta_7$. However, on the basis of an approximate test, there does not seem to be strong evidence that it is of advantage to include θ_5, θ_6, and θ_7. With this in mind, one may decide to select the alternative model having only θ_1, θ_2, θ_3, and θ_4.

Figure 8.1 shows the fitted mean function and the polynomial of degree 7 which is determined by the eight observed means.

EXAMPLE 8.3.2. Consider the monthly average temperatures in Paris which are given by Knoch (1947) for the eight decades between 1850 and 1930 (Table 8.6).

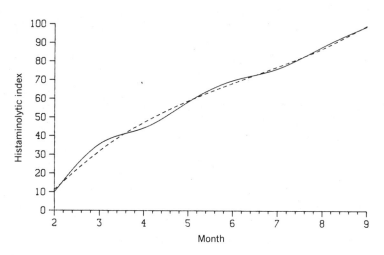

Figure 8.1. Estimators of expected histaminolytic indices using 3rd (– – –) and 7th (———) degree polynomial models.

Table 8.6. Monthly Average Temperatures in Paris in the Decades Between 1850 and 1930

	51–60	61–70	71–80	81–90	91–00	01–10	11–20	21–30
J	3.9	3.0	2.8	2.0	2.1	2.6	3.2	4.5
F	3.2	4.9	4.6	3.3	4.7	3.3	4.3	4.4
M	5.3	6.1	7.2	5.4	6.2	6.1	6.7	7.0
A	9.4	11.0	9.8	9.0	10.3	9.5	9.2	9.7
M	12.5	14.0	12.1	13.3	13.0	13.3	14.8	13.8
J	16.2	16.2	16.2	16.3	17.0	16.2	16.3	16.4
J	18.0	18.4	18.3	17.9	18.5	18.2	17.6	18.9
A	18.0	17.4	18.1	17.2	18.1	17.4	17.7	17.8
S	14.4	15.1	14.8	14.4	15.3	14.5	14.7	15.3
O	10.4	10.1	9.9	9.0	10.2	10.8	9.6	11.2
N	5.0	5.8	5.8	6.4	6.4	5.4	5.0	6.0
D	2.8	3.9	2.1	2.3	3.6	3.5	5.1	3.9

One could assume the operating model

$$y_{ij} = \mu_i + b_j + e_{ij}, \qquad i = 1, 2, \ldots, 12,$$
$$j = 1, 2, \ldots, 8,$$

with independently distributed random variables b_j and e_{ij}, $Eb_j = Ee_{ij} = 0$, Var $b_j = \sigma_b^2$, Var $e_{ij} = \sigma_i^2$. Here μ_i is the effect of the month (A) and b_j the random effect of the decade (B). (The variances in the 12 columns which estimate $J\sigma_{ii} = \sigma_b^2 + \sigma_i^2$ differ markedly and a simplifying assumption $\sigma_i^2 = \sigma^2$ is not advisable here.)

Suppose that we wish to estimate the mean monthly temperatures, μ_i, for the 12 months. Truncated Fourier series are often suitable for this type of data because the sequence of μ_i is likely to be quite smooth. We consider the question of how many harmonics should be used in the approximating model.

The approximating families of mean functions are linear combinations of the orthogonal functions:

$$P_1(x) = (1/I)^{1/2}, \qquad\qquad I = 12,$$
$$P_{2j}(x) = (2/I)^{1/2} \cos \omega_j x, \qquad j = 1, 2, \ldots, 5,$$
$$P_{2j+1}(x) = (2/I)^{1/2} \sin \omega_j x, \qquad j = 1, 2, \ldots, 5,$$
$$P_{12}(x) = (1/I)^{1/2} \cos \omega_6 x, \qquad \omega_j = 2\pi j/12.$$

For convenience, we will again use α_0, α_j, and β_j instead of θ_1, θ_{2j}, and θ_{2j+1}. Then

$$\hat{\alpha}_0 = \frac{\sum_i \bar{y}_{i.}}{I^{1/2}}, \qquad \hat{\alpha}_j = \sum_i \bar{y}_{i.} P_{2j}(x_i),$$

$$\hat{\beta}_j = \sum_i \bar{y}_{i.} P_{2j+1}(x_i),$$

where $x_i = i - 1$, $i = 1, 2, \ldots, I$.

A simplification is possible if one considers the contributions of *pairs* (α_r, β_r) simultaneously because

$$\sum_i (P_{2r}^2(x_i) + P_{2r+1}^2(x_i))\sigma_i^2 = \frac{2 \sum_i \sigma_i^2}{I} \quad \text{for } r = 1, \ldots, 5,$$

and

$$\sum_i P_{12}^2(x_i)\sigma_i^2 = \frac{\sum_i \sigma_i^2}{I}.$$

The contribution to the criterion of (α_r, β_r) is *negative* if

$$2J(\widehat{\text{Var }} \hat{\alpha}_r + \widehat{\text{Var }} \hat{\beta}_r) - J(\hat{\alpha}_r^2 + \hat{\beta}_r^2)$$

$$= 2 \sum_i (P_{2r}^2(x_i) + P_{2r+1}^2(x_i))\hat{\sigma}_i^2 - J(\hat{\alpha}_r^2 + \hat{\beta}_r^2)$$

$$= \frac{4 \sum_i \hat{\sigma}_i^2}{I} - J(\hat{\alpha}_r^2 + \hat{\beta}_r^2) < 0.$$

But $\sum_i \hat{\sigma}_i^2/I = \text{MS}AB$ and (α_r, β_r) should be used in the approximating model if

$$F = \frac{J(\hat{\alpha}_r^2 + \hat{\beta}_r^2)/2}{\text{MS}AB} = \frac{\text{MS}\alpha_r, \beta_r}{\text{MS}AB} > 2.$$

Similarly, it follows that α_6 should be used if

$$F = \frac{J\hat{\alpha}_6^2}{\text{MS}AB} = \frac{\text{MS}\alpha_6}{\text{MS}AB} > 2.$$

This is again a special case in which the methods of the homoskedastic case can simply be applied. The basic analysis of variance is given by

Variation	d.f.	SS	MS
Months (A)	11	2887.612	262.510
Decades (B)	7	9.511	1.359
AB	77	26.093	0.339
Total	95	2923.216	

The estimators of the Fourier parameters are

$$\hat{\delta}_0 = 35.45,$$
$$\hat{\alpha}_1 = -18.91, \qquad \hat{\beta}_1 = -0.64,$$
$$\hat{\alpha}_2 = 0.84, \qquad \hat{\beta}_2 = 1.43,$$
$$\hat{\alpha}_3 = 0.40, \qquad \hat{\beta}_3 = -0.11,$$
$$\hat{\alpha}_4 = 0.11, \qquad \hat{\beta}_4 = 0.02,$$
$$\hat{\alpha}_5 = -0.12, \qquad \hat{\beta}_5 = 0.03,$$
$$\hat{\alpha}_6 = -0.02.$$

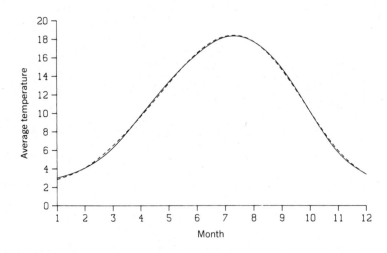

Figure 8.2. Estimated expected temperatures in Paris, 1850–1930, using a finite Fourier series with first and second harmonic (---) and using all six harmonics (—).

The analysis of the criterion is given by

Parameter	d.f.	SS	MS	F
α_1, β_1	2	2864.08	1432.04	4224
α_2, β_2	2	21.94	10.97	32
α_3, β_3	2	1.36	0.68	2.0
α_4, β_4	2	0.11	0.05	0.2
α_5, β_5	2	0.12	0.06	0.2
α_6	1	0.00	0.00	0.0

Only α_1, β_1, α_2, and β_2 are selected for the approximating mean function. The fitted mean function is given in Figure 8.2.

CHAPTER 9

Analysis
of Covariance

9.1. A DISCREPANCY FOR ANALYSIS OF COVARIANCE

Although the particular discrepancy which was used for analysis of variance is applicable to linear models in general, it is not always appropriate for analysis of covariance.

In some applications of analysis of covariance the focus of interest is the expected response for given values of the covariates. However, in other applications covariates are introduced primarily to increase the precision of the estimates for the treatment effects. In the first case the discrepancy mentioned above would be suitable, but in the second case the discrepancy should be sensitive to treatment effects only, and not to the influence of the covariates on the expectation of the observations. Here a particular covariate should be included in the model only if it leads to better estimates of the treatment effects.

Let the operating model be

$$y = A_1\theta_1^* + A_2\theta_2^* + A_3\theta_3^* + e,$$

where y is a column vector of n observations, A_1, A_2, and A_3 are known matrices of dimension $n \times I$, $n \times p$, and $n \times (k - p)$, respectively, $0 \leq p \leq k$, e is a random vector with $Ee = 0$ and $\operatorname{Var} e = \sigma^2 \mathscr{I}$, θ_1^* is a vector of I treatment parameters, and θ_2^* and θ_3^* are vectors of regression coefficients of the covariates. (We use the asterisk here to indicate that the model is fully specified. Later if no confusion can arise the asterisks will be omitted.) We assume that there are equal numbers of observations for each treatment.

Suppose that we wish to decide whether the estimate of θ_1^* is improved if the $k - p$ covariates with coefficients θ_3^* in the operating model are omitted, that is, if the approximating family of models is

$$y = A_1\theta_1 + A_2\theta_2 + e.$$

A simple discrepancy is

$$\Delta(\theta_1) = (\theta_1^* - \theta_1)^{\mathsf{T}}(\theta_1^* - \theta_1).$$

The discrepancy due to approximation is zero, but each different value of p leads to a different estimator $\hat{\theta}_1$ and hence to a different discrepancy due to estimation and a different overall discrepancy. The expected overall discrepancy is given by

$$E(\hat{\theta}_1 - \theta_1^*)^{\mathsf{T}}(\hat{\theta}_1 - \theta_1^*),$$

where $\hat{\theta}_1$ is an estimator of θ_1^* using the approximating family of models.

In the following sections, which are based on Linhart and Zucchini (1982c), a criterion is derived for this discrepancy.

9.2. THE EXPECTED DISCREPANCY

The following notation is used:

$$A = (A_1 A_2 A_3), \qquad A_{12} = (A_1 A_2),$$
$$C_{ij} = A_i^{\mathsf{T}} A_j, \qquad C = \{C_{ij}\}, \qquad C^{-1} = \{C^{ij}\},$$
$$V = \begin{bmatrix} C_{11} & C_{12} \\ C_{21} & C_{22} \end{bmatrix}, \qquad V^{-1} = \begin{bmatrix} V^{11} & V^{12} \\ V^{21} & V^{22} \end{bmatrix},$$
$$C_3 = (C_{31} C_{32}), \qquad W = (V^{11} V^{12}),$$
$$\theta^{\mathsf{T}} = (\theta_1^{\mathsf{T}} \theta_2^{\mathsf{T}} \theta_3^{\mathsf{T}}).$$

The least squares estimator of θ_1 using the (reduced) approximating model is

$$\hat{\theta}_1 = W A_{12}^{\mathsf{T}} y.$$

The expectation and variance of $\hat{\theta}_1$ under the operating model are given by

$$E\hat{\theta}_1 = WA_{12}^\mathsf{T}A\theta \quad \text{and} \quad \text{Var}\ \hat{\theta}_1 = \sigma^2 V^{11}.$$

It is easy to show that

$$(E\hat{\theta}_1 - \theta_1)^\mathsf{T}(E\hat{\theta}_1 - \theta_1) = \theta_3^\mathsf{T}C_3 W^\mathsf{T}WC_3^\mathsf{T}\theta_3$$

and so the expected discrepancy is given by

$$E(\hat{\theta}_1 - \theta_1)^\mathsf{T}(\hat{\theta}_1 - \theta_1) = (E\hat{\theta}_1 - \theta_1)^\mathsf{T}(E\hat{\theta}_1 - \theta_1) + \text{tr Var}\ \hat{\theta}_1$$
$$= \theta_3^\mathsf{T}C_3 W^\mathsf{T}WC_3^\mathsf{T}\theta_3 + \sigma^2\ \text{tr}\ V^{11}.$$

The result holds for all $p = 0, 1, 2, \ldots, k$. If the family containing the operating model is used for fitting (i.e., $p = k$, $V = C$), then $E\hat{\theta}_1 = \theta_1$ and the expected discrepancy reduces to $\sigma^2\ \text{tr}\ C^{11}$.

A family with $p < k$ covariates should be fitted in preference to the operating model if the difference in expected discrepancies is negative, that is, if

$$\sigma^2\ \text{tr}\ (V^{11} - C^{11}) + \theta_3^\mathsf{T}C_3 W^\mathsf{T}WC_3^\mathsf{T}\theta_3 < 0.$$

9.3. THE CRITERION

Since $\theta_3^\mathsf{T}C_3 W^\mathsf{T}WC_3^\mathsf{T}\theta_3$, with $\hat{\theta}_3 = (C^{31}C^{32}C^{33})A^\mathsf{T}y$, has expectation

$$\theta_3^\mathsf{T}C_3 W^\mathsf{T}WC_3^\mathsf{T}\theta_3 + \sigma^2\ \text{tr}\ C_3 W^\mathsf{T}WC_3^\mathsf{T}C^{33},$$

it follows that an unbiased estimator for the difference in expected discrepancies is given by

$$\hat{\sigma}^2\ \text{tr}\ (V^{11} - C^{11}) + \hat{\theta}_3^\mathsf{T}C_3 W^\mathsf{T}WC_3^\mathsf{T}\hat{\theta}_3 - \hat{\sigma}^2\ \text{tr}\ C_3 W^\mathsf{T}WC_3^\mathsf{T}C^{33},$$

where $\hat{\sigma}^2$ is the residual mean square when the operating model is fitted, that is,

$$\hat{\sigma}^2 = \frac{y^\mathsf{T}y - \hat{\theta}^\mathsf{T}C\hat{\theta}}{n - I - p}.$$

One can prove that $\text{tr}\ C_3 W^\mathsf{T}WC_3^\mathsf{T}C^{33} = \text{tr}\ (C^{11} - V^{11})$, and so the criterion for the reduced family of models is smaller if

$$\hat{\theta}_3^\mathsf{T} C_3 W^\mathsf{T} W C_3^\mathsf{T} \hat{\theta}_3 - 2\hat{\sigma}^2 \operatorname{tr}(C^{11} - V^{11}) < 0.$$

Clearly, the above results can also be applied to decide whether p_1 or p_2, where $p_1 < p_2 < k$, covariates should be used. The family leading to the smaller value of the left-hand side of the above inequality should be used. This left-hand side is the difference between the criterion for the approximating model used and the criterion for the operating model. One can compare these differences rather than the criteria themselves.

9.4. SPECIAL CASES

9.4.1. The Case of One Covariate Only ($p = 0$, $k = 1$)

In this case the results can be expressed in terms of the usual quantities which are calculated in analysis of covariance. We use the notation

$$y_{ij} = \mu_i + \theta_3 x_{ij} + e_{ij}, \qquad i = 1, 2, \ldots, I,$$

$$j = 1, 2, \ldots, J.$$

Variation	d.f.	SS_{yy}	SP_{yx}	SS_{xx}
Groups	I	$\sum \bar{y}_{i.}^2$	$\sum \bar{y}_{i.}\bar{x}_{i.}$	$\sum \bar{x}_{i.}^2$
Error	$n - I$	$\sum y_{ij}^2 - \sum \bar{y}_{i.}^2$	$\sum y_{ij}x_{ij} - \sum \bar{y}_{i.}\bar{x}_{i.}$	$\sum x_{ij}^2 - \sum \bar{x}_{i.}^2$
Total	$n = IJ$	$\sum y_{ij}^2$	$\sum y_{ij}x_{ij}$	$\sum x_{ij}^2$

All sums are over i and j; we use G for Groups and E for Error.
   Here $C_3 = C_{31}$, $V = C_{11}$, and therefore

$$\hat{\theta}_3 = \frac{SP_{yx}(E)}{SS_{xx}(E)},$$

$$\hat{\theta}_3^\mathsf{T} C_3 W^\mathsf{T} W C_3^\mathsf{T} \hat{\theta}_3 = \frac{\hat{\theta}_3^2 SS_{xx}(G)}{J},$$

$$\operatorname{tr}(C^{11} - V^{11}) = \frac{SS_{xx}(G)/SS_{xx}(E)}{J}.$$

The criterion is

$$\left(\frac{SS_{xx}(G)}{SS_{xx}(E)}\right) \frac{SP_{yx}^2(E)/SS_{xx}(E) - 2\hat{\sigma}^2}{J}$$

and so the covariate should be omitted if

$$F = \frac{SP_{yx}^2(E)/SS_{xx}(E)}{\hat{\sigma}^2} < 2,$$

where $\hat{\sigma}^2 = (SS_{yy}(E) - SP_{yx}^2(E)/SS_{xx}(E))/(IJ - I - 1)$. This F is the statistic which one uses in analysis of covariance to test the hypothesis that the regression coefficient of the covariate (here θ_3) is zero.

The above result does not depend on the assumption that the errors are normally distributed. If one assumes that this does hold then it is also possible to test the hypothesis that the difference in expected discrepancies is negative, that is, the hypothesis that

$$E(\hat{\theta}_3^T C_3 W^T W C_3^T \hat{\theta}_3 - 2\hat{\sigma}^2 \operatorname{tr}(C^{11} - V^{11})) < 0,$$

or

$$E(SP_{yx}^2(E)/SS_{xx}(E))/\sigma^2 < 2.$$

It is well known that under normality the sum of squares for the regression coefficient of the covariate, $SP_{yx}^2(E)/SS_{xx}(E)$, is distributed as σ^2 times non-central chi-squared with one degree of freedom and noncentrality parameter λ. The hypothesis to be tested is thus

$$\frac{E\sigma^2\chi^2(1, \lambda)}{\sigma^2} < 2, \quad \text{that is,} \quad \lambda < 1.$$

As test statistic one can use F which, under the operating model, has the noncentral F distribution $F(n_1 = 1, n_2 = n - I; \lambda)$.

The upper percentage points of noncentral F with $\lambda = n_1 = 1$ are tabulated in the Appendix (Tables A.4.5 to A.4.8). The hypothesis has to be rejected (and the covariate should be included) if the observed F is larger than the tabulated critical value.

In this case (only) the method which emerges is equivalent to that obtained for analysis of variance. This is in spite of the fact that a different discrepancy is used.

The rule of using a critical value of 2 for F corresponds again roughly to a test of the mentioned hypothesis at a significance level of approximately 0.5.

9.4.2. The Case $p = 0$, $k > 1$

For this case too the results can be somewhat simplified. Let $SS_{x_1 x_1}$, $SP_{x_1 x_2}$, SP_{yx_1}, ... be defined analogously to the sums of squares and products in the previous case and let

$$S = \begin{bmatrix} SS_{x_1 x_1} & \cdots & SP_{x_1 x_p} \\ \vdots & & \\ SP_{x_p x_1} & \cdots & SS_{x_p x_p} \end{bmatrix},$$

$$S_{yx}^{\mathsf{T}} = (SP_{yx_1}, \ldots, SP_{yx_p}).$$

Here one obtains

$$\mathrm{tr}\,(C^{11} - V^{11}) = \frac{\mathrm{tr}\, S(G)S^{-1}(E)}{J},$$

$$C_3 W^{\mathsf{T}} W C_3^{\mathsf{T}} = \frac{S(G)}{J},$$

and

$$\hat{\theta}_3 = S^{-1}(E)S_{yx}(E).$$

The estimated difference of expected discrepancies is

$$\frac{\hat{\theta}_3^{\mathsf{T}} S(G)\hat{\theta}_3 - 2\hat{\sigma}^2\, \mathrm{tr}\, S(G)S^{-1}(E)}{J}$$

which is negative if

$$H = \frac{\hat{\theta}_3^{\mathsf{T}} S(G)\hat{\theta}_3 / \mathrm{tr}\, S(G)S^{-1}(E)}{\hat{\sigma}^2} < 2.$$

An exact test of the hypothesis that the difference in expected discrepancies is negative can not be carried out by means of H. One can prove, in the normal case, that the numerator of H is independently distributed of $\hat{\sigma}^2$ but it does not have a chi-squared distribution. The mean and variance of the quadratic form were calculated using the characteristic function of quadratic forms in normal variables which is given, for example, in Plackett (1960):

$$E\hat{\theta}_3^T S(G)\hat{\theta}_3 = \theta_3^T S(G)\theta_3 + \sigma^2 \operatorname{tr} S(G)S^{-1}(E),$$

$$\operatorname{Var} \hat{\theta}_3^T S(G)\hat{\theta}_3 = 4\sigma^2 \theta_3^T S(G)S^{-1}(E)S(G)\theta_3 + 2\sigma^4 \operatorname{tr} (S(G)S^{-1}(E))^2.$$

9.5. EXAMPLES OF APPLICATION

EXAMPLE 9.5.1. The data in Table 9.1, taken from Snedecor and Cochran (1980), show the initial weight, x_1 (lb), the initial age, x_2 (days), and the

Table 9.1. Initial Weight, Initial Age, and Weight Gain of Pigs Under Four Treatments

Treatment 1			Treatment 2		
Weight	Age	Gain	Weight	Age	Gain
61	78	1.40	74	78	1.61
59	90	1.79	74	99	1.31
76	94	1.72	64	80	1.21
50	71	1.47	48	75	1.35
61	99	1.26	62	94	1.29
54	80	1.28	42	91	1.24
57	83	1.34	52	75	1.29
45	75	1.55	43	63	1.43
41	62	1.57	50	62	1.29
40	67	1.26	40	67	1.26

Treatment 3			Treatment 4		
Weight	Age	Gain	Weight	Age	Gain
80	78	1.67	62	77	1.40
61	83	1.41	55	71	1.47
62	79	1.73	62	78	1.37
47	70	1.23	43	70	1.15
59	85	1.49	57	95	1.22
42	83	1.22	51	96	1.48
47	71	1.39	41	71	1.31
42	66	1.39	40	63	1.27
40	67	1.56	45	62	1.22
40	67	1.36	39	67	1.36

average weight gain, y (lb/day), of 40 pigs subjected to four different treatments. The sums of squares and products are given below:

Variation	d.f.	$SS_{x_1 x_1}$	$SP_{x_1 x_2}$	$SS_{x_2 x_2}$
Groups (G)	4	111386.1	162658.6	237655.8
Error (E)	36	4876.9	2877.4	4548.2
Total	40	116263.0	165536.0	242204.0
Variation	d.f.	SS_{yy}	SP_{yx_1}	SP_{yx_2}
Groups (G)	4	77.2671	2929.141	4279.887
Error (E)	36	0.8452	26.219	5.623
Total	40	78.1123	2955.360	4285.510

A conventional analysis of covariance reveals that the coefficient of covariate x_1 is significant at the 5% level but that the coefficient of x_2 is not. A rule of thumb, apparently due to W. G. Cochran, is to include a covariate in the model if the "correlation coefficient,"

$$\rho = \frac{SP_{yx}(E)}{(SS_{xx}(E)SS_{yy}(E))^{1/2}}$$

is greater than 0.3. In this example the values of ρ for x_1 and x_2 are approximately 0.4 and 0.1, respectively. Although this rule was probably intended for cases where there is a single covariate these results suggest that only x_1, the initial weight, should be retained in the model.

It is easy to show that for the case of a single covariate the rule described in Section 9.4 reduces to: include the covariate in the model if $\rho > (2/(IJ - I + 1))^{1/2}$. (For $I = 4$ and $J = 10$ this critical value is 0.23.)

Now we apply the results derived above to these data, to decide whether it is preferable to omit both covariates from the model, that is, whether $p = 0$ leads to a smaller criterion than the operating model. Using the results of Section 9.4.2 one obtains

$$S(E) = \begin{bmatrix} 4876.9 & 2877.4 \\ 2877.4 & 4548.2 \end{bmatrix}, \quad S^{-1}(E) = 10^{-4} \begin{bmatrix} 3.2717 & -2.0698 \\ -2.0698 & 3.5081 \end{bmatrix},$$

$$S(G) = \begin{bmatrix} 111386.1 & 162658.6 \\ 162658.6 & 237655.8 \end{bmatrix}, \quad S_{yx}(E) = [26.219 \quad 5.623].$$

Hence $\hat{\theta}_3^{\mathsf{T}} = (0.007414 \quad -0.003454)$ and $\hat{\sigma}^2 = 0.019713$. The estimated difference in expected discrepancies is then $(0.62716 - 2 \times 0.01971 \times 52.48)/10 = -0.144$. Since this is *negative* the model having no covariates is preferred to that having both.

It is conceivable, however, that the use of the single covariate x_1 is preferable to using no covariates at all. To decide on this we compare at first the model containing both covariates to that having only x_1. To apply the results of Section 9.3 one needs to compute C^{-1} and V^{-1}. With these inverses one can then compute

$$\text{tr}\,(C^{11} - V^{11}) = 2.964, \qquad C_3 W^{\mathsf{T}} W C_3^{\mathsf{T}} = 7768.7,$$

$$\hat{\theta}_3 = -0.003457, \quad \text{and} \quad \hat{\sigma}^2 = 0.019713.$$

The difference between the two criteria is therefore

$$(-0.003457)^2 \times 7768.7 - 2 \times 2.964 \times 0.01971 = -0.0240.$$

The difference is *negative* and so the model with the single covariate x_1 is to be preferred to that with both covariates.

Finally, since $-0.144 < -0.024$ the model without covariates is preferred to that having x_1 only. It is not really necessary to calculate the criterion for the model using x_2 only since the correlation between y and x_2 (0.1) is much smaller than the correlation between y and x_1 (0.4). The selected model thus has no covariates.

EXAMPLE 9.5.2. The Institute of Medical Statistics in the University of Göttingen investigated the effect of large doses of a certain drug on the offspring of rabbits (see Ahlborn, 1982).

We give, in Table 9.2, the results which were obtained in an experiment with 20 rabbits each in the four groups

1   50-fold normal dose,
2   100-fold normal dose,
3   only dilutant,
4   no treatment (control).

Observed were y, the average fetal length in the litter (mm), x_1, the number of fetuses in the litter, x_2, the weight of the mother (g).

The object of this experiment was the estimation of the treatment effects. As operating model one has

$$y_{ij} = \mu_i + \beta_1(x_{1ij} - \bar{x}_{1..}) + \beta_2(x_{2ij} - \bar{x}_{2..}) + e_{ij},$$
$$i = 1, 2, 3, 4; \quad j = 1, 2, \ldots, 20.$$

Table 9.2   Average Fetal Length, Number of Fetuses, and Weight of Mother of Litters of Rabbits Under Four Treatments

i	1			2			3			4		
j	y	x_1	x_2	y	x_1	x_2	y	x_1	x_2	y	x_1	x_2
1	100.3	4	337	91.2	6	329	97.0	5	326	105.5	2	326
2	94.0	2	353	96.0	8	308	104.5	2	374	95.3	4	289
3	98.0	6	328	93.0	2	377	91.7	7	319	98.0	2	481
4	100.4	5	360	90.0	6	298	93.3	3	323	96.1	9	334
5	98.9	9	337	94.7	9	307	97.0	4	341	96.9	7	347
6	94.5	2	453	84.5	8	358	94.9	8	337	88.5	6	404
7	91.1	9	357	92.0	6	325	100.9	7	308	99.4	5	350
8	89.7	7	418	81.0	7	347	88.9	8	411	88.9	9	339
9	94.8	8	372	97.3	6	368	105.0	5	361	93.8	4	385
10	95.0	4	402	91.3	3	406	104.8	4	350	86.0	4	303
11	96.6	7	348	92.3	8	360	94.5	4	335	101.5	2	342
12	95.7	6	353	96.0	2	306	95.0	3	398	92.0	9	419
13	95.4	9	347	96.3	4	337	94.5	4	356	104.0	4	350
14	103.3	3	342	85.2	5	387	94.8	8	350	92.9	8	393
15	92.5	4	404	88.4	8	308	93.5	6	366	89.6	7	395
16	89.0	6	368	95.0	6	374	93.7	6	362	100.3	3	348
17	87.1	9	445	89.6	5	358	88.7	3	436	98.7	6	438
18	89.0	7	384	88.0	6	336	90.2	6	365	91.8	9	403
19	95.3	6	383	100.4	12	314	86.3	12	386	87.6	9	455
20	87.2	11	315	94.9	11	329	94.8	10	340	89.8	9	374

The question which we consider is whether the estimation of μ_i should be based on *this* model (two covariates), or on a simplified model with one covariate (either x_1 or x_2), or no covariates.

We consider first the criterion for the case in which both variables are omitted. (One could invoke the special results of Section 9.4.2, but we use here the general results of Section 9.3.) One obtains

$$C_3 W^\mathsf{T} W C_3^\mathsf{T} = C_{31} C_{11}^{-1} C_{11}^{-1} C_{31}^\mathsf{T} = \frac{S(G)}{J}$$

$$= \begin{bmatrix} \sum_i (\bar{x}_{1i.} - \bar{x}_{1..})^2 & \sum_i (\bar{x}_{1i.} - \bar{x}_{1..})(\bar{x}_{2i.} - \bar{x}_{2..}) \\ & \sum_i (\bar{x}_{2i.} - \bar{x}_{2..})^2 \end{bmatrix}$$

$$= \begin{bmatrix} 0.2569 & -5.809 \\ -5.809 & 692.15 \end{bmatrix}.$$

One also gets $\hat{\theta}_3^{\mathsf{T}} = (-0.751, -0.041)$, $\hat{\sigma}^2 = 20.653$, and tr $(C^{11} - C_{11}^{-1})$ = 6.688×10^{-3}. The criterion has the value

$$0.951 - 2 \times 20.653 \times 6.618 \times 10^{-3} = 0.678.$$

This is *positive*, so the estimator based on a covariance analysis with two covariates is preferred to the estimator based on a simple analysis of variance model (without covariates). To decide whether a single variate should be used, one calculates the criterion for a model with x_1 as covariate:

$$\hat{\theta}_3^{\mathsf{T}} C_3 W^{\mathsf{T}} W C_3^{\mathsf{T}} \hat{\theta}_3 - 2\hat{\sigma}^2 \text{ tr } (C^{11} - V^{11})$$

$$= 1.1366 - 2 \times 20.653 \times 6.116 \times 10^{-3} = 0.884.$$

This is also positive, and it is preferable to use *both* covariates.

The criterion for a model with x_2 only becomes 0.102. Again a model with both variates should be preferred.

The rank order of the various possible estimation methods is the following:

1 (best)   use both covariates,
2   use x_2 only (0.102),
3   use neither x_1 nor x_2 (0.678),
4   use x_1 only (0.884).

EXAMPLE 9.5.3.   We give now an example in which the covariate is not used as "blocking" variable only. Here the effect of the covariate is as important as the treatment effects, and the discrepancy of this chapter which uses only the deviation between true and fitted approximating *treatment* effects is not suitable. One would use the discrepancy suggested in the chapter on analysis of variance (Chapter 8). The results which were obtained there are immediately applicable.

The table below gives the monthly index (minus 100) of net industrial production in the Federal Republic of Germany from July 1969 to June 1973 (*Statistisches Bundesamt*, 1976):

j \ i	1	2	3	4	5	6	7	8	9	10	11	12
1	48	37	58	71	60	52	43	44	54	66	49	61
2	58	44	64	73	66	60	48	51	72	63	57	63
3	58	46	68	72	70	60	52	57	75	62	67	71
4	56	56	69	84	83	67	70	69	85	75	86	76

The figures were rounded to the nearest whole number. A graphical representation of the data is given in Figure 9.1.

Inspection of the data suggests an operating model of the type

$$y_{ij} = \pi_i + \tau(z_{ij}) + e_{ij}, \qquad i = 1, \ldots, 12; \quad j = 1, 2, 3, 4,$$

where z_{ij} is the time, $z_{ij} = 12(j - 1) + i$, and $\tau(z_{ij})$, the trend, is some polynomial in z_{ij}, and π_i is the effect of the month (the seasonal variation).

From the data one can see that at most a polynomial of the third degree could be plausible for τ. It is convenient to use the orthogonal polynomials of Hald (1948):

$$\xi_{1ij} = z_{ij} - \bar{z}_{..},$$

$$\xi_{2ij} = (z_{ij} - \bar{z}_{..})^2 - \frac{(I^2 J^2 - 1)}{12},$$

$$\xi_{3ij} = (z_{ij} - \bar{z}_{..})^3 - \frac{(3I^2 J^2 - 2I^2 - 5)(z_{ij} - \bar{z}_{..})}{20}.$$

The operating model is

$$y_{ij} = \mu + \pi_i + \gamma_1 \xi_{1ij} + \gamma_2 \xi_{2ij} + \gamma_3 \xi_{3ij} + e_{ij},$$

$$\sum \pi_i = 0, \qquad E e_{ij} = 0, \qquad \text{Var } e_{ij} = \sigma_{ij}^2.$$

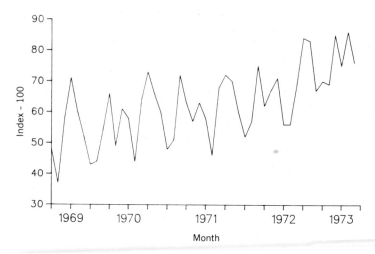

Figure 9.1. Index of net industrial production in the Federal Republic of Germany.

This is an analysis of covariance model with three covariates, ξ_1, ξ_2, and ξ_3. We reparametrize as follows:

$$\pi_i = \sum_{k=1}^{5} (\alpha_k \cos \omega_k i + \beta_k \sin \omega_k i) + (-1)^i \alpha_6,$$

where $\omega_k = 2\pi k/12$; that is, we replace π_i by its Fourier series representation. This leads, on inversion, to the 11 orthogonal contrasts

$$\alpha_k = \frac{1}{6} \sum \pi_i \cos \omega_i k,$$

$$\beta_k = \frac{1}{6} \sum \pi_i \sin \omega_i k, \qquad k = 1, 2, \ldots, 5,$$

$$\alpha_6 = \frac{1}{12} \sum \pi_i \cos \omega_i 6.$$

The problem is to decide how many covariates should be used (i.e., the choice of the order of the polynomial for the approximating trend) and how many of the harmonics should be used to explain the approximating seasonal variation.

The usual orthogonality of balanced analysis of variance is missing here; the sums of squares for the Fourier coefficients depend on the chosen trend. Strictly speaking, the criteria of Section 7.5 should be calculated for each possible model.

It seems reasonable (as an approximate method) to first concentrate on the degree of the polynomial to be used and then to select the appropriate harmonics. One then obtains the following analysis of covariance:

Variation	d.f.	yy	$y\xi_1$	$y\xi_2$	$y\xi_3$
Months	11	3303.67	512.00	−576.33	122326.6
Error	33	3104.00	4428.00	12971.01	110390.4
Total	44	6407.67	4940.00	12394.68	232717.0

Variation	d.f.	$\xi_1\xi_1$	$\xi_2\xi_2$	$\xi_3\xi_3$
Months	11	572.00	5338.67	30384118.7
Error	33	8640.00	1407168.01	179894823.5
Total	44	9212.00	1412506.68	210278942.2

From this one can calculate

$$\text{SS Regr } \xi_r = \frac{(\text{SS}_{y\xi_r}(E))^2}{\text{SS}_{\xi_r\xi_r}(E)} :$$

r	SS Regr ξ_r	F
1	2269.35	116
2	119.56	6.1
3	67.74	3.5

The values of F (with one and 33 degrees of freedom) are obtained by dividing SS Regr by

$$\text{MS Res} = \frac{\text{SS}_{yy}(E) - \sum\limits_{r=1}^{3} \text{SS Regr } \xi_r}{33}$$

$$= 19.617$$

The hypothesis that the inclusion of ξ_r results in an increase of the expected overall discrepancy can be rejected at the 0.10 level if $F > 5.6$ (see Table A.4.6). One would therefore decide here to use a second degree polynomial, that is, ξ_1 and ξ_2, only. (If, however, the *criterion* is applied, one would have to use ξ_1, ξ_2, *and* ξ_3, that is, a polynomial of degree three.)

With the estimators $\hat{\gamma}_1 = 0.5125$ and $\hat{\gamma}_2 = 0.009219$ an approximate covariance analysis can now be carried out. (The approximate analysis, originally suggested by W. G. Cochran, amounts to carrying out an analysis of variance for $y_{ij} - \bar{y}_{..} - \hat{\gamma}_1\xi_{1ij} - \hat{\gamma}_2\xi_{2ij}$. This usually results in values of F which are practically the same as those obtained in the *exact* analysis of covariance.) One gets the following estimators for π_i and for the 11 orthogonal contrasts in the Fourier analysis:

i	1	2	3	4	5	6
$\hat{\pi}_i$	-4.767	-14.437	4.124	13.917	8.191	-2.303

i	7	8	9	10	11	12
$\hat{\pi}_i$	-9.315	-7.846	7.854	2.287	-0.050	2.346

k	1	2	3	4	5	6
$\hat{\alpha}_k$	-1.277	-1.978	3.812	3.005	-0.211	-1.006
$\hat{\beta}_k$	1.171	-8.763	2.753	2.347	-0.283	

The approximate analysis of covariance is given in Table 9.3.

Only the F values of $\alpha_2, \alpha_3, \alpha_4, \beta_2, \beta_3$, and β_4 are substantially larger than 2. Of these, all except that for α_2 are larger than the tabulated critical value for a significance level of 0.10. At this level of significance one would thus use only the second, third, and fourth harmonics to estimate the seasonal variation.

Finally, on simultaneous estimation of the selected parameters one obtains

$$\hat{\gamma}_1 = \quad 0.4945, \qquad \hat{\gamma}_2 = \quad 0.0090,$$
$$\hat{\alpha}_2 = -1.959, \qquad \hat{\beta}_2 = -8.794,$$
$$\hat{\alpha}_3 = \quad 3.830, \qquad \hat{\beta}_3 = \quad 2.735,$$
$$\hat{\alpha}_4 = \quad 3.023, \qquad \hat{\beta}_4 = \quad 2.337.$$

Table 9.3. Approximate Analysis of Covariance for Index of Net Industrial Production in the Federal Republic of Germany

Variation	d.f.	SS	F
α_1	1	39.12	1.9
α_2	1	93.87	4.5
α_3	1	348.69	16.6
α_4	1	216.76	10.3
α_5	1	1.07	0.1
α_6	1	48.60	2.3
β_1	1	32.93	1.6
β_2	1	1842.88	87.6
β_3	1	181.93	8.7
β_4	1	132.24	6.3
β_5	1	1.93	0.1
Months	11	2940.02	
Residual	34	715.09	MS = 21.03
Total	45	3655.11	

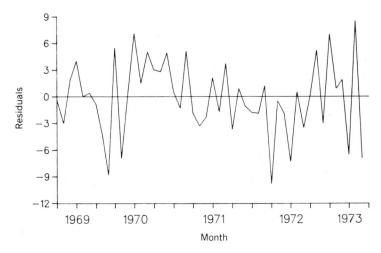

Figure 9.2. Residuals in the analysis of index of net industrial production in the Federal Republic of Germany.

The residuals

$$y_{ij} - \bar{y}.. - \hat{\gamma}_1 \xi_{1ij} - \hat{\gamma}_2 \xi_{2ij} - \sum_{k=2}^{4} (\hat{\alpha}_k \cos \omega_k i + \hat{\beta}_k \sin \omega_k i)$$

were analyzed and showed no evidence for dependence, heteroskedasticity, or deviation from normality. They are illustrated in Figure 9.2.

CHAPTER 10

The Analysis
of Proportions:
The Binomial Case

Suppose that we wish to estimate the "probability of success," π_i, for each of I independent Bernoulli random variables associated with I categories represented by x_i, $i = 1, 2, \ldots, I$, and that in N_i trials in the ith category n_i successes were observed.

If the x_i represent purely nominal categories then we have little alternative but to use the proportions $p_i = n_i/N_i$ to estimate the corresponding π_i. In many situations, however, there is some natural ordering of the categories and we have some information about the behavior of the probabilities π_i, $i = 1, 2, \ldots, I$.

In Example 10.2.2 the x_i, $i = 1, 2, \ldots, 5$, represent increasing doses of radiation and π_i the corresponding probabilities of mutations of Drosophila. Here one would expect the probability of mutation to be an increasing and (in this range) *smooth* function of the dose level. Consequently, it may be preferable to fit a smooth function $h(x, \theta)$, for example, a low degree polynomial, to the π_i and estimate the parameters of this function, namely, θ, rather than the individual π_i. In this way we can reduce the number of parameters that need to be estimated. There is the additional advantage that we can use the model to estimate the probability of mutation for those dosage levels which fall between the observed x_i.

The operating model in this chapter is described by I independently binomially distributed random variables, n_i, having parameters π_i and N_i, $i = 1, 2, \ldots, I$.

Useful candidates for the $h(x, \theta)$ are the polynomials. By using a polynomial of degree $I - 1$ we are simply reparametrizing the operating model in terms of

174

the polynomial coefficients, θ. Approximating models are obtained by setting some of these coefficients (usually the last few) to zero. We discuss methods which can be used to decide which particular approximating model is estimated to lead to the best fit. Polynomials are not the only approximating families which can be used. They have the disadvantage of sometimes leading to inadmissible estimates of the probabilities, that is, estimates which are less than zero or greater than one. We will therefore not restrict our attention to polynomials.

In the sequel the approximating functions are only needed for the arguments x_i, $i = 1, 2, \ldots, I$, and to simplify the notation we will again use $h_i(\theta)$ for $h(x_i, \theta)$.

We will derive methods for three different discrepancies, the Kullback–Leibler, Gauss, and Mahalanobis discrepancies. The proofs of asymptotic results are given in the Appendix, as are regularity conditions which we will assume to hold.

For discrepancies which are not covered in this chapter and for which results are not available we recommend the use of Bootstrap methods. These require additional computing but are easy to apply.

In this chapter $p = (p_1, p_2, \ldots, p_I)^{\mathsf{T}}$ is used to denote the vector of relative frequencies. We will also continue to use p to denote the number of free parameters in the approximating model. The context should make it clear which of these two quantities is meant.

10.1. KULLBACK–LEIBLER DISCREPANCY

If π_i is the expectation of p_i under the operating model and $h_i(\theta)$ the expectation under the approximating model the essential part of the Kullback–Leibler discrepancy is

$$\Delta(\theta) = - \sum_i N_i \pi_i \log h_i(\theta) - \sum_i N_i (1 - \pi_i) \log (1 - h_i(\theta)).$$

The corresponding empirical discrepancy is

$$\Delta_n(\theta) = - \sum_i n_i \log h_i(\theta) - \sum_i (N_i - n_i) \log (1 - h_i(\theta)).$$

The minimum discrepancy estimator, $\hat{\theta}$, is the maximum likelihood estimator. It is computed by solving the system of p equations

$$\sum_{i=1}^{I} \frac{(N_i h_i(\theta) - n_i) h_i'(\theta)}{h_i(\theta)(1 - h_i(\theta))} = 0.$$

Here $h_i'(\theta)$ is the column vector with elements $\partial h_i(\theta)/\partial\theta_r$, $r = 1, 2, \ldots, p$. The second derivative $h_i''(\theta)$, which is needed later, is the $p \times p$ matrix with elements $\partial^2 h_i(\theta)/\partial\theta_r\partial\theta_s$, $r, s = 1, 2, \ldots, p$. These equations can be solved using numerical methods.

The propositions of Section 2.4 hold with

$$\Sigma = \sum_{i=1}^{I} \frac{N_i\pi_i(1 - \pi_i)h_i'(\theta_0)h_i'(\theta_0)^{\mathsf{T}}}{h_i^2(\theta_0)(1 - h_i(\theta_0))^2}$$

and

$$\Omega = \sum_{i=1}^{I} \frac{N_i\pi_i[h_i'(\theta_0)h_i'(\theta_0)^{\mathsf{T}} - h_i''(\theta_0)h_i(\theta_0)]}{h_i^2(\theta_0)}$$
$$+ \sum_{i=1}^{I} \frac{N_i(1 - \pi_i)[h_i''(\theta_0)(1 - h_i(\theta_0)) + h_i'(\theta_0)h_i'(\theta_0)^{\mathsf{T}}]}{(1 - h_i(\theta_0))^2}.$$

It is not difficult to find Ω_n and Σ_n, estimators of Ω and Σ, by replacing $N_i\pi_i$ by n_i and θ_0 by $\hat{\theta}$ in Ω and Σ.

The estimated expected discrepancy, the criterion, is then

$$\Delta_n(\hat{\theta}) + \operatorname{tr} \Omega_n^{-1}\Sigma_n.$$

If the operating model is a member of the approximating family then $\Omega = \Sigma$. Even if this is not the case, p is usually a good approximation to $\operatorname{tr} \Omega^{-1}\Sigma$. This justifies the simpler criterion,

$$\Delta_n(\hat{\theta}) + p.$$

In most applications the simpler criterion (Akaike's Information Criterion) leads to comparable, if not better, results.

EXAMPLE 10.1. The rate of arrival of storms at different times of the year is of interest to engineers and hydrologists. Table 10.1 gives, for each week of the year, the number of times that at least one storm occurred at Botanic Gardens, Durban, for the two periods 1.6.1932–31.12.1979 (Period I) and 1.1.1964–31.12.1979 (Period II). Here a "storm" is defined as a rainfall event of at least 30 mm in 24 hours. In Period I, $N_i = 47$ for $i \leqslant 22$, $N_i = 48$ for $i > 22$; in Period II, $N_i = 16$ for all i.

The data record in this example is atypically long for many parts of the world—one often has to estimate the rate of arrival from less than 20 years of record. Consequently, for purposes of illustration the analysis is repeated

Table 10.1 Number of Weeks with at Least One Storm at Botanic Gardens, Durban

Week	Begin	Period I n_i	Period II n_i	Week	Begin	Period I n_i	Period II n_i
1	1 Jan	6	1	27	2 Jul	4	1
2	8 Jan	8	3	28	9 Jul	0	0
3	15 Jan	7	5	29	16 Jul	2	0
4	22 Jan	6	2	30	23 Jul	0	0
5	29 Jan	9	3	31	30 Jul	3	1
6	5 Feb	15	5	32	6 Aug	1	1
7	12 Feb	6	1	33	13 Aug	1	0
8	19 Feb	12	5	34	20 Aug	5	1
9[a]	26 Feb	16	7	35	27 Aug	4	2
10	5 Mar	7	2	36	3 Sep	3	1
11	12 Mar	9	2	37	10 Sep	6	0
12	19 Mar	6	1	38	17 Sep	1	0
13	26 Mar	8	4	39	24 Sep	8	5
14	2 Apr	2	1	40[b]	1 Oct	3	1
15	9 Apr	7	2	41	9 Oct	4	2
16	16 Apr	4	1	42	16 Oct	6	4
17	23 Apr	4	1	43	23 Oct	9	2
18	30 Apr	3	1	44	30 Oct	5	3
19	7 May	3	2	45	6 Nov	8	1
20	14 May	10	4	46	13 Nov	6	1
21	21 May	3	1	47	20 Nov	5	0
22	28 May	3	2	48	27 Nov	7	2
23	4 Jun	0	0	49	4 Dec	5	0
24	11 Jun	5	1	50	11 Dec	8	1
25	18 Jun	1	0	51	18 Dec	5	1
26	25 Jun	2	2	52	25 Dec	4	1

[a]Eight days on leap years.
[b]Eight days.
Note: In Period I, $N_i = 47$ for $i \leqslant 22$, $N_i = 48$ for $i > 22$; in Period II, $N = 16$ for all i.

using the last 16 years of observation (Period II). But even with the full sample a simple relative frequency estimate is unsatisfactory because of the high variation from week to week. There are meteorological grounds to believe that the rate of arrival should be smooth and this can be verified for locations at which long data records exist. This information can be used to improve the estimate.

It is well known that finite Fourier series are particularly suitable to estimate this type of meteorological parameter. We will use maximum likelihood estimation of the mean rate of arrival for a finite Fourier series model. The question about how many terms to include can be answered as follows, Zucchini and Adamson (1984a).

Suppose that the year is divided into I time intervals denoted by $i = 1, 2, \ldots, I$, that N_i observations were made at time i, and that n_i successes were observed (a success here is at least one storm in the week in question).

The expectation of n_i/N_i under the approximating family with p parameters is

$$h_i(\theta) = \sum_{j=1}^{p} \theta_j \psi_{ji}, \qquad i = 1, 2, \ldots, I,$$

where $\psi_{1i} = 1$, $\psi_{2i} = \cos 2\pi(i-1)/I$, $\psi_{3i} = \sin 2\pi(i-1)/I$, $\psi_{4i} = \cos 4\pi(i-1)/I$, $\psi_{5i} = \sin 4\pi(i-1)/I, \ldots$.

The maximum likelihood estimators are the solutions to the equations:

$$\sum_{i=1}^{I} \frac{(N_i h_i(\theta) - n_i)\psi_{ri}}{h_i(\theta)(1 - h_i(\theta))} = 0, \qquad r = 1, 2, \ldots, p.$$

These equations are concave in the parameters and this makes Newton's iterative method particularly suitable for solving them. For the examples given here very few iterations are required for convergence. Suitable starting values can be obtained by the usual least-squares method of fitting finite Fourier series.

The matrix of second derivatives of the log-likelihood has elements given by

$$\frac{\partial^2 \log L}{\partial \theta_r \partial \theta_s} = -\sum_{i=1}^{I} \frac{[n_i(1 - h_i(\theta))^2 + (N_i - n_i)h_i^2(\theta)]\psi_{ri}\psi_{si}}{h_i^2(\theta)(1 - h_i(\theta))^2}, \qquad r, s = 1, 2, \ldots, p.$$

The *negative* of this, taken at $\theta = \hat{\theta}$, gives the elements of the matrix Ω_n, which is thus available at no extra cost.

There is also no difficulty in computing Σ_n, whose elements are given by

$$\sum_{i=1}^{I} \frac{n_i(1 - n_i/N_i)\psi_{ri}\psi_{si}}{h_i^2(\hat{\theta})(1 - h_i(\hat{\theta}))^2}, \qquad r, s = 1, 2, \ldots, p.$$

Table 10.2 gives the computed values for $\Delta_n(\hat{\theta})$, $\mathrm{tr}\,\Omega_n^{-1}\Sigma_n$, and the criterion for each of the two periods and for increasing numbers of parameters. It can be seen that the trace term is approximately equal to the number of parameters, p. Using p instead of the trace term amounts to using Akaike's Information Criterion.

Table 10.2. Criteria for Fourier Series with p Parameters Approximating the Probabilities of Storms in Durban

Period	p	$\Delta_n(\hat{\theta})$	$\operatorname{tr}\Omega_n^{-1}\Sigma_n$	Criterion
	1	863.24	0.95	864.19
	3	832.47	2.91	835.38
	5	829.94	4.86	834.80
I	7	826.87	6.82	833.69
	9	824.41	8.79	<u>833.20</u>
	11	824.04	10.75	834.79
	13	823.17	12.71	835.88
	1	285.11	0.90	286.01
	3	276.91	2.77	279.68
	5	274.10	4.65	278.75
II	7^a			
	9	267.66	8.51	<u>276.17</u>
	11	267.40	10.39	277.79
	13	267.08	12.24	279.32

^{a}Model leads to negative probabilities.

Note that for Period II no estimates are available when seven parameters are used. Here the solution of the normal equations leads to "negative probabilities." This problem occurs frequently in practice, particularly in arid regions, if one uses polynomial models.

The estimated parameters corresponding to the models leading to the smallest value of the criterion are given below and the estimated probabilities $h_i(\hat{\theta})$ are given in Figure 10.1.

Estimator	Period I	Period II
$\hat{\theta}_1$	0.1115	0.1081
$\hat{\theta}_2$	0.0625	0.0511
$\hat{\theta}_3$	0.0249	0.0339
$\hat{\theta}_4$	−0.0159	−0.0247
$\hat{\theta}_5$	0.0158	0.0290
$\hat{\theta}_6$	−0.0167	−0.0124
$\hat{\theta}_7$	0.0193	0.0501
$\hat{\theta}_8$	−0.0183	−0.0126
$\hat{\theta}_9$	−0.0028	0.0149

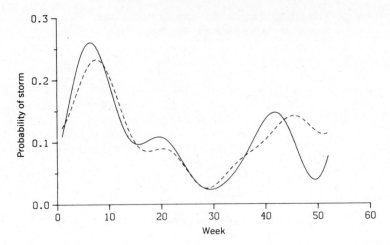

Figure 10.1. Estimated probabilities of storms in Durban. Probabilities approximated by Fourier series. (Period I: – – –; Period II: ——.)

To overcome the problem of obtaining negative estimates for the probabilities one can fit a finite Fourier series to the *logits* rather than to the probabilities. This amounts to using a different set of approximating families, namely, those with probabilities

$$h_i(\theta) = \frac{e^{Q_i(\theta)}}{1 + e^{Q_i(\theta)}},$$

where

$$Q_i(\theta) = \sum_{j=1}^{p} \theta_j \psi_{ji}, \qquad i = 1, 2, \ldots, I.$$

Apart from a constant, the log-likelihood under this model is given by

$$\log L(\theta) = \sum_{i=1}^{I} (n_i Q_i(\theta) - N_i \log (1 + e^{Q_i(\theta)})).$$

The Kullback–Leibler discrepancy is the negative expectation of this expression and the minimum discrepancy (maximum likelihood) estimators are the solution to the p equations:

$$\sum_{i=1}^{I} \left(n_i - \frac{N_i e^{Q_i(\theta)}}{1 + e^{Q_i(\theta)}} \right) \psi_{ri} = 0, \qquad r = 1, 2, \ldots, p.$$

As before these equations can be solved using Newton's iterative method or alternatively using the method of iteratively reweighted least squares. The matrix of second derivatives required to apply the method is given by

$$\frac{\partial^2 \log L}{\partial \theta_r \partial \theta_s} = - \sum_{i=1}^{I} \left(\frac{N_i e^{Q_i(\theta)}}{(1 + e^{Q_i(\theta)})^2} \right) \psi_{ri} \psi_{si}, \qquad r, s = 1, 2, \ldots, p.$$

The matrix Ω_n is simply the negative of this matrix taken at $\theta = \hat{\theta}$ and is also in this case available at no extra computational cost. The matrix Σ_n has elements

$$\sum_i n_i \left(1 - \frac{n_i}{N_i} \right) \psi_{ri} \psi_{si}, \qquad r, s = 1, 2, \ldots, p.$$

Table 10.3 gives the values for $\Delta_n(\hat{\theta})$, $\mathrm{tr}\,\Omega_n^{-1}\Sigma_n$, and the criterion. Again the trace term is approximately equal to the number of parameters, p.

The problem of obtaining negative estimates for probabilities does not arise because logits are not restricted to any specific interval. Note that for Period II the smallest value of the criterion is achieved for $p = 7$; precisely the case which failed to yield estimates using the original method. It can

Table 10.3. Criteria for Fourier Series with p Parameters Approximating the Logits of Storms in Durban

Period	p	$\Delta_n(\hat{\theta})$	$\mathrm{tr}\,\Omega_n^{-1}\Sigma_n$	Criterion
	1	863.24	0.95	864.19
	3	833.83	2.91	836.74
	5	829.17	4.87	834.04
I	7	826.37	6.83	833.20
	9	823.91	8.80	832.91
	11	823.89	10.76	834.65
	13	823.40	12.70	836.10
	1	285.11	0.90	286.01
	3	277.03	2.75	279.78
	5	273.98	4.63	278.61
II	7	268.10	6.61	274.71
	9	267.86	8.47	276.33
	11	266.86	10.41	277.27
	13	266.69	12.31	279.00

also be seen that the logit models lead to slightly smaller criteria than the models based on (untransformed) probabilities.

The estimated parameters corresponding to the models leading to the smallest value of the criterion are given below and the estimated probabilities

$$h_i(\hat{\theta}) = \frac{e^{Q_i(\theta)}}{1 + e^{Q_i(\theta)}}, \qquad i = 1, 2, \ldots, I,$$

are given in Figure 10.2.

Estimator	Period I	Period II
$\hat{\theta}_1$	-2.2161	-2.2955
$\hat{\theta}_2$	0.7419	0.5964
$\hat{\theta}_3$	0.2291	0.3778
$\hat{\theta}_4$	-0.2912	-0.4099
$\hat{\theta}_5$	0.0417	0.0918
$\hat{\theta}_6$	-0.0375	-0.0592
$\hat{\theta}_7$	0.2070	0.5868
$\hat{\theta}_8$	-0.2051	
$\hat{\theta}_9$	-0.0650	

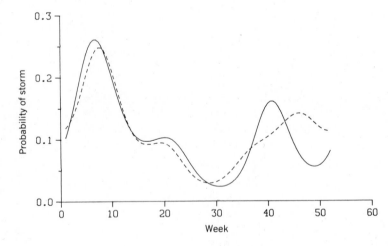

Figure 10.2. Estimated probabilities of storms in Durban. Logits approximated by Fourier series. (Period I: – – –; Period II: ——.)

10.2. GAUSS DISCREPANCY

10.2.1. The General Case

Here the discrepancy is defined by

$$\Delta(\theta) = \sum_i (\pi_i - h_i(\theta))^2$$

and the empirical discrepancy by

$$\Delta_n(\theta) = \sum_i (p_i - h_i(\theta))^2.$$

The minimum discrepancy estimator is the ordinary least-squares estimator, the solution of the p equations:

$$\sum_i (p_i - h_i(\theta))h_i'(\theta) = 0.$$

In this case

$$\Sigma = 4 \sum_i \frac{h_i'(\theta_0)h_i'(\theta_0)^{\mathsf{T}}\pi_i(1 - \pi_i)}{N_i}$$

and

$$\Omega = \Delta''(\theta_0) = 2 \sum_i [(h_i(\theta_0) - \pi_i)h_i''(\theta_0) + h_i'(\theta_0)h_i'(\theta_0)^{\mathsf{T}}].$$

Estimators of Ω and Σ are

$$\Omega_n = 2 \sum_i [(h_i(\hat{\theta}) - p_i)h_i''(\hat{\theta}) + h_i'(\hat{\theta})h_i'(\hat{\theta})^{\mathsf{T}})]$$

and

$$\Sigma_n = 4 \sum_i \frac{h_i'(\hat{\theta})h_i'(\hat{\theta})^{\mathsf{T}}p_i(1 - p_i)}{N_i - 1}.$$

If the operating model is a member of the approximating family then $\pi_i \equiv h_i(\theta_0)$. In this case an estimator of Ω is

$$\Omega_n^* = 2 \sum_i h_i'(\hat{\theta})h_i'(\hat{\theta})^{\mathsf{T}}.$$

This estimator can be used as an approximation to Ω_n. The criterion is

$$\Delta_n(\hat{\theta}) + \operatorname{tr} \Omega_n^{-1} \Sigma_n$$

and the simpler criterion is obtained by replacing Ω_n by Ω_n^*.

10.2.2. Linear Models

Criteria for linear models which are based on the Gauss discrepancy were derived in Chapters 6, 7, and 8. Since homoskedasticity was not required for the operating models in these chapters one would expect these methods to be immediately applicable to proportions. This is in fact the case but special features arise and need explanations.

We use the operating model

$$p = \pi + e,$$

where $p = (p_1, p_2, \ldots, p_I)^{\mathsf{T}}$, $Ee = 0$, and $Eee^{\mathsf{T}} = \operatorname{Diag}\{\pi_i(1 - \pi_i)/N_i\} = \Sigma$. The approximating models have expectations

$$h(\theta) = A\theta,$$

where A is a given $p \times I$ matrix of full rank p.

The least-squares estimator is here

$$\hat{\theta} = C^{-1}A^{\mathsf{T}}p,$$

where $C = A^{\mathsf{T}}A$. The expected discrepancy becomes

$$E\Delta(\hat{\theta}) = \pi^{\mathsf{T}}\pi - \pi^{\mathsf{T}}AC^{-1}A^{\mathsf{T}}\pi + \operatorname{tr} AC^{-1}A^{\mathsf{T}}\Sigma.$$

For the operating family C is the unit matrix and the expected discrepancy reduces to $\operatorname{tr} \Sigma$. Therefore, the difference between the expected discrepancies of operating and approximating families is

$$-(\pi^{\mathsf{T}}\pi - \pi^{\mathsf{T}}AC^{-1}A^{\mathsf{T}}\pi) + \operatorname{tr}(\Sigma - AC^{-1}A^{\mathsf{T}}\Sigma).$$

This difference can be estimated by

$$-(p^{\mathsf{T}}p - p^{\mathsf{T}}AC^{-1}A^{\mathsf{T}}p) + 2\operatorname{tr}(\hat{\Sigma} - AC^{-1}A^{\mathsf{T}}\hat{\Sigma}),$$

where

$$\hat{\Sigma} = \text{Diag}\{p_i(1 - p_i)/(N_i - 1)\}.$$

The difference is negative and the approximating model should *not* be used if

$$F = \frac{(p^{\mathsf{T}}p - p^{\mathsf{T}}AC^{-1}A^{\mathsf{T}}p)/(I - p)}{\text{tr}\,(\hat{\Sigma} - AC^{-1}A^{\mathsf{T}}\hat{\Sigma})/(I - p)} > 2.$$

The numerator is the mean square for residuals in a regression analysis under the approximating model. The denominator can be calculated by replacing in the formula for the numerator $p_i p_j$ by $\delta_{ij}p_i(1 - p_i)/(N_i - 1)$, where δ_{ij} is the Kronecker delta.

The method becomes very convenient for factorial analysis of variance models. The operating model is then the saturated model and the approximating models are obtained by omitting certain effects or interactions. Because of orthogonality one can decide for each effect (or interaction) separately whether it should stay in the model.

To calculate the criterion one sets up an analysis of variance table. Although there is only one replication (one observation for each treatment combination), one must also calculate a sum of squares for the highest order interaction. There is no sum of squares for residuals.

The above "F ratio", now used to decide whether a certain effect (or interaction) should be retained in an approximating model, becomes in this case

$$F = \frac{\text{MS (effect)}}{\hat{V}},$$

where $\hat{V}$ is an estimator of the average variance of the proportions.

Take, for example, the case of two factors (A and B) with observed proportions p_{ij}, $i = 1, 2, \ldots, I, j = 1, 2, \ldots, J$. The analysis of variance table is then

Variation	d.f.	SS
Mean	1	$IJ\bar{p}_{..}^2$
A	$I - 1$	$J\sum_i(\bar{p}_{i.} - \bar{p}_{..})^2$
B	$J - 1$	$I\sum_j(\bar{p}_{.j} - \bar{p}_{..})^2$
AB	$(I - 1)(J - 1)$	$\sum_{ij}(p_{ij} - \bar{p}_{i.} - \bar{p}_{.j} + \bar{p}_{..})^2$
Total	IJ	$\sum_{ij}p_{ij}^2$

and

$$\hat{V} = \frac{1}{IJ} \sum_{ij} \frac{p_{ij}(1 - p_{ij})}{N_{ij} - 1}$$

Note, however, that this simple algorithm does not automatically hold for other subdivisions of SS Total. This can be seen from Example 10.2.2 to follow. There one is interested in a set of orthogonal contrasts of the π_i, the linear, quadratic, ... contrasts with coefficients λ_i as given in this example. Let the contrast in question be $\sum_i \lambda_i \pi_i$. Then the numerator of F is as before the mean square for the contrast:

$$\frac{(\sum_i \lambda_i p_i)^2}{\sum_i \lambda_i^2},$$

but the denominator is no longer $\hat{V}$. One has to return to the original formula and calculate the denominator of F by replacing $p_i p_j$ by $\delta_{ij} p_i (1 - p_i)/(N_i - 1)$ in the formula for the numerator:

$$F = \frac{(\sum_i \lambda_i p_i)^2/\sum_i \lambda_i^2}{[\sum_i \lambda_i^2 p_i (1 - p_i)/(N_i - 1)]/\sum_i \lambda_i^2}.$$

EXAMPLE 10.2.2.   Linder and Berchtold (1976, p. 150) analyze data of Traut and Scheid. Observed were the numbers of mutations (n_i) of (N_i) Drosophila under different doses of radiation $[x_i$ (roentgen)$]$:

i	x_i	n_i	N_i	$100p_i$
1	0	2	2000	0.10
2	100	18	3118	0.58
3	200	29	2603	1.11
4	300	133	7347	1.81
5	400	278	8891	3.13

The mean and the four orthogonal contrasts have the following coefficients (λ_i):

i	1	2	3	4	5	$\sum \lambda_i^2$
Mean	1	1	1	1	1	5

i	1	2	3	4	5	$\sum \lambda_i^2$
Linear	-2	-1	0	1	2	10
Quadratic	2	-1	-2	-1	2	14
Cubic	-1	2	0	-2	1	10
Quartic	1	-4	6	-4	1	70

One obtains the following analysis:

Variation	d.f.	SS $\times 10^4$	Denominator $\times 10^4$	F
Mean	1	9.05	0.0248	365
Linear	1	5.31	0.0199	267
Quadratic	1	0.241	0.0263	9.2
Cubic	1	0.0315	0.0210	1.5
Quartic	1	0.0019	0.0321	0.1
Total	5	14.63	0.1241	

The selected model is a parabola. This selection agrees with that of Linder and Berchtold.

10.3. DISCREPANCIES BASED ON MAHALANOBIS' GENERALIZED DISTANCE

A classical measure of the distance between two multivariate distributions with expectations μ_1 and μ_2 and common nonsingular covariance matrix Λ is Mahalanobis' generalized distance:

$$(\mu_2 - \mu_1)^\mathsf{T} \Lambda^{-1} (\mu_2 - \mu_1).$$

Consider the operating model for p with

$$Ep = \pi \quad \text{and} \quad E(p - \pi)(p - \pi)^\mathsf{T} = \Lambda$$

and an approximating model for which one only wants to specify

$$Ep = h(\theta).$$

A possible discrepancy, which we call the Mahalanobis discrepancy, is

$$(\pi - h(\theta))^T \Lambda^{-1} (\pi - h(\theta)).$$

If the approximating model also specifies the variance covariance matrix of p, say $\Lambda(\theta)$, then an alternative discrepancy is available, namely,

$$(\pi - h(\theta))^T \Lambda^{-1}(\theta)(\pi - h(\theta)),$$

which we call the modified Mahalanobis discrepancy.

In the case with which we deal in this chapter these two discrepancies are given by

$$\sum_{i=1}^{I} \frac{N_i(\pi_i - h_i(\theta))^2}{\pi_i(1 - \pi_i)}, \qquad 0 < \pi_i < 1,$$

and

$$\sum_{i=1}^{I} \frac{N_i(\pi_i - h_i(\theta))^2}{h_i(\theta)(1 - h_i(\theta))}, \qquad 0 < h_i(\theta) < 1,$$

and we call them the Neyman and Pearson chi-squared discrepancies, respectively.

10.4. NEYMAN CHI-SQUARED DISCREPANCY

In this case we define

$$\Delta(\theta) = \sum_{i=1}^{I} \frac{N_i(\pi_i - h_i(\theta))^2}{\pi_i(1 - \pi_i)}$$

and the empirical discrepancy by

$$\Delta_n(\theta) = \sum_{i=1}^{I} \frac{N_i(p_i - h_i(\theta))^2}{p_i(1 - p_i)}.$$

If p_i is zero or one, the denominator should be replaced by $h_i(\theta)(1 - h_i(\theta))$ for the corresponding i.

The minimum discrepancy estimator which is obtained by minimizing $\Delta_n(\theta)$ is a minimum Neyman χ^2 estimator.

One can show that the propositions of Section 2.4 hold with

$$\Sigma = 4 \sum_{i=1}^{I} \frac{N_i h_i'(\theta_0) h_i'(\theta_0)^{\mathsf{T}} [\pi_i^2 + h_i(\theta_0)(1 - 2\pi_i)]^2}{\pi_i^3 (1 - \pi_i)^3}$$

and

$$\Omega = 2 \sum_{i=1}^{I} \frac{N_i [(h_i(\theta_0) - \pi_i) h_i''(\theta_0) + h_i'(\theta_0) h_i'(\theta_0)^{\mathsf{T}}]}{\pi_i (1 - \pi_i)}.$$

If the operating model is a member of the approximating family then $\pi_i = h_i(\theta_0)$ and consequently $\Sigma = 2\Omega$. It would not be difficult to give unbiased estimators of Ω and Σ, but it seems preferable to use the simple criterion

$$\Delta_n(\hat{\theta}) + 2p.$$

10.5. PEARSON CHI-SQUARED DISCREPANCY

The discrepancy here is

$$\Delta(\theta) = \sum_{i=1}^{I} \frac{N_i (\pi_i - h_i(\theta))^2}{h_i(\theta)(1 - h_i(\theta))},$$

and the empirical discrepancy is

$$\Delta_n(\theta) = \sum_{i=1}^{I} \frac{N_i (p_i - h_i(\theta))^2}{h_i(\theta)(1 - h_i(\theta))}.$$

The minimum discrepancy estimator is a minimum Pearson χ^2 estimator. One can show that the results of Section 2.4 hold with

$$\Omega = \Delta''(\theta_0)$$

and

$$\Sigma = 4 \sum_{i=1}^{I} \frac{h_i'(\theta_0) h_i'(\theta_0)^{\mathsf{T}} N_i \pi_i (1 - \pi_i)[h_i^2(\theta_0) + \pi_i(1 - 2h_i(\theta_0))]^2}{h_i^4(\theta_0)(1 - h_i(\theta_0))^4}.$$

If $\pi_i = h_i(\theta_0)$, that is, if the approximating family contains the operating model, then again $\Sigma = 2\Omega$. The simple criterion is

$$\Delta_n(\hat{\theta}) + 2p.$$

EXAMPLE 10.5.1. We consider the data of Pearson and Hartley (1972, p. 95). The experiment investigated the effect of various doses of antipneumococcus serum on the survival of infected mice. Five different doses (working scale $x_i = -2, -1, 0, 1, 2$) were applied to five groups of 40 mice each, of which n_i survived. The results were:

i	x_i	n_i	N_i	p_i
1	-2	7	40	0.175
2	-1	18	40	0.450
3	0	32	40	0.800
4	1	35	40	0.875
5	2	38	40	0.950

We wish to decide whether the logit or the normit model should be used. For the logit model

$$h_i(\theta) = \frac{e^{\alpha + \beta x_i}}{1 + e^{\alpha + \beta x_i}}$$

and for the normit model

$$h_i(\theta) = \Phi(\alpha + \beta x_i),$$

where $\theta = (\alpha, \beta)^{\mathsf{T}}$ and Φ is the standard normal distribution function.

We use both the Neyman and the Pearson chi-squared discrepancies. Since in all cases the number of parameters, p, is equal to 2 it suffices, for the simple criterion, to calculate $\Delta_n(\hat{\theta})$ in each case. The results are the following:

Discrepancy	Approximating Family	$\hat{\alpha}$	$\hat{\beta}$	$\Delta_n(\hat{\theta})$
Neyman χ^2	Logit	0.973	1.167	1.735
Neyman χ^2	Normit	0.538	0.662	2.641
Pearson χ^2	Logit	1.170	1.354	0.225
Pearson χ^2	Normit	0.695	0.810	0.322

The logit model is selected in both cases.

EXAMPLE 10.5.2. A given weight of charge was detonated at a distance x_i (ft) from a disk of cardboard and it was recorded whether the disk was perforated or not. This was done $N_i = 16$ times at each of five distances (Pearson and Hartley, 1976, p. 8.) The observations were:

i	x_i	Working Scale	n_i	N_i	p_i
1	53	0	0	16	0.0000
2	49	1	9	16	0.5625
3	45	2	9	16	0.5625
4	41	3	12	16	0.7500
5	37	4	16	16	1.0000

We wish to decide whether the logit or the normit model should be used. Here one of the p_i is zero and the Neyman chi-squared discrepancy leads to difficulties. We therefore use the Pearson chi-squared discrepancy.

For the logit model one gets $\hat{\alpha} = -1.546$, $\hat{\beta} = 1.022$, and $\Delta_n(\hat{\theta}) = 7.938$. In the normit model $\hat{\alpha} = -0.965$, $\hat{\beta} = 0.629$, and $\Delta_n(\hat{\theta}) = 7.583$. Since in the two cases the number of parameters is equal (2) it suffices to consider $\Delta_n(\hat{\theta})$. The normit model has a smaller criterion and is therefore estimated to result in a better fit.

10.6. ANALYSIS IN THE TRANSFORMED DOMAIN

It is established practice in biometry to fit models, in particular linear models, to *transformed* proportions. Well-known examples are logits $[z_i = z(p_i) = \log p_i/(1 - p_i)]$ and normits $[z_i = \Phi^{-1}(p_i)$, where Φ is the standard normal distribution function]. A list of other possible transformations can be found in Linder and Berchtold (1976). In such situations discrepancies can be defined in the transformed domain and methods such as Berkson's minimum logit and minimum normit estimation (e.g., see Pearson and Hartley, 1972, pp. 91–95) can be interpreted as minimum discrepancy estimation.

10.6.1. Mahalanobis Discrepancy: The General Case

We analyze the problem in general. We write z_i for $z(p_i)$, ξ_i for $z(\pi_i)$, assume that $z(x)$ is *differentiable* for $0 < x < 1$, and write z_i' for $z'(p_i)$ and ξ_i' for $z'(\pi_i)$. Using an $O(1/n^2)$ approximation to the variance of $z(p_i)$, the Mahalanobis discrepancy becomes

$$\Delta(\theta) = \sum_{i=1}^{I} \frac{N_i(\xi_i - h_i(\theta))^2}{\xi_i'^2 \pi_i(1 - \pi_i)},$$

where $h_i(\theta)$ is now the approximation to $z(\pi_i)$, and an empirical discrepancy is

$$\Delta_n(\theta) = \sum_{i=1}^{I} \frac{N_i(z_i - h_i(\theta))^2}{z_i'^2 p_i(1 - p_i)}.$$

For the minimum discrepancy estimator the results of Section 2.4 hold with

$$\Sigma = 4 \sum_{i=1}^{I} \frac{N_i h_i'(\theta_0) h_i'(\theta_0)^{\mathsf{T}}}{\xi_i'^2 \pi_i(1 - \pi_i)} \left[1 - (\xi_i - h_i(\theta_0)) \left(\frac{2\xi_i''}{\xi_i'^2} + \frac{1 - 2\pi_i}{\xi_i' \pi_i(1 - \pi_i)} \right) \right]^2$$

and

$$\Omega = 2 \sum_{i=1}^{I} \frac{N_i[(h_i(\theta_0) - \xi_i) h_i''(\theta_0) + h_i'(\theta_0) h_i'(\theta_0)^{\mathsf{T}}]}{\xi_i'^2 \pi_i(1 - \pi_i)}.$$

If $\xi_i = h_i(\theta_0)$ then $\Sigma = 2\Omega$ and the simpler criterion is

$$\Delta_n(\hat{\theta}) + 2p.$$

10.6.2. Mahalanobis Discrepancy: Linear Models

The analysis becomes particularly convenient in the case of linear models

$$h(\theta) = A\theta,$$

where A is a nonsingular $I \times p$ matrix of rank p. Here

$$\Delta_n(\theta) = (z - A\theta)^{\mathsf{T}} U(z - A\theta),$$

where

$$U = \text{Diag}\{N_i/z_i'^2 p_i(1 - p_i)\},$$
$$\hat{\theta} = (A^{\mathsf{T}} U A)^{-1} A^{\mathsf{T}} U z,$$

and

$$\Delta_n(\hat{\theta}) = z^{\mathsf{T}} U z - z^{\mathsf{T}} U A (A^{\mathsf{T}} U A)^{-1} A^{\mathsf{T}} U z = z^{\mathsf{T}} U z - z^{\mathsf{T}} U A \hat{\theta}.$$

If one calls the quadratic forms, as in analysis of variance, SS Total and SS Regr and their difference SS Res then

$$\Delta_n(\hat{\theta}) = \text{SS Total} - \text{SS Regr} = \text{SS Res}$$

and the simpler criterion is

$$\text{SS Res} + 2p.$$

The standard regression analysis calculations have to be carried out with the proviso that all sums of squares and products are now *weighted* sums of squares and products. For example, let

$$A = \{a_{ij}: i = 1, \ldots, I, \quad j = 1, \ldots, p\}$$

and

$$a_j = (a_{1j}, a_{2j}, \ldots, a_{Ij})^\mathsf{T}, \quad j = 1, 2, \ldots, p.$$

Then $z^\mathsf{T} A = (\text{SP}_{za_1}, \ldots, \text{SP}_{za_p})$ with $\text{SP}_{za_j} = \sum_i z_i a_{ij}$ is now to be replaced by $z^\mathsf{T} U A = (\text{SP}u_{za_1}, \ldots, \text{SP}u_{za_p})$ with $\text{SP}u_{za_j} = \sum_i u_i z_i a_{ij}$, and similarly $\text{SP}_{a_j a_k} = \sum_i a_{ij} a_{ik}$ has to be replaced by $\text{SP}u_{a_j a_k} = \sum_i u_i a_{ij} a_{ik}$.

Unfortunately, the above discrepancy does not preserve the convenient orthogonality property in analysis of variance models. The criterion has to be computed for each approximating model.

EXAMPLE 10.6.2.1.   We consider once more the data of Example 10.5.1 to see whether the logit model is still selected if one uses the Mahalanobis discrepancy in the transformed domain. Both models are *linear* in the transformed domain and the method of this section may be used.

For the logit model $z_i = \log p_i/(1 - p_i)$ and $u_i = N_i p_i (1 - p_i)$, and for the normit model $z_i = \Phi^{-1}(p_i)$ and $u_i = N_i \varphi^2(z_i)/p_i(1 - p_i)$. Here Φ^{-1} is the inverse of the normal distribution function and φ the unit normal density function.

One gets:

Approximating Model	$\hat{\alpha}$	$\hat{\beta}$	$\Delta_n(\hat{\theta})$
Logit	0.971	1.167	1.716
Normit	0.546	0.669	2.664

The criterion is again smaller for the logit model.

EXAMPLE 10.6.2.2.   Consider Example 10.2.2. We compare the quadratic model with the saturated model because we want to see whether the simpler model, with the cubic and quartic contrasts equal to zero, leads to a smaller criterion.

Here $z(x)$ is the identity $z(p_i) = p_i$, and thus $U = \mathrm{Diag}\{N_i/p_i(1 - p_i)\}$. Under the saturated model $A = \mathcal{I}$ and SS Res $= 0$. Under the approximating model one has the restrictions

$$-\pi_1 + 2\pi_2 - 2\pi_4 + \pi_5 = 0,$$

$$\pi_1 - 4\pi_2 + 6\pi_3 - 4\pi_4 + \pi_5 = 0,$$

which leads to

$$\pi_4 = \pi_1 - 3\pi_2 + 3\pi_3,$$

$$\pi_5 = 3\pi_1 - 8\pi_2 + 6\pi_3.$$

The approximating model, $Ep = A\theta$, has

$$A^{\mathsf{T}} = \begin{bmatrix} 1 & 0 & 0 & 1 & 3 \\ 0 & 1 & 0 & -3 & -8 \\ 0 & 0 & 1 & 3 & 6 \end{bmatrix}, \qquad \theta = \begin{bmatrix} \pi_1 \\ \pi_2 \\ \pi_3 \end{bmatrix}.$$

One gets

$$z^{\mathsf{T}}Uz - z^{\mathsf{T}}UA(A^{\mathsf{T}}UA)^{-1}A^{\mathsf{T}}Uz = 1.513.$$

The criterion is equal to 10 (twice the number of parameters) for the saturated model and is $1.513 + 6 = 7.513$ for the restricted model. The latter is therefore selected.

CHAPTER 11

The Analysis
of Proportions:
Contingency Tables

An $I \times J \times K$ contingency table gives the frequencies of certain events. There are discrete variables x, y, z which have I, J, K possible values. Assume, for simplicity, that these values are $1, 2, \ldots, I; 1, 2, \ldots, J; 1, 2, \ldots, K$. The integers in the table, n_{ijk}, give the frequencies of the event that $x = i$ and $y = j$ and $z = k$.

The immediate object of the statistical analysis—to whatever use it is put in the end—is the study of the joint distribution of the random variables x, y, z. Often *simpler* approximating models lead to better estimates than the operating model

$$P(x = i, y = j, z = k) = \pi_{ijk}.$$

There is a rich literature on contingency tables, which is concerned with simpler structures for discrete multivariate distributions and associated methods of analysis. Particularly successful are loglinear models in which the logarithm of π_{ijk} is expressed as a sum of certain effects and interactions, analogous to factorial analysis of variance models.

Our main concern in this chapter is model selection methods for loglinear models. These are discussed in Sections 11.2 and 11.3. As an introduction we consider initially, in Section 11.1, a simpler problem for two-dimensional contingency tables.

11.1. INDEPENDENCE IN $I \times J$ TABLES

We consider an $I \times J$ contingency table in which the frequencies n_{ij} were obtained in a single sample of size $n_{..} = n$ by simultaneous observations of x and y. In this situation the operating model for n_{ij} is a multinomial distribution with parameters π_{ij} and n.

In traditional analyses the hypothesis that $\pi_{ij} = \pi_{i.}\pi_{.j}$ is tested. The operating model is very rarely *exactly* of this form but it will sometimes almost have this structure and it might be better to estimate π_{ij} by $p_{i.}p_{.j} = (n_{i.}/n)(n_{.j}/n)$ rather than by $p_{ij} = n_{ij}/n$.

We use the Gauss discrepancy. Denote the probabilities in the approximating model by $h_{ij}(\theta)$, then

$$\Delta(\theta) = -2 \sum_{ij} \pi_{ij} h_{ij}(\theta) + \sum_{ij} h_{ij}^2(\theta).$$

One could ask whether the analysis should be based on the operating model or on the model

$$h_{ij}(\theta) = \theta_{1i}\theta_{2j},$$

where all $\theta_{kl} \geqslant 0$ and $\theta_{1.} = \theta_{2.} = 1$. One can show that $\theta^{0\mathsf{T}} = (\theta_1^{0\mathsf{T}}, \theta_2^{0\mathsf{T}})$, the vector minimizing $\Delta(\theta)$, is equal to $(\pi_{1.}, \ldots, \pi_{I.}, \pi_{.1}, \ldots, \pi_{.J})$, which is estimated by $(p_{1.}, \ldots, p_{I.}, p_{.1}, \ldots, p_{.J})$.

The expected discrepancy is

$$E\Delta(\hat{\theta}) = -2E \sum_{ij} \pi_{ij} p_{i.} p_{.j} + E \sum_{ij} p_{i.}^2 p_{.j}^2$$

$$= -\frac{2}{n}\left(\sum_{ij} \pi_{ij}^2 + (n-1) \sum_{ij} \pi_{ij}\pi_{i.}\pi_{.j}\right) + E \sum_{ij} p_{i.}^2 p_{.j}^2.$$

By a straightforward but tedious calculation an unbiased estimator (a criterion) can be calculated. It is

$$n\hat{E}\Delta(\hat{\theta}) = \frac{2n}{n-2}\left[\frac{\sum_{ij} p_{ij}^2}{n-1} + \sum_i p_{i.}^2 + \sum_j p_{.j}^2 - n \sum_{ij} p_{ij}p_{i.}p_{.j} - \frac{1}{n}\right]$$

$$+ n \sum_{ij} p_{i.}^2 p_{.j}^2.$$

This must then be compared with an estimator of the expected discrepancy for the operating model, that is, if π_{ij} is estimated by p_{ij}. One gets

$$-2E \sum_{ij} \pi_{ij}p_{ij} + E \sum_{ij} p_{ij}^2 = -2 \sum_{ij}\pi_{ij}^2 + E \sum_{ij} p_{ij}^2$$

and the criterion, an estimator of n times this expectation, is

$$\frac{n}{n-1} [2 - (n+1) \sum_{ij} p_{ij}^2].$$

EXAMPLE 11.1.1. Lukatis (1972) studied the employment of university graduates in industry and commerce. The study was based on a random sample from the population of all industrial firms with more than 200 employees and all commercial firms with more than 500 employees in Württemberg and Bavaria.

The following table gives the number of graduates employed in relation to the distance of the employing firm from the nearest university:

Number of Graduates	Distance to Nearest University			Total
	< 50 km	50–100 km	> 100 km	
0	31	40	17	88
1	32	27	10	69
2–4	44	46	15	105
5–9	29	22	7	58
10–25	31	9	4	44
26 or more	37	3	4	44
Total	204	147	57	408

For the calculation of the criteria one needs

$$\sum_{ij} p_{ij}^2 = 0.07693, \qquad \sum_i p_{i.}^2 = 0.18482,$$
$$\sum_j p_{.j}^2 = 0.399330, \qquad \sum_{ij} p_{ij}p_{i.}p_{.j} = 0.0725441,$$

and obtains

$$\frac{408}{407} (2 - 409 \times 0.07693) = -29.53$$

for the family of the operating model, and

$$\frac{816}{406}\left(\frac{0.07693}{407} + 0.1848 + 0.3993 - 408 \times 0.0725441 - \frac{1}{407}\right)$$

$$+ 408 \times 0.1848 \times 0.3993 = -28.21$$

for the simpler model, based on independence. The simpler model does not fit sufficiently well to warrant its use.

EXAMPLE 11.1.2. The numbers of graduates employed versus the firms' forecasts of employment, observed in the same study, are given below:

Number of Graduates	Expected Increase of Number of Employees				
	Negative	Zero	$1-20\%$	$>20\%$	Total
0	7	17	34	30	88
1	10	17	23	17	67
2–4	14	21	36	33	104
5–9	5	17	20	16	58
10–25	2	6	21	13	42
26 or more	2	8	23	11	44
Total	40	86	157	120	403

Here the criterion has a value of -20.34 for the family of the operating model and a value of -20.97 for the simpler model. The simpler model for which π_{ij} is estimated by $p_{i.}p_{.j}$ is selected by the criterion.

11.2. THREE GENERAL ASYMPTOTIC METHODS

We discuss methods for contingency tables with *multinomial* operating models. Such models are appropriate when the observed frequencies n_{ijk} are the result of a *single* sample from a population. There is no difficulty in developing methods for the case where one has product–multinomial operating models. The latter arise when some of the marginal totals are fixed, that is, where samples are taken from subpopulations.

We will consider the Kullback–Leibler, the Neyman and the Pearson chi-squared discrepancies. The corresponding methods have in fact been introduced in Section 4.1.2 which deals with multinomial operating and approximating models. (Data on a single discrete variable, or grouped data on a continuous variable, can be regarded as a one-dimensional contingency table.)

The operating model admits arbitrary probabilities π_{ijk} ($\pi_{...} = 1$) and the approximating model uses restricted probabilities $h_{ijk}(\theta)$ ($h_{...}(\theta) = 1$), which are a function of p free parameters.

The three resulting criteria are

$$G^2(\hat{\theta}) + 2p,$$

$$\chi_N^2(\hat{\theta}) + 2p,$$

and

$$\chi_P^2(\hat{\theta}) + 2p,$$

where

$$G^2(\hat{\theta}) = 2 \sum_{ijk} n_{ijk} (\log n_{ijk} - \log nh_{ijk}(\hat{\theta})),$$

$$\chi_N^2(\hat{\theta}) = \sum_{ijk} \frac{(n_{ijk} - nh_{ijk}(\hat{\theta}))^2}{n_{ijk}},$$

and

$$\chi_P^2(\hat{\theta}) = \sum_{ijk} \frac{(n_{ijk} - nh_{ijk}(\hat{\theta}))^2}{nh_{ijk}(\hat{\theta})},$$

where n_{ijk} is replaced by $nh_{ijk}(\hat{\theta})$ in the denominator of $\chi_N^2(\hat{\theta})$ whenever $n_{ijk} = 0$.

In the above criteria it is not necessary to use the corresponding minimum discrepancy estimator as $\hat{\theta}$; any of the three possible estimators can be used in each case.

These criteria are particularly suitable for the analysis of loglinear models. These models and the associated methodology emerged during the last decades and provide very important tools for the analysis of multi-dimensional contingency tables. A good introduction to these methods is the short book by Fienberg (1980) which also gives some historical background and references to further literature. A more detailed treatment can be found in Plackett (1974) and in Bishop et al. (1975). Model selection methods based on G^2, that is, methods using Akaike's Information Criterion, are described in Sakamoto and Akaike (1978a, 1978b).

Subject to some restrictions, the selection criteria for loglinear models developed for the *multinomial* case are directly applicable to the *product-multinomial* case. The restrictions are that the parameters must be estimated by maximum likelihood and that selection is restricted to approximating models which contain all the main effects and interactions which correspond to the fixed marginal totals.

Consider, for example, a four-way contingency table in which the marginal totals $n_{ij..}$ are fixed, that is, the data constitute IJ samples taken from IJ subpopulations. In this case all approximating models must contain the main effects, α_i and β_j, as well as the interactions $(\alpha\beta)_{ij}$. If only such approximating models are admitted then the maximum likelihood estimators of the expected frequencies are the same as those for the multinomial case, and consequently $G^2(\hat\theta)$ and $\chi^2(\hat\theta)$ can be computed as before.

The number of free parameters in the product-multinomial case is lower than in the corresponding multinomial case. In a four-way table with fixed totals $n_{ij..}$, for example, there are $IJ - 1$ fewer free parameters. This might suggest that an adjustment is necessary to the criteria $G^2(\hat\theta) + 2p$ and $\chi^2(\hat\theta) + 2p$, but this is in fact not the case. The reason is that by restricting the permissible approximating models as outlined above, the identical adjustment applies to all these approximating models. It is therefore an inessential component of the criteria. In other words, in the criteria $G^2(\hat\theta) + 2p$ and $\chi^2(\hat\theta) + 2p$ the same p can be used in both multinomial and product-multinomial situations.

EXAMPLE 11.2.1.   The Town Council of Göttingen conducted a survey in 1985 to gauge the inhabitants' opinions on the desirability of supplementing the existing refuse collection service by introducing separate bins for organic waste (Schäfer, 1985). Four areas were selected: Geismar, Hetjershausen, Göttingen-East, and Grone. These represent four housing types: clustered housing, central village housing, suburban housing, and high-rise housing (flats). The inhabitants were asked to state whether they were in favor of the proposed scheme, and also to indicate their opinion (in one of four categories) on the existing refuse collection service. The observed frequencies are given in Table 11.1.

The operating model is the saturated model

$$\log n\pi_{ijk} = \mu + \alpha_i + \beta_j + \gamma_k + (\alpha\beta)_{ij} + (\alpha\gamma)_{ik} + (\beta\gamma)_{jk} + (\alpha\beta\gamma)_{ijk},$$

where i, j, and k refer to attitude toward proposed scheme, opinion on existing service, and area:

$$i = 1: \text{against}, \qquad\qquad i = 2: \text{in favor};$$

Table 11.1. Observed Frequencies in Survey on a Proposed Refuse Collection Scheme

Area	Opinion on Existing Service	Proposed Scheme	
		Against	In Favor
Geismar	Very satisfied	18	61
	Satisfied	43	203
	Indifferent	16	36
	Not satisfied	1	10
Hetjershausen	Very satisfied	13	56
	Satisfied	16	78
	Indifferent	6	12
	Not satisfied	5	2
Göttingen-East	Very satisfied	29	135
	Satisfied	53	203
	Indifferent	15	30
	Not satisfied	9	17
Grone	Very satisfied	15	28
	Satisfied	33	125
	Indifferent	17	57
	Not satisfied	8	16

Total number of observations: 1366

$j = 1$: very satisfied, $\quad j = 2$: satisfied,

$j = 3$: indifferent, $\quad j = 4$: not satisfied;

$k = 1$: Geismar, $\quad k = 2$: Hetjershausen,

$k = 3$: Göttingen-East, $\quad k = 4$: Grone

The Akaike Information Criterion (AIC) corresponding to each of the possible hierarchical models is given in Table 11.2. In the table the models are identified by their sufficient configurations. For example, model 6 has sufficient configurations $\{n_{i.k}\}$ and $\{n_{.j.}\}$. The corresponding model,

$$\log nh_{ijk}(\theta) = \mu + \alpha_i + \beta_j + \gamma_k + (\alpha\beta)_{ik},$$

has $1 + (I - 1) + (J - 1) + (K - 1) + (I - 1)(K - 1) - 1 = 10$ free parameters. The last "1" is subtracted because the sum of the probabilities is one.

Tables 11.2. Models and Criteria for Refuse Collection Data

Model	Sufficient Configurations	d.f.	p	G^2	AIC
1	$(ij)(ik)(jk)$	9	22	16.45	60.45
2	$(ij)(jk)$	12	19	17.14	55.14
3	$(ik)(jk)$	12	19	28.31	66.31
4	$(ij)(ik)$	18	13	104.50	130.50
5	$(ij)(k)$	21	10	106.38	126.38
6	$(ik)(j)$	21	10	117.54	137.54
7	$(jk)(i)$	15	16	30.18	62.18
8	(jk)	16	15	493.31	523.31
9	(ik)	24	7	896.37	910.37
10	(ij)	24	7	258.09	272.09
11	$(i)(j)(k)$	24	7	119.42	133.42
12	$(i)(k)$	27	4	898.24	906.24
13	$(j)(k)$	25	6	582.55	594.55
14	$(i)(j)$	27	4	271.13	279.13
15	(k)	28	3	1361.37	1367.37
16	(j)	28	3	734.26	740.26
17	(i)	30	1	1049.95	1051.95

The criterion selects model 2:

$$\log nh_{ijk}(\theta) = \mu + \alpha_i + \beta_j + \gamma_k + (\alpha\beta)_{ij} + (\beta\gamma)_{jk}.$$

The estimates of the loglinear parameters are

$$\hat{\alpha}_1 = -0.542, \quad \hat{\alpha}_2 = 0.542;$$

$$\hat{\beta}_1 = 0.322, \quad \hat{\beta}_2 = 1.085, \quad \hat{\beta}_3 = -0.205, \quad \hat{\beta}_4 = -1.202;$$

$$\hat{\gamma}_1 = 0.065, \quad \hat{\gamma}_2 = -0.587, \quad \hat{\gamma}_3 = 0.437, \quad \hat{\gamma}_4 = 0.086;$$

$(\widehat{\beta\gamma})_{jk}$ k	j	1	2	3	4
1		-0.062	0.276	0.148	-0.362
2		0.455	-0.033	-0.260	-0.162
3		0.297	-0.056	-0.368	0.126
4		-0.691	-0.187	0.480	0.397

$(\widehat{\alpha\beta})_{ij}$	i	j	1	2	3	4
	1		-0.116	-0.175	0.084	0.207
	2		0.116	0.175	-0.084	-0.207

The frequencies $nh_{ijk}(\hat{\theta})$ under the fitted model are in Table 11.3.

Most inhabitants are in favor of the new proposal and, on the whole, opinion on the existing service is favorable. The estimated $(\alpha\beta)$ interactions indicate that those who are less satisfied with the existing service are also against the new proposals. Opinion on the existing service varies from area to area.

EXAMPLE 11.2.2. It becomes increasingly difficult to scan all possible models when the dimension of a contingency table increases. Although this example, which deals with a four-way classification, is still manageable, we will also use it to illustrate a technique to reduce the number of approximating models to be considered. A different method is described by Brown (1976). Bishop et al. (1975, pp. 142–146, 158–159) analyzed data on detergent

Table 11.3. Expected Frequencies in Selected Model for Refuse Collection Data

Area	Opinion on Existing Service	Proposed Scheme	
		Against	In Favor
Geismar	Very satisfied	16.7	62.3
	Satisfied	47.3	198.7
	Indifferent	14.9	37.1
	Not satisfied	3.7	7.3
Hetjershausen	Very satisfied	14.6	54.4
	Satisfied	18.1	75.9
	Indifferent	5.1	12.9
	Not satisfied	2.4	4.6
Göttingen-East	Very satisfied	34.6	129.4
	Satisfied	49.2	206.8
	Indifferent	12.9	32.1
	Not satisfied	8.8	17.2
Grone	Very satisfied	9.1	33.9
	Satisfied	30.4	127.6
	Indifferent	21.1	52.9
	Not satisfied	8.1	15.9

preferences. The reported frequencies refer to:

Water softness	Soft, medium, hard,	$(i = 1, 2, 3)$
Previous use of brand M	Yes, no,	$(j = 1, 2)$
Water temperature	Low, high,	$(k = 1, 2)$
Brand preference	$X, M,$	$(l = 1, 2)$

The data are reproduced below:

i	l	k	j 1 / 1	2	2 / 1	2	
1	1	1	19	57	29	63	168
		2	29	49	27	53	158
	2	1	23	47	33	66	169
		2	47	55	23	50	175
	3	1	24	37	42	68	171
		2	43	52	30	42	167
			185	297	184	342	1008

For this case the saturated loglinear model is

$$\log n\pi_{ijkl} = \mu + \alpha_i + \beta_j + \gamma_k + \delta_l + (\alpha\beta)_{ij} + (\alpha\gamma)_{ik} + (\alpha\delta)_{il} + (\beta\gamma)_{jk} + (\beta\delta)_{jl}$$
$$+ (\gamma\delta)_{kl} + (\alpha\beta\gamma)_{ijk} + (\alpha\beta\delta)_{ijl} + (\alpha\gamma\delta)_{ikl} + (\beta\gamma\delta)_{jkl} + (\alpha\beta\gamma\delta)_{ijkl},$$

with the usual restrictions of analysis of variance ($\sum_i \alpha_i = 0$, $\sum_j \beta_j = 0$, ...) and the additional restriction that $\sum_{ijkl} \pi_{ijkl} = 1$. Thus there are $3 \times 2 \times 2 \times 2 - 1 = 23$ free parameters in this operating family. Approximating models are obtained by eliminating some of the terms in the saturated model. We consider hierarchical models only.

There are nevertheless very many such models. To compute a criterion for each one is a costly computational task for high-dimensional tables. There is no orthogonality and the criterion for a model *can not* be obtained as a sum of components belonging to the individual effects or interactions. In such situations one may be forced to apply a screening procedure in order to preselect those models which are likely to be appropriate. Such a procedure was applied here.

To arrive at hierarchical loglinear models with small values of the criterion we analyzed the logarithms of the frequencies. A factorial analysis of variance model was assumed for log n_{ijkl} and the model selection methods of Chapter 8 were applied. Of course, a loglinear model does *not* imply an analysis of variance model for log n_{ijkl}. Also, the methods of Chapter 8 are based on the Gauss discrepancy. The results of this analysis are used to preselect models. The analysis of variance of Table 11.4 resulted.

The 90% point of F with one and two degrees of freedom is given as 17.5 in Table A.4.6. Only γ and $(\beta\delta)$ would pass the corresponding test; a final model should contain these two effects. The $F > 2$ rule suggests that apart from β, which belongs in any case to a hierarchical model containing γ and $(\beta\delta)$, also $(\gamma\delta)$, $(\alpha\gamma)$, $(\beta\gamma\delta)$ and $(\alpha\beta\delta)$ should be included in the model. We consider the hierarchical models generated by successively including these interactions in the given order. The values of the three criteria are given below. Model 2 is selected by all criteria:

Model	Sufficient Configurations	p	$G^2 + 2p$	$\chi_P^2 + 2p$	$\chi_N^2 + 2p$
1	$(k)(jl)$	4	30.85	31.54	29.32
2	$(jl)(kl)$	5	28.49	28.74	28.38
3	$(ik)(jl)(kl)$	9	29.89	29.92	30.04
4	$(ik)(jkl)$	11	30.41	30.44	30.42

Table 11.4. Analysis of Variance for Logarithms of Frequencies in Study on Detergent Preferences

Variation	d.f.	MS	F
α	2	0.013	1.4
β	1	0.043	4.4
γ	1	1.959	202.5
δ	1	0.006	0.6
$(\alpha\beta)$	2	0.019	2.0
$(\alpha\gamma)$	2	0.084	8.7
$(\alpha\delta)$	2	0.001	0.1
$(\beta\gamma)$	1	0.016	1.7
$(\beta\delta)$	1	0.592	61.2
$(\gamma\delta)$	1	0.099	10.2
$(\alpha\beta\gamma)$	2	0.019	1.9
$(\alpha\beta\delta)$	2	0.053	5.5
$(\alpha\gamma\delta)$	2	0.003	0.3
$(\beta\gamma\delta)$	1	0.061	6.3
e	2	0.010	
Total	23		

This set of data has been analyzed by several authors and is discussed in detail in Goodman (1971) and in Bishop et al. (1975). The latter make use of standardized parameters to make a preliminary selection of effects and interactions which should be included in the model. Their minimal model in this sense is model 1 above. To extend the model they suggest that the main effect α should be considered for inclusion, and possibly also the two-way interactions $(\alpha\gamma)$ and $(\gamma\delta)$. The corresponding values of $G^2 + 2p$ for these four models are 30.85, 34.35, 36.25, and 29.89.

Naturally, it is neither necessary nor always advisable to rely solely on data-based methods to preselect candidate models for final selection.

The model which we selected is

$$\log nh_{ijk}(\hat{\theta}) = \mu + \beta_j + \gamma_k + \delta_l + (\beta\delta)_{jl} + (\gamma\delta)_{kl}.$$

The corresponding estimated expected frequencies are:

i	l	j k	1 1	2	2 1	2
1	1		23.1	45.9	33.6	66.8
	2		36.5	55.2	29.9	45.2
2	1		23.1	45.9	33.6	66.8
	2		36.5	55.2	29.9	45.2
3	1		23.1	45.9	33.6	66.8
	2		36.5	55.2	29.9	45.2

11.3.　CROSS-VALIDATORY METHODS

In the two preceding sections we gave a finite sample criterion for the Gauss discrepancy (for two-dimensional tables only) and asymptotic criteria for the Kullback–Leibler, the Pearson chi-squared, and the Neyman chi-squared discrepancy. For discrepancies and situations not covered so far, Bootstrap or cross-validatory methods could be used. We make a few remarks on cross-validatory methods.

In the data situation of this chapter the "items" (compare Section 2.6) are realizations of *identically* distributed random vectors and the crossvalidatory criteria are unbiased estimators of the expected discrepancy belonging to a sample of size $n - 1$. (See the Proposition of Section 2.6.)

For the Gauss discrepancy, the cross-validatory criterion is (e.g., in the three-dimensional case)

$$CV = -\frac{2}{n}\sum_{ijk} n_{ijk}\, m_{ijk}^{(ijk)} + \frac{1}{n^2}\sum_{ijk} n_{ijk}\sum_{rst} (m_{ijk}^{(rst)})^2,$$

where $m_{ijk}^{(rst)}$ is the estimator of the expected frequency of cell (ijk) under the considered approximating model if a single observation is removed from cell (rst) (Stone, 1974b). The method of estimation need not be minimum discrepancy, it could for instance be maximum likelihood. This criterion and formulae for criteria based on other discrepancies can be found in Section 2.6.

For purposes of illustration we calculated the cross-validatory Gauss criteria for the data and the models considered in Examples 11.1.1, 11.1.2, 11.2.1, and 11.2.2:

Example	Sufficient Configurations	Cross-validatory Gauss criterion
11.1.1	(ij)	-29.54
	$(i)(j)$	-28.21
11.1.2	(ij)	-20.34
	$(i)(j)$	-20.98
11.2.1	$(ij)(ik)(jk)$	-104.24
	$(ij)(jk)$	-104.39
	$(i)(jk)$	-104.29
11.2.2	$(k)(jl)$	-45.57
	$(jl)(kl)$	-45.67
	$(ik)(jl)(kl)$	-45.54
	$(ik)(jkl)$	-45.54

It can be seen that in all four examples this criterion selects the same models which were selected in Sections 11.1 and 11.2. In Examples 11.2.1 and 11.2.2 the order of merit in the group of the best models changes if compared with the AIC results. (Akaike's Information Criterion uses the same method of estimation but with another basic discrepancy.)

CHAPTER 12

Time Series

The usefulness of parametric models, in particular autoregressive integrated moving average (ARIMA) processes, for analyzing and forecasting time series was demonstrated by Box and Jenkins (1970). One of the more difficult steps in applying the methodology is that of selecting the order of the model, that is, the number of parameters in the autoregressive and moving average components. This step requires judgment and expertise. It is therefore not surprising that in recent years some automatic selection procedures have been proposed to supplement existing "model identification" methods. Initially these were mainly concerned with the *order* of autoregressive processes.

Stationary solutions of the stochastic difference equation

$$x_t + \alpha_1 x_{t-1} + \cdots + \alpha_p x_{t-p} = a_t,$$

where the a_t are independently and identically distributed with $Ea_t = 0$ and Var $a_t = \sigma_a^2$, are called autoregressive processes of order p. Before one can fit such a model to data one has to decide which p should be used.

Akaike (1969, 1970) developed a criterion (FPE) which uses the mean squared error of prediction as discrepancy. He seems to assume in his analysis (1969, Section 3) that the operating model is contained in the approximating family, in other words, that there is no misspecification. Later (1973) he introduced a general model selection criterion (AIC) which is based on the Kullback–Leibler discrepancy. Ogata (1980) derived the AIC criterion for the autoregressive case.

Apart from autoregressive processes, there are of course other families of models for time series, and general model selection methods are needed. In this chapter we discuss a method (Linhart and Volkers, 1985), which uses the same discrepancy as Akaike's FPE. Misspecification is admitted and in the autoregressive case the criterion coincides with Akaike's criterion if a certain approximation is made.

In Section 12.4 we discuss a method due to Parzen (1974, 1977) who approximates an arbitrary stationary time series by a suitable AR(p) process and uses a discrepancy which is the negative reciprocal of the one described in Section 12.1.

A review of model selection methods for stationary time series is given in Anděl (1982).

The derivation of model selection criteria for nonstationary time series is more difficult than for stationary series. In the absence of additional information there is usually no obvious operating model which can be postulated. It is particularly difficult to specify an operating model having a deterministic trend. In Section 12.5 we will assume that the parametric family to which the trend belongs is specified. We derive a criterion for selecting a suitable residual process. This criterion can also be applied to select a residual process in regression models that possibly have correlated residuals.

12.1. A GENERAL CRITERION

Assume that the operating model is a stationary Gaussian process with zero mean. Such a process is described completely by its covariance function $\gamma(v) = E(x_t x_{t+v})$, $v = 0, \pm 1, \pm 2, \ldots$, or equivalently by its spectral density function

$$f(\omega) = \frac{1}{2\pi} \sum_{v=-\infty}^{\infty} \gamma(v)e^{-iv\omega}, \qquad -\pi \leq \omega \leq \pi.$$

Suppose that the observations $x_1, x_2, \ldots, x_T$ are available and that we wish to fit a model having spectral density g_θ, where θ is a vector of parameters. We will take it that the purpose of fitting this model is to forecast future values of the series and therefore that a suitable discrepancy is the (one step ahead) mean squared forecast error. Under f (which is unknown) the best linear predictor of x_{T+1}, given $x_T, x_{T-1}, \ldots$, based on the model with spectral density g_θ has mean squared forecast error given by

$$\Delta(g_\theta, f) = \int_{-\pi}^{\pi} \frac{f(\omega)}{k(\omega, \theta)} \, d\omega,$$

where

$$k(\omega, \theta) = \frac{2\pi g_\theta(\omega)}{\sigma_g^2}$$

and σ_g^2 is the innovation variance of the approximating model:

$$\sigma_g^2 = 2\pi \exp\left(\frac{1}{2\pi} \int_{-\pi}^{\pi} \log g_\theta(\omega)d\omega\right)$$

(Grenander and Rosenblatt, 1957, p. 261). This result holds under the assumption that both the operating and fitted processes are purely nondeterministic and that the discrepancy exists for all θ.

Let

$$I_T(\omega_t) = \frac{1}{2\pi T} \left|\sum_{s=0}^{T-1} x_{s+1} e^{i\omega_t s}\right|^2, \qquad t = 0, 1, \ldots, T-1,$$

$$\omega_t = \frac{2\pi t}{T},$$

be the periodogram. Then an empirical discrepancy is

$$\Delta_T(\theta) = \frac{2\pi}{T} \sum_{t=0}^{T-1} I_T(\omega_t)k^{-1}(\omega_t, \theta).$$

The minimum discrepancy estimator

$$\hat{\theta} = \arg\min\{\Delta_T(\theta): \theta \in \Theta\}$$

is a generalized least-squares estimator in the frequency domain, which is described in Hannan (1973b). It can be shown that for autoregressive approximating processes this is the Yule–Walker estimator (Box and Jenkins, 1970, p. 278) in the time domain.

Under certain regularity conditions one can prove (see Appendix A.2.6), using essentially the methods of Hannan (1973b) and Robinson (1978) that $\sqrt{T}(\hat{\theta} - \theta_0)$ converges in distribution to $N(0, \Omega^{-1}\Sigma\Omega^{-1})$, where

$$\Omega = \int_{-\pi}^{\pi} \psi(\omega, \theta_0)f(\omega)d\omega = \frac{\partial^2\Delta(\theta_0)}{\partial\theta^2},$$

$$\Sigma = 4\pi \int_{-\pi}^{\pi} \phi(\omega, \theta_0)\phi^{\mathsf{T}}(\omega, \theta_0)f^2(\omega)d\omega,$$

$$\phi(\omega, \theta) = \frac{\partial k^{-1}(\omega, \theta)}{\partial\theta}$$

and

$$\psi(\omega, \theta) = \frac{\partial^2 k^{-1}(\omega, \theta)}{\partial \theta^2},$$

and one gets again

$$E\Delta(\hat{\theta}) \approx \Delta(\theta_0) + \left(\frac{1}{2T}\right) \operatorname{tr} \Omega^{-1}\Sigma.$$

The criterion is

$$\hat{E}\Delta(\hat{\theta}) = \Delta_T(\hat{\theta}) + \left(\frac{1}{T}\right) \operatorname{tr} \hat{\Omega}^{-1}\hat{\Sigma},$$

where

$$\hat{\Omega} = \frac{2\pi}{T} \sum_{t=0}^{T-1} \psi(\omega_t, \hat{\theta}) I_T(\omega_t)$$

and

$$\hat{\Sigma} = \frac{4\pi^2}{T} \sum_{t=0}^{T-1} \phi(\omega_t, \hat{\theta}) \phi^{\mathsf{T}}(\omega_t, \hat{\theta}) I_T^2(\omega_t).$$

If the approximating family contains the operating model, that is, if $f \equiv g_{\theta_0}$,

$$\Delta(\theta_0) = \sigma_f^2,$$

and

$$\Sigma = 2\sigma_f^2 \Omega,$$

in which case

$$\operatorname{tr} \Omega^{-1}\Sigma = 2p\sigma_f^2$$

and

$$E\Delta(\hat{\theta}) \approx \sigma_f^2 \left(1 + \frac{p}{T}\right).$$

Even if $f \not\equiv g_{\theta_0}$, $\sigma_f^2(1 + p/T)$ or $\Delta(\theta_0)(1 + p/T)$ can be used as an approximation to the expected discrepancy and a simpler criterion is

$$\Delta_T(\hat{\theta})\left(1 + \frac{2p}{T}\right).$$

This last criterion is very similar to that proposed by Akaike (1969, 1970) for *autoregressive* model fitting. He uses Yule–Walker estimators of $\theta = (\alpha_1, \alpha_2, \ldots, \alpha_p)^\mathsf{T}$ and arrives at the same expected discrepancy:

$$E\Delta(\hat{\theta}) = \sigma_f^2\left(1 + \frac{p}{T}\right).$$

Akaike estimates σ_f^2 by

$$\frac{1}{T - p} \sum_{t=1}^{T} (x_t + \hat{\alpha}_1 x_{t-1} + \cdots + \hat{\alpha}_p x_{t-p})^2,$$

where $x_0, x_{-1}, \ldots, x_{1-p}$ are set to zero.

12.2. THE DERIVATIVES ϕ AND ψ FOR TWO IMPORTANT CLASSES OF MODELS

We recommend that as a rule the simple criterion

$$\Delta_T(\hat{\theta})\left(1 + \frac{2p}{T}\right)$$

be used, but for completeness we give the matrices Ω and Σ for two important classes of models, namely, ARMA models and Bloomfield's (1973) models.

12.2.1. Autoregressive Moving Average Models

A detailed discussion of the class of ARMA models is given in Box and Jenkins (1970). An ARMA (p, q) process with autoregressive parameters $\alpha_1, \alpha_2, \ldots, \alpha_p$ and moving average parameters $\beta_1, \beta_2, \ldots, \beta_q$ has spectral density

$$g(\omega, \theta) = \frac{\sigma_g^2}{2\pi} \frac{B(\omega)}{A(\omega)},$$

where

$$A(\omega) = \left| \sum_{j=0}^{p} \alpha_j e^{ij\omega} \right|^2,$$

$$B(\omega) = \left| \sum_{j=0}^{q} \beta_j e^{ij\omega} \right|^2,$$

and $\alpha_0 = \beta_0 = 1$. Here

$$\theta = (\alpha_1, \alpha_2, \ldots, \alpha_p, \beta_1, \beta_2, \ldots, \beta_q)^{\mathsf{T}}.$$

The assumption that the fitted model is purely nondeterministic implies that these processes satisfy the stationarity and invertibility conditions (Box and Jenkins, 1970, p. 49).

Here

$$k^{-1}(\omega, \theta) = \frac{A(\omega)}{B(\omega)}$$

and the derivatives ϕ and ψ which are required for the criterion are given by

$$\phi(\omega, \theta) = (\phi_1(\omega, \theta), \ldots, \phi_{p+q}(\omega, \theta))^{\mathsf{T}}$$

where

$$\phi_s(\omega, \theta) = \frac{2A_s(\omega)}{B(\omega)} \quad \text{for } s = 1, 2, \ldots, p$$

$$= -\frac{2k^{-1}(\omega, \theta)B_s(\omega)}{B(\omega)} \quad \text{for } s = p + 1, p + 2, \ldots, p + q,$$

$$A_s(\omega) = \sum_{j=0}^{p} \alpha_j \cos (j - s)\omega \quad \text{for } s = 1, 2, \ldots, p,$$

$$B_s(\omega) = \sum_{j=0}^{q} \beta_j \cos (j - (s - p))\omega \quad \text{for } s = p + 1, \ldots, p + q.$$

Also

$$\psi_{uv}(\omega, \theta) = \frac{2 \cos (u - v)\omega}{B(\omega)} \qquad \text{for } 1 \leqslant u, v \leqslant p$$

$$= -\frac{2\phi_u(\omega, \theta)B_v(\omega)}{B(\omega)} \qquad \begin{aligned} &\text{for } 1 \leqslant u \leqslant p \\ &\quad p + 1 \leqslant v \leqslant p + q \end{aligned}$$

$$= -\frac{2\phi_v(\omega, \theta)B_u(\omega)}{B(\omega)} \qquad \begin{aligned} &\text{for } 1 \leqslant v \leqslant p \\ &\quad p + 1 \leqslant u \leqslant p + q \end{aligned}$$

$$= k^{-1}(\omega, \theta) \left(\frac{8B_u(\omega)B_v(\omega)}{B^2(\omega)} - \frac{2 \cos (u - v)\omega}{B(\omega)} \right)$$

$$\text{for } p + 1 \leqslant u, v \leqslant p + q.$$

12.2.2. Bloomfield's Models

Bloomfield (1973) proposed families of models with spectral densities

$$g(\omega, \theta) = \frac{\sigma_g^2}{2\pi} \exp\left(2 \sum_{j=1}^{p} \theta_j \cos \omega j \right),$$

so that

$$k^{-1}(\omega, \theta) = \exp\left(-2 \sum_{j=1}^{p} \theta_j \cos \omega j \right).$$

One gets

$$\phi_s(\omega, \theta) = -2(\cos \omega s)k^{-1}(\omega, \theta), \qquad\qquad s = 1, 2, \ldots, p,$$

$$\psi_{uv}(\omega, \theta) = 4(\cos \omega u)(\cos \omega v)k^{-1}(\omega, \theta), \qquad u, v = 1, 2, \ldots, p.$$

12.3. EXAMPLES OF APPLICATIONS

EXAMPLE 12.3.1.   Zucchini and Hiemstra (1983) investigated the relationship between tree-ring indices of Widdingtonia Cedarbergensis and the annual rainfall totals at a neighboring location. (The tree-ring indices are given in the Appendix in Table A.3.5.) The logarithms of index/100 (they are displayed in Figure 12.1) were identified as an AR(2) process.

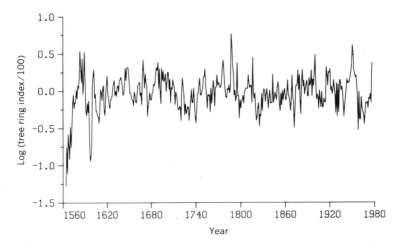

Figure 12.1. Logarithms of tree ring indices minus log 100 of Widdingtonia Cedarbergensis, 1564 – 1970.

The table below gives the values of the criterion for a number of ARMA (p, q) and Bloomfield approximating models. The criterion also selects an AR(2) process. The estimated parameters are $\hat{\alpha}_1 = -0.51$, $\hat{\alpha}_2 = -0.18$, and $\hat{\sigma}_a^2 = 0.03584$.

ARMA(p, q)						Bloomfield (p)	
p	q	Criterion	p	q	Criterion	p	Criterion
1	0	372	0	1	452	1	409
2	0	362	0	2	391	2	370
3	0	363	0	3	378	3	370
4	0	364	0	4	379	4	372
5	0	365	1	1	394	5	365
6	0	366	1	2	377	6	366
7	0	368	2	1	364	7	367

EXAMPLE 12.3.2. The model selection criteria of Section 12.1 are applicable to stationary time series. For nonstationary series one can apply these criteria to select a model for the series of first (or higher order) differences of the series. It is first necessary to decide how many times the original series should be differenced to achieve stationarity. This can be done by inspecting the correlation function of the series and its successive differences.

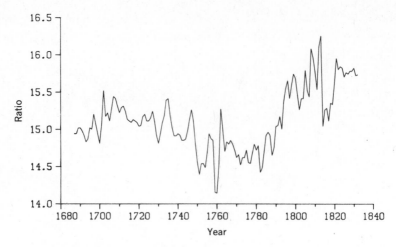

Figure 12.2. Ratio of price of gold to price of silver, 1687 – 1832.

The ratios of the price of gold to that of silver for the years 1687 to 1832 (from Soetbeer, 1879) are given in Table A.3.6 and displayed in Figure 12.2. The series of first differences of the ratios is stationary. The criterion is applied to select an ARMA model for these differences, so one is in effect selecting the orders p and q of an ARIMA(p, 1, q) series. The values of the criterion (as multiples of 10^{-4}) for $1 \leqslant p + q \leqslant 4$ are given below.

p	q	0	1	2	3	4
0			465	399	402	406
1		460	417	400	403	
2		422	401	405		
3		411	402			
4		410				

The corresponding values of the simple criterion are given by

p	q	0	1	2	3	4
0			462	400	404	409
1		463	422	404	409	
2		423	404	408		
3		415	408			
4		414				

Although the two criteria have different numerical values the overall pattern is the same. Both criteria select an ARIMA(0, 1, 2) approximating model.

EXAMPLE 12.3.3. The proposed selection procedure was applied to series A to F given in Box and Jenkins (1970, p. 524). (In the case of series B and C the first differences were analyzed.) The purpose of this exercise was to check whether the models which are selected by the given criterion are the same as, or at least similar to, those selected "by hand." The results are summarized in the table below.

	Criterion Selections		Box–Jenkins Selections	
Data	Model	Criterion	Model	Criterion
A	(1, 1)	0.110	(1, 1)	0.110
B	(1, 0)	53.15	(0, 1)	53.16
C	(1, 0)	0.0195	(1, 0)	0.0195
D	(1, 0)	0.0942	(1, 0)	0.0942
E	(2, 1)	230	(3, 0)	232
F	(0, 2)	122	(2, 0)	123

For series A, C, and D the criterion selects the same models as are identified by Box and Jenkins. The selection differs for series A, E, and F but the differences are minor.

In the case of series B the criterion selects an AR(1) model whereas Box and Jenkins identified an MA(1) model. First, the values of the criteria for these two models are almost equal, and second, the fitted models are practically the same for the relevant parameter values. The fitted MA(1) process $x_t = a_t - 0.09\, a_{t-1}$ has autoregressive representation

$$x_t + 0.09x_{t-1} + (0.09)^2 x_{t-2} + \cdots = a_t$$

which is approximately the AR(1) process

$$x_t + 0.09x_{t-1} = a_t.$$

For series E the criterion selects an ARMA(2, 1) model [closely followed by an AR(3) model]. Box and Jenkins recommend either an AR(3) or alternatively an AR(2) model. The fitted models, that is, ARMA(2, 1) and AR(3), differ very little for this series. Similarly, for series F the criterion selects an MA(2) process closely followed by the AR(2) model identified by Box and Jenkins, but again these two models are quite similar for this series.

It can be seen that the models selected by the criterion are either the same or practically equivalent to those selected by Box and Jenkins. We note also that the criterion selects parsimonious models for each of these series.

12.4. PARZEN'S AUTOREGRESSIVE ORDER DETERMINATION

Parzen (1974, 1977) describes methods to approximate stationary time series by means of autoregressive processes of finite order p (AR(p)). The order p is determined by minimizing a criterion denoted by CAT(p).

There are two version of CAT(p). They are based on two different discrepancies, but both discrepancies are closely related to the mean squared error of prediction. We describe the later version which is given in Parzen (1977).

Let

$$g_p(B) = (1 + \alpha_1^{(p)}B + \cdots + \alpha_p^{(p)}B^p)$$

be the approximating autoregressive transfer function (ARTF). It is assumed that the operating process has an autoregressive representation

$$g_\infty(B)x_t = a_t,$$

where

$$g_\infty(B) = (1 + \alpha_1 B + \alpha_2 B^2 + \cdots)$$

is the operating ARTF and a_t is a sequence of independently and identically distributed random variables with zero mean and variance $\sigma_\infty^2 > 0$.

The criterion is based on the discrepancy

$$\Delta(\theta^{(p)}) = \int_{-\pi}^{\pi} \left| \frac{g_p(e^{i\omega})}{\eta(\theta^{(p)})} - \frac{g_\infty(e^{i\omega})}{\sigma_\infty^2} \right|^2 f(\omega)d\omega$$

$$= \frac{1}{\sigma_\infty^2} - \frac{1}{\eta(\theta^{(p)})},$$

where

$$\eta(\theta^{(p)}) = \frac{1}{2\pi} \int_{-\pi}^{\pi} k^{-1}(\omega, \theta^{(p)}) f(\omega)d\omega$$

$$= \frac{1}{2\pi} \int_{-\pi}^{\pi} \left| g_p(e^{i\omega}) \right|^2 f(\omega)d\omega$$

is the discrepancy of Section 12.1 and

$$\theta^{(p)} = (\alpha_1^{(p)}, \alpha_2^{(p)}, \ldots, \alpha_p^{(p)})^{\mathsf{T}}.$$

Since σ_∞^2 does not depend on the approximating model, $\Delta(\theta^{(p)})$ is simply the negative reciprocal of the discrepancy of Section 12.1.

The parameters $\alpha_j^{(p)}$ are estimated by solving the Yule–Walker equations:

$$\sum_{j=0}^{p} \hat{\alpha}_j^{(p)} c(j - k) = 0, \qquad k = 1, 2, \ldots, p,$$

where $\hat{\alpha}_0^{(p)} = 1$ and $c(v)$ is the estimated covariance of lag v:

$$c(v) = \frac{1}{T} \sum_{t=1}^{T-|v|} x_t x_{t+|v|}.$$

The resulting criterion is an estimator of the essential part of a large sample approximation to the expected discrepancy:

$$\mathrm{CAT}(p) = \frac{1}{T} \sum_{j=1}^{p} \left(\frac{1}{\hat{\sigma}_j^2} \right) - \frac{1}{\hat{\sigma}_p^2},$$

where

$$\hat{\sigma}_i^2 = \frac{T}{T-i} \sum_{j=0}^{i} \hat{\alpha}_j^{(i)} c(j).$$

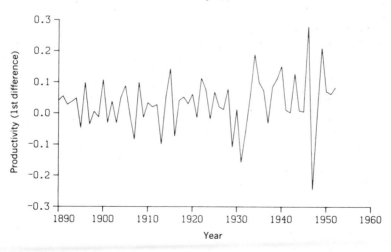

Figure 12.3. First difference of productivity in the United States, 1890–1953.

Once p is chosen by minimizing $\mathrm{CAT}(p)$ the corresponding approximating ARTF:

$$\hat{g}_p(B) = (1 + \hat{\alpha}_1^{(p)} B + \cdots + \hat{\alpha}_p^{(p)} B^p)$$

is used to compute the forecasts.

EXAMPLE 12.4.   The first difference of the productivity in the United States ($1000, 1929) from 1890 to 1953 is illustrated in Figure 12.3 (U.S. Department of Commerce, 1966). (The original data are given in Table A.3.7.) Parzen's autoregressive model fitting leads to the results:

p	$\hat{\sigma}_p^2 \times 10^8$	$\mathrm{CAT}(p)$	p	$\hat{\sigma}_p^2 \times 10^8$	$\mathrm{CAT}(p)$
1	669079	-147.1	7	690497	-128.4
2	661571	-146.4	8	701221	-123.9
3	666912	-142.8	9	671077	-127.9
4	676223	-138.4	10	678374	-124.0
5	687659	-133.6	15	729589	-102.5
6	680829	-132.7	20	795892	-80.7

The most appropriate autoregressive process is AR(1). The parameter is $\hat{\alpha}_1 = 0.169$.

12.5.   SELECTING THE RESIDUAL PROCESS OF A NONSTATIONARY TIME SERIES

12.5.1.   Introduction

One of the standard ways to model a nonstationary time series is to describe it as the sum of a deterministic mean function, the *trend*, and a stationary stochastic component, the *residual process:*

$$y_t = \mu_t + x_t, \qquad t = 1, 2, \ldots .$$

For theoretical reasons we will restrict our attention to models whose trend components are *linear* functions of their parameters.

The main difficulty in modeling nonstationary time series is that it is not easy to decide on an operating family. The object of fitting a model to a given series may be to summarize the data in terms of a few descriptive constants (parameters). It is then a matter of convenience which operating family is assumed. If, however, the object is to forecast future values of the series then the choice of operating family can be crucial.

There are usually several different models which could have given rise to the observations but which lead to very different forecasts. Within the class of models considered here—trend plus stationary residuals—one could combine a relatively smooth trend with highly correlated residuals, or alternatively a very flexible trend with uncorrelated residuals. There is no obvious way to decide how much should be modeled by the trend and how much by the residuals. Furthermore, there are other classes of models which may be compatible with the data. The most important examples are the autoregressive *integrated* moving average processes of Box and Jenkins (1970).

The forecasts depend strongly on the operating family which is assumed. This presents particular difficulties in the class of processes considered in this chapter, because a trend must be specified and this often dominates the forecasts.

Criteria for selecting models for nonstationary time series are discussed in Volkers (1983). We will consider the problem of selecting a model for the residual process when the family of mean functions in the operating process is specified. We will take it that the mean function of the approximating process is to be of the same form as that of the operating process.

Econometric theory sometimes leads to certain regression relationships which effectively determine the form and number of parameters in the mean function, but not the properties of the residual process. Durbin and Watson (1951) proposed a test which is now in general use to test the hypothesis that the residual process is serially uncorrelated against the alternative hypothesis that it has an autoregressive structure of lag 1, AR(1). The methods given here are concerned with selection rather than testing. They can be used to select a residual process, which could be white noise, AR(1), or some other process. Selection is based on a discrepancy which is the mean squared error of prediction in the residual process.

12.5.2. The Criterion

Suppose that $y = (y_1, y_2, \ldots, y_T)^{\mathsf{T}}$ are observations from an operating model of the form

$$y = Z\alpha + x,$$

where Z is a known $T \times k$ matrix of rank k, $x = (x_1, x_2, \ldots, x_T)^{\mathsf{T}}$, and $x_1, x_2, \ldots$ is a stationary Gaussian process with mean zero and spectral density $f(\omega)$.

Both Z and x are assumed to satisfy the regularity conditions given in Appendix A.2.7. One of these conditions, namely, Grenander's condition for Z, excludes the case in which the trend is periodic. Alternative criteria are available for this case (Volkers, 1983) but they are rather complicated and inconvenient to apply. For the purposes of this chapter we will assume that the mentioned regularity conditions are met.

In this section we will consider approximating models having a stationary Gaussian residual process with spectral density

$$g_\theta(\omega) = \left(\frac{\sigma_g^2}{2\pi}\right) k(\omega, \theta), \quad \dim(\theta) = p.$$

The discrepancy is that of Section 12.1:

$$\Delta(\theta) = \int_{-\pi}^{\pi} \frac{f(\omega)}{k(\omega, \theta)} \, d\omega.$$

If α were known, then $x = y - Z\alpha$ could be treated using the methods of Section 12.1 and in particular a suitable empirical discrepancy would be

$$\Delta_T^{(x)}(\theta) = \frac{2\pi}{T} \sum_{t=0}^{T-1} \frac{I_{T;x}(\omega_t)}{k(\omega_t, \theta)},$$

where $I_{T;x}$ denotes the periodogram of the x series. Since α is unknown it is estimated by the method of ordinary least squares:

$$\hat{\alpha} = (Z^\mathsf{T} Z)^{-1} Z^\mathsf{T} y.$$

This leads to the empirical discrepancy

$$\Delta_T(\theta) = \Delta_T^{(\hat{x})}(\theta),$$

where $\hat{x} = y - Z\hat{\alpha}$ are the residuals.

Under the regularity conditions given in the Appendix, a minimum discrepancy estimator $\hat{\theta}$ [minimizing $\Delta_T(\theta)$] exists and is asymptotically normally distributed, independently of $\hat{\alpha}$.

One obtains a simple criterion, an estimator of the expected discrepancy for the case in which the operating model is a member of the approximating family:

$$\Delta_T^{(\hat{x})}(\hat{\theta}) \, \frac{T + p}{T - p - k}.$$

If the approximating model is white noise ($p = 0$) the criterion reduces to

$$\frac{Tc_T(0)}{T - k},$$

where

$$c_T(0) = \frac{1}{T}(y - Z\hat{\alpha})^{\mathsf{T}}(y - Z\hat{\alpha}).$$

12.5.3. Examples of Application

EXAMPLE 12.5.3.1. The price of sugar [mills (tenths of a cent), coded by subtracting 40] in the United States from 1875 to 1936 is given in Table A.3.8 and displayed in Figure 12.4. Durbin and Watson (1951) fitted polynomial trends of degree 3 and 4, and in both cases rejected the hypothesis that the residual process is uncorrelated.

The method of ordinary least squares results in the following estimate of the cubic trend function

$$h_t = 90.78 - 8.5245t + 0.2731t^2 - 0.002587t^3,$$

where $t = 1$ corresponds to 1875.

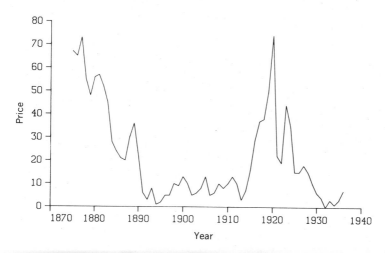

Figure 12.4. Price of sugar in the United States, 1875 – 1936.

The criterion for fitting AR(p), $p = 0, 1, 2, \ldots$ ($p = 0$ corresponds to white noise), is given in Figure 12.5. Clearly, $p = 0$ (white noise) should not be used for the residual process. This conclusion is consistent with the result of the test by Durbin and Watson. The criterion achieves its minimum for an AR(4) process.

As can be seen in Figure 12.5 the selected order, $p = 4$, is unchanged if a trend of degree 4 is fitted.

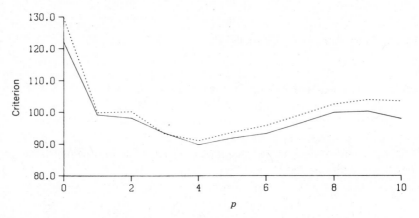

Figure 12.5. Price of sugar: criterion for approximating AR(p) residual process for polynomial trends of degree 3 (...) and 4 (—).

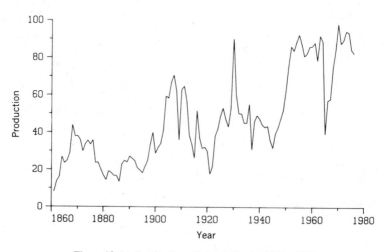

Figure 12.6. Production of zinc in Spain, 1861 – 1976.

EXAMPLE 12.5.3.2.   The annual production of zinc (10^3 tons) in Spain from 1861 to 1976 (Schmitz, 1979) is illustrated in Figure 12.6. (For the data see Table A.3.9.)

A linear trend

$$h_t = 13.792 + 0.566t$$

was subtracted, where $t = 1$ corresponds to 1861.

The criterion (see Figure 12.7) selects an AR(1) residual process. The Durbin–Watson test rejects the hypothesis of an uncorrelated residual process.

EXAMPLE 12.5.3.3.   The position is very similar for the price of magnesium (pounds/ton) in Great Britain, 1800–1911 (Schmitz, 1979) (see Table A.3.10 and Figure 12.8), for which the estimated trend is

$$h_t = 24.343 + 0.1078t,$$

where $t = 1$ corresponds to 1800.

The analysis resulted in the criterion values given in Figure 12.9. An AR(1) residual process is selected. Here too, the Durbin–Watson test rejects the hypothesis that the residual process is white noise.

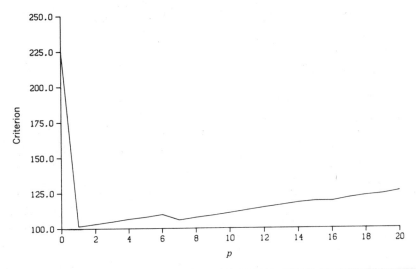

Figure 12.7. Production of zinc: criterion for approximating AR(p) residual process if a linear operating trend is assumed.

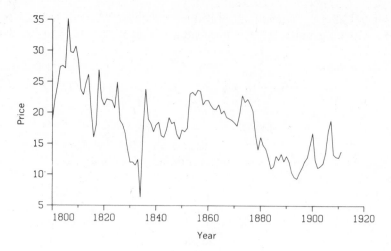

Figure 12.8.   Price of magnesium in Great Britain, 1800 – 1911.

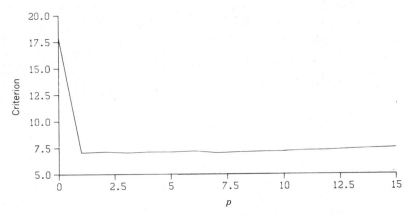

Figure 12.9.   Price of magnesium: criterion for approximating AR(p) residual process if linear operating trend is assumed.

EXAMPLE 12.5.3.4.   Table A.3.11 and Figure 12.10 give the average yield of wheat (cwt/acre) in Great Britain from 1885 to 1944 (Ministry of Agriculture, 1968). A linear trend ($h_t = 16.204 + 0.0408t$) is fitted by least squares where $t = 1$ corresponds to 1885.

The criterion for AR(p) approximating models can be seen in Figure 12.11. The selected residual process is white noise. The Durbin–Watson test does not reject the hypothesis of white noise.

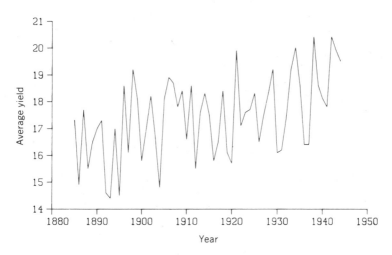

Figure 12.10. Yield of wheat in Great Britain, 1885 – 1944.

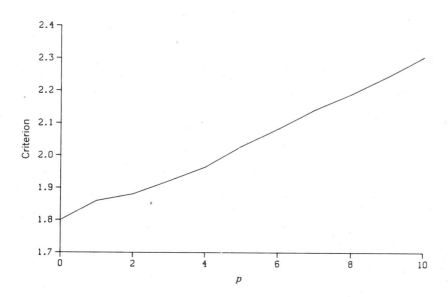

Figure 12.11. Yield of wheat: criterion for approximating AR(p) residual process if a linear operating trend is assumed.

227

CHAPTER 13

Spectral Analysis

The estimation of the spectral density $f(\omega)$ of a stationary time series is a problem which has attracted considerable attention in the statistical literature. The central difficulty is that one is obliged to use estimators $\hat{f}(\omega)$ which are biased. In general, $E\hat{f}(\omega)$ is not equal to $f(\omega)$, but to some weighted average of f in the neighborhood of ω. The magnitude of the bias depends on the extent of this averaging, which is characterized by the *bandwidth* associated with the estimator. A precise definition of bandwidth, for example, is given in Jenkins and Watts (1968). By reducing the bandwidth one can reduce the bias but this increases the variance of $\hat{f}(\omega)$. This dilemma is summarized in the well-known relationship

$$\text{variance} \times \text{bandwidth} \approx \frac{\text{constant}}{n}.$$

The problem of finding a suitable compromise between bias and variance is simply one more manifestation of the problem described in Chapter 1, namely, that of balancing the discrepancy due to approximation against the discrepancy due to estimation in order to minimize their combined effect.

Several criteria to facilitate objective decisions in this respect are in use, and many of them can be interpreted in the general framework described here. As an example we will consider, in Section 13.1, the estimator of the log spectral density proposed by Wahba (1980).

In Section 13.2 we demonstrate, using a particular class of approximating models, how one can derive methods to select a *parametric* model for the spectral density. The models discussed are orthogonal with respect to the

228

Gauss discrepancy and lead to a selection procedure which is particularly easy to implement.

Naturally, one can also apply the methods of Chapter 12 to estimate the spectral density. The estimate is simply the spectral density which characterizes the fitted stationary process.

13.1. WAHBA'S OPTIMALLY SMOOTHED SPLINE ESTIMATOR OF THE LOG SPECTRAL DENSITY

Some operating models are not parametric, that is, they can not be described by a finite number of parameters. One can approximate them by parametric models, as we did in Chapters 3 and 4, but it is also possible to use non-parametric methods. It would be awkward to refer to *parametric* nonparametric methods and *nonparametric* nonparametric methods, respectively, and we propose to distinguish between these two situations by referring to parametric and nonparametric distribution-free methods (although "distribution-free" is not entirely satisfactory for the case in which mean functions are in the focus of interest, as in Chapter 5).

This book deals almost exclusively with parametric methods (many of them are distribution-free), where the focus of interest is the selection of a parametric family of models. The essential features of these methods are also present if one applies nonparametric methods. For example, smoothing methods usually depend on a smoothing constant which has to be selected. The selection can be based on criteria which estimate an expected discrepancy. This section describes an example of a nonparametric method in the context of spectral analysis, the estimation of the logarithm of the spectral density as proposed by Wahba (1980).

Suppose that the observations $x_1, x_2, \ldots, x_T$ can be considered to have been generated by some stationary Gaussian operating model. Since such a model is fully specified by its spectral density $f(\omega)$, an estimate of $f(\omega)$, or equivalently $g(\omega/2\pi) = \log 2\pi f(\omega)$, $-\pi \leqslant \omega \leqslant \pi$, provides a fitted model. (Note that in the original article ω denotes frequency rather than radial frequency and $-\frac{1}{2} \leqslant \omega \leqslant \frac{1}{2}$.)

Wahba (1980) proposed (for T even) an estimator of $g(\omega/2\pi)$ which has the following form:

$$\hat{g}_{T,m,\lambda}\left(\frac{\omega}{2\pi}\right) = \sum_{v=-T/2+1}^{T/2} \tilde{g}_v (1 + \lambda(2\pi v)^{2m})^{-1} \exp(iv\omega),$$

$$\lambda > 0, \quad m = 2, 3, 4, \ldots,$$

where

$$\tilde{g}_v = \frac{1}{T} \sum_{k=-T/2+1}^{T/2} y_k \exp\left(-\frac{2\pi i v k}{T}\right),$$

$$y_k = \log 2\pi I_T(\omega_k) + \gamma \qquad \text{if } k \neq 0, T/2$$

$$= \log 2\pi I_T(\omega_k) + \frac{\log 2 + \gamma}{\pi} \qquad \text{if } k = 0, T/2,$$

$\gamma = 0.57721\ldots$ is Euler's constant, and $I_T(\omega)$ is the periodogram:

$$I_T(\omega_j) = \frac{1}{2\pi} \sum_{v=-(T-1)}^{(T-1)} c(v) \cos \omega_j v,$$

$$\omega_j = \frac{2\pi j}{T},$$

$$c(v) = \frac{1}{T} \sum_{t=1}^{T-|v|} x_t x_{t+|v|}.$$

The properties of the estimators are controlled by the constants λ and m which determine the bandwidth and shape of the window, respectively. The method proposed by Wahba for selecting approximate values of λ and m is based on the Gauss discrepancy. The overall discrepancy for a given (λ, m) is then

$$R_T(\lambda, m) = \frac{1}{2\pi} \int_{-\pi}^{\pi} \left| \hat{g}_{T,m,\lambda}\left(\frac{\omega}{2\pi}\right) - g\left(\frac{\omega}{2\pi}\right) \right|^2 d\omega.$$

The expectation is approximately

$$ER_T(\lambda, m) \approx \sum_{v=-T/2+1}^{T/2} |g_v|^2 (1 - (1 + \lambda(2\pi v)^{2m})^{-1})^2$$

$$+ \frac{\pi^2}{6T} \sum_{v=-T/2+1}^{T/2} (1 + \lambda(2\pi v)^{2m})^{-2},$$

where g_v are the Fourier coefficients of g:

$$g\left(\frac{\omega}{2\pi}\right) = \sum_{v=-\infty}^{\infty} g_v \exp(iv\omega).$$

An approximately unbiased estimator of the expected overall discrepancy, a criterion, is

$$\hat{R}_T(\lambda, m) = \sum_{v = -T/2 + 1}^{T/2} \left(|\tilde{g}_v|^2 - \frac{\pi^2}{6T} \right) (1 - (1 + \lambda(2\pi v)^{2m})^{-1})^2$$

$$+ \frac{\pi^2}{6T} \sum_{v = -T/2 + 1}^{T/2} (1 + \lambda(2\pi v)^{2m})^{-2}.$$

The pair (λ, m) that minimizes this criterion is selected. The method is illustrated in Wahba (1980).

13.2. A FAMILY OF APPROXIMATING SPECTRAL DENSITIES

One of the more difficult aspects of applying autoregressive moving average (ARMA) models is that of model selection. Objective selection procedures require considerable computing effort, so one relies mainly on subjective judgment which requires expertise and experience. The objective procedures [see Chapter 12, Akaike (1969, 1973), and Parzen (1974, 1977)] require that criteria be calculated for each competing model and the absence of orthogonality makes it necessary to fit each model individually. For example, if one wants to select the order p of an AR model one has to re-estimate the parameters for each order p to compute the corresponding criterion. This is analogous to nonorthogonal polynomial regression, where one has to re-estimate the regression coefficients if the order of the polynomial is changed.

The class of models which is used in this section is *orthogonal*, and consequently the above difficulties do not arise. One can look at each parameter separately and decide whether it should be retained in the model. The expected discrepancy is simply the sum of components due to the individual parameters. See also Linhart and Zucchini (1984a).

The models which we consider are given by spectral density

$$g_{A,\theta}(\omega) = \frac{1}{2\pi} (\theta_0 + 2 \sum_{v \in A} \theta_v \cos \omega v),$$

where A is a set of p natural numbers. The parameters are restricted by the requirement that the covariance function corresponding to $g_{A,\theta}$ is positive definite. Each A determines an approximating family. We consider again the problem of model selection and spectral estimation using the Gauss discrepancy:

$$\Delta(\theta) = -2 \int_{-\pi}^{\pi} f(\omega) g_{A,\theta}(\omega) d\omega + \int_{-\pi}^{\pi} g_{A,\theta}^2(\omega) d\omega.$$

The set of functions $\{\cos \omega v\}$ is orthogonal with respect to this discrepancy.

We assume that the operating model is a Gaussian linear process with finite innovation variance, a spectral density which satisfies a Lipschitz condition of order ξ, and a covariance function $\gamma(v)$ with $\sum |v\gamma(v)| < \infty$.

It is well known that the covariances $\gamma(v)$ are the optimal parameter values $\theta_0 = \{\theta_{0v}\}$ minimizing $\Delta(\theta)$. The discrepancy due to approximation is therefore

$$\Delta(\theta_0) = -\frac{1}{2\pi}(\gamma^2(0) + 2\sum_{v \in A} \gamma^2(v)).$$

We use as empirical discrepancy

$$\Delta_T(\theta) = \frac{2\pi}{T} \sum_{j=0}^{T-1} (-2I_T(\omega_j) g_{A,\theta}(\omega_j) + g_{A,\theta}^2(\omega_j)),$$

where $I_T(\omega_j)$ is the periodogram defined in Section 13.1.

The minimum discrepancy estimator

$$\hat{\theta}_T = \arg\min\{\Delta_T(\theta) : \theta \in \Theta\}$$

is the vector whose components are the estimated covariances

$$c(v) = \frac{1}{T} \sum_{t=1}^{T-|v|} x_t x_{t+|v|}, \qquad v \in A^*.$$

The minimum of $\Delta_T(\theta)$ is

$$\Delta_T(\hat{\theta}_T) = -\frac{1}{2\pi}(c^2(0) + 2\sum_{v \in A} c^2(v)).$$

We now calculate the expected discrepancy $E\Delta(\hat{\theta}_T)$. One has

$$T\Delta(\hat{\theta}_T) = T\Delta(\theta_0) + \frac{T(\hat{\theta}_T - \theta_0)^\mathsf{T} \Omega(\hat{\theta}_T - \theta_0)}{2},$$

where

$$\Omega = \Delta''(\theta_0)$$

and (Anderson, 1971, Theorem 8.4.2)

$$\lim_{T \to \infty} ET(\hat{\theta}_T - \theta_0)(\hat{\theta}_T - \theta_0)^{\mathsf{T}} = \Sigma,$$

where

$$\{\Sigma\}_{ij} = 4\pi \int_{-\pi}^{\pi} f^2(\omega) \cos i\omega \cos j\omega d\omega$$

Hence

$$E\Delta(\hat{\theta}_T) = \Delta(\theta_0) + \frac{\operatorname{tr} \Omega\Sigma}{2T} + o\left(\frac{1}{T}\right).$$

To obtain an estimator of this expectation we consider

$$T\Delta_T(\hat{\theta}_T) = T\Delta_T(\theta_0) - \frac{T(\hat{\theta}_T - \theta_0)^{\mathsf{T}}\Omega(\hat{\theta}_T - \theta_0)}{2},$$

$$E\Delta_T(\hat{\theta}_T) = E\Delta_T(\theta_0) - \frac{\operatorname{tr} \Omega\Sigma}{2T} + o\left(\frac{1}{T}\right).$$

By Brillinger (1975, Theorems 5.2.2 and 5.2.4) and Robinson (1978, p. 147)

$$E\Delta_T(\theta_0) = \Delta(\theta_0) + O(T^{-\min(1,\xi)})$$

and, with

$$\widehat{\operatorname{Var}} \, c(v) = \frac{4\pi^2}{T^2} \sum_{j=0}^{T-1} \cos^2 (\omega_j v) I_T^2(\omega_j),$$

and

$$\widehat{\operatorname{tr} \Omega\Sigma} = \frac{T}{\pi} (\widehat{\operatorname{Var}} \, c(0) + 2\sum_{v \in A} \widehat{\operatorname{Var}} \, c(v)),$$

$$E(\widehat{\operatorname{tr} \Omega\Sigma}) = \operatorname{tr} \Omega\Sigma + O(T^{-\min(1,\xi)}).$$

An estimator of the expected discrepancy, which is asymptotically unbiased to the order $O(T^{-\min(1,\xi)})$, is therefore

$$\hat{E}\Delta(\hat{\theta}_T) = \Delta_T(\hat{\theta}_T) + \frac{\widehat{\operatorname{tr}\Omega\Sigma}}{T}$$

$$= -\frac{1}{2\pi}\left(c^2(0) + 2\sum_{v\in A} c^2(v)\right) + \frac{1}{\pi}\left(\widehat{\operatorname{Var}} c(0) + 2\sum_{v\in A} \widehat{\operatorname{Var}} c(v)\right).$$

By this criterion a parameter θ_v should remain in the approximating family if

$$\frac{c^2(v)}{\widehat{\operatorname{Var}} c(v)} > 2,$$

Table 13.1. Criterion for Monthly Number of Passengers of an Airline

v	$c(v) \times 10^3$	$\widehat{\operatorname{Var}} c(v) \times 10^{10}$	$c^2(v)/\widehat{\operatorname{Var}} c(v)$
0	2.086	948	45.9
1	−0.712	568	8.9
2	0.219	496	1.0
3	−0.422	451	3.9
4	0.045	505	0.0
5	0.116	417	0.3
6	0.064	174	0.2
7	−0.116	397	0.3
8	−0.002	389	0.0
9	0.368	556	2.4
10	−0.159	361	0.7
11	0.134	416	0.4
12	−0.807	585	11.1
13	0.316	584	1.7
14	−0.120	532	0.3
15	0.312	363	2.7
16	−0.290	576	1.5
17	0.147	492	0.4
18	0.032	418	0.0
19	−0.022	315	0.0
20	−0.244	391	1.5

This selection procedure results in the choice of a particular A (and consequently a particular approximating family). The corresponding estimator for the spectral density is simply

$$\hat{f}(\omega) = \frac{1}{2\pi}\left(c(0) + 2\sum_{v \in A} c(v) \cos \omega v\right).$$

EXAMPLE 13.2. The methods derived above were applied to the airline passenger data given in Box and Jenkins (1970). In particular, a model was

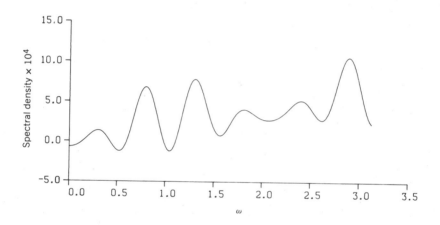

Figure 13.1.   Estimated spectral density for airline data.

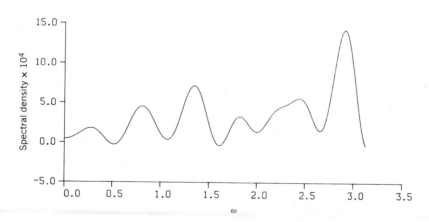

Figure 13.2.   Bartlett's spectral density estimate for airline data.

selected for $\nabla\nabla_{12} \log x_t$, where x_t is the number of passengers in month t, $t = 1, 2, \ldots, 144$, and the difference operator is defined for any sequence z_t by $\nabla_s z_t = z_t - z_{t-s}$. The values of $c(v)$ and $\widehat{\text{Var}}\ c(v)$ are given in Table 13.1.

The ratio $c^2(v)/\widehat{\text{Var}}\ c(v)$ is larger than 2 for $v \in A^* = \{0, 1, 3, 9, 12, 15\}$. This A^* defines the model to be used, the corresponding $c(v)$ are the minimum discrepancy estimators of the parameters θ_v, $v \in A^*$.

The spectral estimator is given in Figure 13.1. For comparison we give also, in Figure 13.2, Barlett's (modified) estimator for $m = 17$. (See Jenkins, 1961, p. 146, window 2.)

Appendix

A.1. ASYMPTOTIC CRITERIA

In this section we prove the results which were summarized in Section 2.4 and which provide the basis for the derivation of asymptotic criteria. The key reference is Linhart and Volkers (1984).

A.1.1. The Asymptotic Properties of Minimum Discrepancy Estimators

Let x_t, $t \in \mathbb{N}$, $\mathbb{N} = \{1, 2, \dots \}$, be a k-dimensional stochastic process on a measurable space $(\Omega, \mathfrak{A})$. All *models* considered here are distributions on the sample space $(\mathbb{R}^T, \mathfrak{P}_T)$. Let M be the set of all such models. The *operating model* $v \in M$ belongs to a subset of M, the *operating family*, which is known. The *approximating family* is a subset of M with elements μ_θ, where $\theta \in \Theta$ is a real p-dimensional vector of parameters. A *discrepancy* is a functional Δ on $M \times M$ with the property

$$\Delta(\mu, v) \geq \Delta(v, v) \quad \text{for all } \mu, v \in M.$$

Assume a discrepancy has been chosen and that Δ, μ_θ, and v are such that

$$\theta_0 = \arg \inf \{\Delta(\mu_\theta, v) \colon \theta \in \Theta\}$$

exists. The minimal value

$$\Delta(\theta_0) = \Delta(\mu_{\theta_0}, v)$$

is called the *discrepancy due to approximation*. A consistent estimator

$$\Delta_n(\omega, \theta) = \Delta_n(x_1(\omega), \ldots, x_n(\omega); \theta)$$

of $\Delta(\theta)$ is called an *empirical discrepancy*. If

$$\hat{\theta}_n(\omega) = \arg\inf\{\Delta_n(\omega, \theta): \theta \in \Theta\}$$

exists a.s. and is measurable as a function of ω it is called a *minimum discrepancy estimator*.

We give now a list of conditions which will be needed in the theorems to follow:

B1. Θ is compact.

B2. θ_0 is unique and in the interior of Θ.

B3. $\Delta''(\theta_0) = \Omega$ is positive definite.

B4. $\Delta_n(\omega, \theta)$ is a.s. continuous on Θ $\forall n \in \mathbb{N}$.

B5. $\Delta_n(\omega, \theta) \to \Delta(\theta)$ a.s. uniformly on Θ.

B6. $\Delta(\theta)$ and a.s. $\Delta_n(\omega, \theta)$ ($\forall n \in \mathbb{N}$) are twice continuously differentiable on Θ.

B7. $\Delta_n'(\omega, \theta)$ and $\Delta_n''(\omega, \theta)$ converge almost surely uniformly on Θ to $\Delta'(\theta)$ and $\Delta''(\theta)$.

B8. $\sqrt{n}\Delta_n'(\omega, \theta_0) \overset{d}{\to} X$, where $EX = 0$ and $\operatorname{Var} X = \Sigma$.

B8'. $\sqrt{n}\Delta_n'(\omega, \theta_0) \overset{d}{\to} N(0, \Sigma)$.

The following lemma gives sufficient conditions for the existence of a minimum discrepancy estimator.

Lemma.   Under B1 and B4 there exists for every $n \in \mathbb{N}$ a minimum discrepancy estimator $\hat{\theta}_n$ of θ_0.

Proof: This is Lemma 2 of Jennrich (1969).   ∎

We now prove a law of large numbers.

Theorem 1.   Let θ_0 be unique and $(\hat{\theta}_n)_{n\in\mathbb{N}}$ be a sequence of minimum discrepancy estimators. If B1, B4, and B5 hold

$$\hat{\theta}_n \to \theta_0 \qquad \text{a.s.,}$$

$$\Delta_n(\hat{\theta}_n) \to \Delta(\theta_0) \quad \text{a.s.}$$

Proof: Because of B1 and the continuity of Δ (which follows from B1, B4, and B5) there exists for each sphere $K_\rho(\theta_0)$ (with $\rho > 0$) centered at θ_0 a $\gamma(\rho) > 0$ such that

$$\Delta(\theta) \geqslant \Delta(\theta_0) + \gamma(\rho) \qquad \forall \, \theta \notin K_\rho(\theta_0).$$

Now for fixed ω, if $\hat{\theta}_n(\omega) \nrightarrow \theta_0$, there is a subsequence $\{m\} \subset \{n\}$ such that, for some $\gamma(\omega) > 0$,

$$\lim_{m \to \infty} \Delta_m(\omega, \hat{\theta}_m(\omega)) \geqslant \Delta(\theta_0) + \gamma(\omega).$$

Therefore, if $\hat{\theta}_n(\omega)$ does not converge a.s. to θ_0,

$$P(\varliminf_{n \to \infty} \Delta_n(\omega, \hat{\theta}_n(\omega)) > \Delta(\theta_0)) > 0.$$

This contradicts

$$\varlimsup_{n \to \infty} \Delta_n(\omega, \hat{\theta}_n(\omega)) \leqslant \varlimsup_{n \to \infty} \Delta_n(\omega, \theta_0) = \Delta(\theta_0) \quad \text{a.s.} \qquad \blacksquare$$

The following is a central limit theorem.

Theorem 2. Let $\hat{\theta}_n$ be a sequence of minimum discrepancy estimators. Then under B1–B8

$$\sqrt{n}(\hat{\theta}_n - \theta_0) \overset{d}{\to} X,$$

with

$$EX = 0 \quad \text{and} \quad EXX^\mathsf{T} = \Omega^{-1}\Sigma\Omega^{-1}.$$

If one replaces B8 by B8' then X is normally distributed.

Proof: By Theorem 1, $\hat{\theta}_n \to \theta_0$ a.s. and by B2 there exists a tail equivalent sequence $(\tilde{\theta}_n)$ such that each $\tilde{\theta}_n$ takes its values in a convex compact neighborhood of θ_0 in int Θ. By B6 and Lemma 3 of Jennrich (1969) one can establish the existence of a measurable function $\bar{\theta}_n$ such that

$$\sqrt{n}(\Delta_n'(\omega, \tilde{\theta}_n) - \Delta_n'(\omega, \theta_0)) = \Delta_n''(\omega, \bar{\theta}_n)\sqrt{n}(\tilde{\theta}_n - \theta_0),$$

where

$$\|\bar{\theta}_n - \theta_0\| \leq \|\tilde{\theta}_n - \theta_0\| \quad \text{a.s.} \qquad \forall n \in T.$$

($\|x\|$ is the usual Euclidean norm of $x \in R^p$.)

From the tail equivalence of $(\hat{\theta}_n)$ and $(\tilde{\theta}_n)$ it follows that $\Delta'_n(\tilde{\theta}_n)$ vanishes a.s. for n sufficiently large. Since $\bar{\theta}_n \to \theta_0$ a.s. it follows from B7 that $\Delta''_n(\omega, \bar{\theta}_n) \to \Delta''(\theta_0)$ a.s. and so for n large $\Delta''_n(\omega, \bar{\theta}_n)$ is a.s. nonsingular. Since the inversion is a continuous operation the above implies $\Delta''_n(\omega, \bar{\theta}_n)^{-1} \to \Delta''(\theta_0)^{-1}$ a.s., and therefore $\sqrt{n}(\tilde{\theta}_n - \theta_0)$ has asymptotically the same distribution as $\Delta''(\theta_0)^{-1}\sqrt{n}\Delta'_n(\omega, \theta_0)$. By B8 this converges to a random variable with mean zero and covariance matrix $\Omega^{-1}\Sigma\Omega^{-1}$. This also holds for $\sqrt{n}(\hat{\theta}_n - \theta_0)$, because $(\hat{\theta}_n)$ and $(\tilde{\theta}_n)$ are tail equivalent. ∎

A.1.2. The Expected Discrepancy and Its Estimation

We now consider minimum discrepancy estimators and give a large sample approximation to the expectation and variance of the overall discrepancy $\Delta(\hat{\theta}_n)$ (Proposition 1), and then consider estimators of $\Delta(\theta_0)$, $E\Delta(\hat{\theta}_n)$, and $\text{Var}\,\Delta(\hat{\theta}_n)$ (Proposition 2).

Proposition 1.   Let $\hat{\theta}_n$ be a sequence of minimum discrepancy estimators and assume B1–B8. Then for large n

$$E\Delta(\hat{\theta}_n) \approx \Delta(\theta_0) + \frac{\text{tr}\,\Omega^{-1}\Sigma}{2n}$$

and under B8′

$$\text{Var}\,\Delta(\hat{\theta}_n) \approx \frac{\text{tr}\,\Omega^{-1}\Sigma\Omega^{-1}\Sigma}{2n^2}.$$

If the sequence $n(\hat{\theta}_n - \theta_0)(\hat{\theta}_n - \theta_0)^{\mathsf{T}}$ is uniformly integrable the error in the approximation of $E\Delta(\hat{\theta}_n)$ is $o(1/n)$.

Proof: Without loss of generality we assume that $\hat{\theta}_n$ is a.s. in a compact convex neighborhood of θ_0. Then by B6 and Lemma 3 of Jennrich (1969) there exists a measurable function $\bar{\theta}_n$ with

$$n\Delta(\hat{\theta}_n) = n\Delta(\theta_0) + \tfrac{1}{2}n(\hat{\theta}_n - \theta_0)^{\mathsf{T}}\Omega(\hat{\theta}_n - \theta_0) + R_n,$$

where

$$R_n = \tfrac{1}{2}n(\hat{\theta}_n - \theta_0)^\mathsf{T}(\Delta''(\bar{\theta}_n) - \Omega)(\hat{\theta}_n - \theta_0)$$

and

$$\|\bar{\theta}_n - \theta_0\| \leqslant \|\hat{\theta}_n - \theta_0\| \quad \text{a.s.} \quad \text{for all } n.$$

From the assumption follows the existence of $E\Delta(\hat{\theta}_n)$, $\text{Var } \Delta(\hat{\theta}_n)$ and $\text{Var }(\sqrt{n}(\hat{\theta}_n - \theta_0))$ for all n and by Theorem 1, Theorem 2, and B6 one has $R_n = o_p(1)$.

The approximation of $E\Delta(\hat{\theta}_n)$ is obtained by omitting the expectation of the $o_p(1/n)$-term R_n/n and replacing $\text{Var }(\sqrt{n}(\hat{\theta}_n - \theta_0))$ by the asymptotic variance $\Omega^{-1}\Sigma\Omega^{-1}$. If $n(\hat{\theta}_n - \theta_0)(\hat{\theta}_n - \theta_0)^\mathsf{T}$ is *uniformly* integrable, one has

$$ER_n = o(1) \quad \text{and} \quad nE(\hat{\theta}_n - \theta_0)^\mathsf{T}\Omega(\hat{\theta}_n - \theta_0) = \text{tr } \Omega^{-1}\Sigma + o(1).$$

Thus

$$E\Delta(\hat{\theta}_n) = \Delta(\theta_0) + \frac{\text{tr } \Omega^{-1}\Sigma}{2n} + o\left(\frac{1}{n}\right).$$

The approximation of $\text{Var } \Delta(\hat{\theta}_n)$ is obtained by omitting the $o_p(1/n)$-term and using the asymptotic normality and variance of $\sqrt{n}(\hat{\theta}_n - \theta_0)$. ∎

Proposition 2. Let $\hat{\theta}_n$ be a sequence of minimum discrepancy estimators and $E\Delta_n(\omega, \theta) = \Delta(\theta) + o(1/n)$ for all $\theta \in \Theta$. Then under B1–B8

$$E\Delta_n(\omega, \hat{\theta}_n) + \frac{\text{tr } \Omega^{-1}\Sigma}{2n} \approx \Delta(\theta_0),$$

$$E\Delta_n(\omega, \hat{\theta}_n) + \frac{\text{tr } \Omega^{-1}\Sigma}{n} \approx E\Delta(\hat{\theta}_n).$$

If the sequence $n(\hat{\theta}_n - \theta_0)(\hat{\theta}_n - \theta_0)^\mathsf{T}$ is uniformly integrable and $\Delta_n''(\omega, \theta)$ is bounded for all $\theta \in \Theta$ and $n \in T$, the error in this approximation is $o(1/n)$.

Proof: For n large enough $\hat{\theta}_n$ lies a.s. in a convex compact neighborhood of θ_0 in the interior of Θ, and by B6 and Lemma 3 of Jennrich (1969) there exists a measurable function $\bar{\theta}_n$ such that

$$n\Delta_n(\omega, \hat{\theta}_n) = n\Delta_n(\omega, \theta_0) + (\hat{\theta}_n - \theta_0)^\mathsf{T}n\Delta_n'(\omega, \theta_0)$$

$$+ \tfrac{1}{2}n(\hat{\theta}_n - \theta_0)^\mathsf{T}\Delta_n''(\omega, \bar{\theta}_n)(\hat{\theta}_n - \theta_0)$$

with

$$\|\bar{\theta}_n - \theta_0\| \leqslant \|\hat{\theta}_n - \theta_0\| \quad \text{a.s.}$$

Also

$$-\Delta_n'(\omega, \theta_0) = \Delta_n''(\omega, \tilde{\theta}_n)(\hat{\theta}_n - \theta_0),$$

where

$$\|\tilde{\theta}_n - \theta_0\| \leqslant \|\hat{\theta}_n - \theta_0\| \quad \text{a.s.}$$

Thus

$$n\Delta_n(\omega, \hat{\theta}_n) = n\Delta_n(\omega, \theta_0) - \tfrac{1}{2}n(\hat{\theta}_n - \theta_0)^\mathsf{T}\Omega(\hat{\theta}_n - \theta_0) + R_n,$$

where

$$R_n = \tfrac{1}{2}n(\hat{\theta}_n - \theta_0)^\mathsf{T}(\Delta_n''(\omega, \bar{\theta}_n) + \Omega - 2\Delta_n''(\omega, \tilde{\theta}_n))(\hat{\theta}_n - \theta_0).$$

It is easily seen, using Theorem 1, Theorem 2, and B7, that $R_n = o_p(1)$ and the approximation is obtained by omitting the term $ER_n/n = E(o_p(1/n))$ and replacing $\mathrm{Var}(\sqrt{n}(\hat{\theta}_n - \theta_0))$ by the variance in the asymptotic distribution. Under the additional assumption of Proposition 2

$$En(\hat{\theta}_n - \theta_0)^\mathsf{T}\Omega(\hat{\theta}_n - \theta_0) = \mathrm{tr}\,\Omega^{-1}\Sigma + o\left(\frac{1}{n}\right)$$

and R_n is uniformly integrable with $ER_n = o(1)$.   ■

Remark 1.   If $\mathrm{tr}\,\Omega^{-1}\Sigma$ is not known it has to be estimated, usually by $\mathrm{tr}\,\Omega_n^{-1}\Sigma_n$, where Ω_n and Σ_n are consistent estimators of Ω and Σ. In such cases the following consistent estimators are available:

$$\hat{\Delta}(\theta_0) = \Delta_n(\omega, \hat{\theta}_n) + \frac{\mathrm{tr}\,\Omega_n^{-1}\Sigma_n}{2n},$$

$$\hat{E}\Delta(\hat{\theta}_n) = \Delta_n(\omega, \hat{\theta}_n) + \frac{\mathrm{tr}\,\Omega_n^{-1}\Sigma_n}{n},$$

$$\widehat{\mathrm{Var}}\,\Delta(\hat{\theta}_n) = \frac{\mathrm{tr}\,(\Omega_n^{-1}\Sigma_n)^2}{2n^2}.$$

If the approximating family contains the operating model, then for some discrepancies $\Sigma = c\Omega$. In such cases it is not necessary to estimate the above trace terms and one has

$$\hat{\Delta}(\theta_0) = \Delta_n(\hat{\theta}_n) + \frac{cp}{2n},$$

$$\hat{E}\Delta(\hat{\theta}_n) = \Delta_n(\hat{\theta}_n) + \frac{cp}{n},$$

$$\widehat{\text{Var}}\,\Delta(\hat{\theta}_n) = \frac{c^2 p}{2n^2}. \qquad\blacksquare$$

Remark 2.   In practice, it is usually not possible—apart from special cases—to establish that $n(\hat{\theta}_n - \theta_0)(\hat{\theta}_n - \theta_0)^{\mathsf{T}}$ is uniformly integrable and that $\Delta_n''(\omega, \theta)$ is bounded a.s. But Propositions 1 and 2 are only arguments supporting the use of the estimators of Remark 1 for finite samples. Only special investigations can clarify properties of the estimators for finite samples. $\qquad\blacksquare$

Remark 3.   It can happen that the bias in $\Delta_n(\theta)$ as estimator of $\Delta(\theta)$ (see Proposition 2) is of larger order than $o(1/n)$. To obtain a criterion one must then attempt to correct for this bias by subtracting an estimator of $E\Delta_n(\theta) - \Delta(\theta)$ from the estimators in Remark 1. $\qquad\blacksquare$

A.2.   ASYMPTOTIC CRITERIA FOR A FEW IMPORTANT CASES

We now consider regularity conditions B1–B8 and derive formulas for Ω and Σ for each of the discrepancies which were discussed in the main text. The necessary regularity conditions will be stated at the beginning of each subsection. Throughout we assume:

C_1: Θ is compact, θ_0 is unique and in the interior of Θ, $\Delta''(\theta_0)$ is positive definite.

This is in fact a combination of B1, B2, and B3. [If necessary, conditions C_0 to ensure the existence of $\Delta''(\theta_0)$ will also be assumed.]

A.2.1.   The Kullback–Leibler Criterion

Akaike (1973) and White (1982) are key references relating to the results in this section.

We will deal here with the case of n independently and identically distributed random vectors of k variables whose common operating d.f. F is such that the second moments exist. We denote the approximating d.f. by G_θ and assume that it has a density g_θ. The discrepancy is

$$\Delta(\theta) = - E_F \log g_\theta(x)$$

and the empirical discrepancy is

$$\Delta_n(\theta) = - \frac{1}{n} \sum_{i=1}^{n} \log g_\theta(x_i).$$

We need a condition which ensures the existence of $\Delta(\theta)$:

$C_0^{(K)}$: G_θ has a density $g_\theta(x)$, $\theta \in \Theta$, $x \in \mathbb{R}^k$, such that $\log g_\theta(x)$ is integrable
w.r.t. F for all $\theta \in \Theta$.

In addition to C_1 we need

$C_2^{(K)}$: $\log g_\theta(x)$ is for all $x \in \mathbb{R}^k$ twice continuously differentiable in Θ and
the moduli of the derivatives and their products are dominated by
functions which are integrable w.r.t. F.

From $C_2^{(K)}$ follows the existence of $\Delta'(\theta)$ and $\Delta''(\theta)$, also that

$$\Delta'(\theta) = - E_F \frac{\partial \log g_\theta(x)}{\partial \theta}$$

and

$$\Delta''(\theta) = - E_F \frac{\partial^2 \log g_\theta(x)}{\partial \theta^2}$$

are continuous and that $\Delta_n(\theta)$ is a.s. twice continuously differentiable (conditions B4 and B6). Conditions B5 and B7 follow from the strong law of large numbers (e.g., see Serfling, 1980, Theorem B on p. 27); the *uniform* convergence follows from the convergence on a compactum. The validity of B8′ follows from the central limit theorem (e.g., see Serfling, 1980, Theorem B on p. 28). One has

$$\Delta_n'(\theta_0) = - \frac{1}{n} \sum_{i=1}^{n} \frac{\partial \log g_{\theta_0}(x_i)}{\partial \theta}$$

and the mean vector of $\partial \log g_{\theta_0}(x)/\partial \theta$ is $-\Delta'(\theta_0) = 0$ and its covariance matrix is

$$\Sigma = \{E_F(\partial \log g_{\theta_0}(x)/\partial \theta_r)(\partial \log g_{\theta_0}(x)/\partial \theta_s): r, s = 1, 2, \ldots, p\}.$$

The matrix $\Omega = \Delta''(\theta_0)$ is given by

$$\Omega = \{-E_F \, \partial^2 \log g_{\theta_0}(x)/\partial \theta_r \partial \theta_s: r, s = 1, 2, \ldots, p\}.$$

The criterion is then

$$-\frac{1}{n} \sum_{i=1}^{n} \log g_{\hat{\theta}_n}(x_i) + \frac{\operatorname{tr} \Omega_n^{-1} \Sigma_n}{n},$$

where

$$\Sigma_n = \{(1/n) \sum_{i=1}^{n} (\partial \log g_{\hat{\theta}_n}(x_i)/\partial \theta_r)(\partial \log g_{\hat{\theta}_n}(x_i)/\partial \theta_s): r, s = 1, 2, \ldots, p\}$$
$$n > 1,$$
$$\Omega_n = -\{(1/n) \sum_{i=1}^{n} \partial^2 \log g_{\hat{\theta}_n}(x_i)/\partial \theta_r \partial \theta_s: r, s = 1, 2, \ldots, p\}$$

and $\hat{\theta}_n$ is the maximum likelihood estimator which is also the minimum discrepancy estimator.

If there is no misspecification ($f \equiv g_{\theta_0}$) then

$$\Omega = \Sigma,$$

which follows immediately since $\int g_\theta(x)dx = 1$ for all $\theta \in \Theta$. In this case $\operatorname{tr} \Omega^{-1}\Sigma = p$ and the simpler criterion

$$-\frac{1}{n} \sum_{i=1}^{n} \log g_{\hat{\theta}_n}(x_i) + \frac{p}{n}$$

is Akaike's Information Criterion.

Remark.   The fitting of univariate ($k = 1$) and multivariate distributions to n independent observations (Chapter 4) is obviously covered by these results. The same holds for a regression analysis (Chapters 5 and 6) with n replications (where n could be 1). In this case x_i comprises the k values of the dependent variable (belonging to k values of the independent variables) in one replication.   ∎

A.2.2. The Neyman Chi-squared Criterion

For this section and A.2.3 see also Linhart and Zucchini (1985).

We deal with the case in which a frequency distribution for a discrete variable or a grouped continuous variable is observed. Suppose the observed frequencies $n_1, n_2, \ldots, n_I$ are from a *single* sample of size $n = \sum_i n_i$. Then the operating distribution is a *multinomial* distribution with parameters n and π_i. The approximating distributions considered are multinomial with parameters n and $h_i(\theta)$ $[\sum h_i(\theta) = 1]$, where θ is a vector of p independent parameters.

We use the Neyman chi-squared discrepancy:

$$\Delta(\theta) = \sum_{i=1}^{I} \frac{(\pi_i - h_i(\theta))^2}{\pi_i}, \qquad \pi_i \neq 0.$$

It is possible to derive the discrepancies of this and the next section from Mahalanobis' generalized distance.

An empirical discrepancy is

$$\Delta_n(\theta) = \sum_{i=1}^{I} \frac{(p_i^{(n)} - h_i(\theta))^2}{p_i^{(n)}},$$

where for $p_i^{(n)} = 0$ the term $(p_i^{(n)} - h_i(\theta))^2 / p_i^{(n)}$ is to be replaced by $h_i(\theta)$. The minimum discrepancy estimator $\hat{\theta}_n$ is a minimum Neyman chi-squared estimator and $n\Delta_n(\hat{\theta}_n)$ is Neyman's chi-squared statistic $\chi_N^2(\hat{\theta}_n)$.

We need regularity condition C_1 and in addition

$C_2^{(\chi^2)}$: $0 < \pi_i,\ h_i(\theta) < 1$, and $h_i(\theta)$ is twice continuously differentiable on Θ.

Conditions B4 and B6 follow immediately from $C_2^{(\chi^2)}$ and B5 and B7 follow by the theorem in Serfling (1980, p. 24) since $p_i^{(n)} \xrightarrow{\text{a.s.}} \pi_i$. That

$$\sqrt{n}\Delta_n'(\theta_0) = \sqrt{n}(\Delta_n'(\theta_0) - \Delta'(\theta_0)) \xrightarrow{d} N(0, \Sigma)$$

with

$$\Sigma = 4 \sum_{i=1}^{I} \frac{h_i^2(\theta_0)h_i'(\theta_0)h_i'(\theta_0)^{\mathsf{T}}}{\pi_i^3}$$

follows from the asymptotic normality of $(p_1^{(n)}, p_2^{(n)}, \ldots, p_I^{(n)})$ by Theorem A in Serfling (1980, p. 122).

Now

$$\Omega = 2 \sum_{i=1}^{I} \frac{h_i'(\theta_0)h_i'(\theta_0)^\top + h_i(\theta_0)h_i''(\theta_0)}{\pi_i}$$

and if the operating model belongs to the approximating family $[\pi_i \equiv h_i(\theta_0)]$

$$\Sigma = 2\Omega.$$

Proposition 2 of Section A.1.2 can not be directly applied. There the bias of $\Delta_n(\theta)$ as estimator of $\Delta(\theta)$ should be of order $o(1/n)$; in the case considered here it is $O(1/n)$. Now it is easy to verify that [to the order $o(1/n)$]

$$E\left[\Delta_n(\theta) + \sum_i h_i^2(\theta) \frac{(p_i/n - (n+1)/n^2)}{p_i^2} \right] = \Delta(\theta).$$

If Δ_n were redefined in this way Proposition 2 would hold, but the minimum discrepancy estimator would then no longer be the minimum Neyman chi-squared estimator. We therefore prefer to leave the definition of Δ_n (and therefore also that of $\hat{\theta}_n$) as it is and to replace, in Proposition 2 and in Remark 1, $\Delta_n(\omega, \hat{\theta}_n)$ by

$$\Delta_n(\omega, \hat{\theta}_n) + \sum_i h_i^2(\hat{\theta}_n) \frac{(p_i/n - (n+1)/n^2)}{p_i^2} .$$

It is convenient here to use an estimator of $nE\Delta(\hat{\theta}_n)$ as a criterion:

$$\chi_N^2(\hat{\theta}_n) + C_N(\hat{\theta}_n) + \mathrm{tr}\, \Omega_n^{-1}\Sigma_n,$$

where

$$C_N(\hat{\theta}_n) = \sum_i h_i^2(\hat{\theta}_n) \frac{(p_i - 1 - 1/n)}{p_i^2}$$

and Ω_n and Σ_n are obtained by replacing, in Ω and Σ, θ_0 by $\hat{\theta}_n$ and π_i by p_i. The simple criterion, using $\Sigma = 2\Omega$, is

$$\chi_N^2(\hat{\theta}_n) + C_N(\hat{\theta}_n) + 2p.$$

If one is prepared to use this last approximation, which is based on the argument that $\Sigma = 2\Omega$ if the operating model belongs to the approximating family, then one should also omit $C_N(\hat{\theta}_n)$. [If there is no misspecification the correction term C_N does not depend on the approximating family.] The criterion is then simply

$$\chi_N^2(\hat{\theta}_n) + 2p.$$

A.2.3.   The Pearson Chi-squared Criterion

The basic situation is as in Section A.2.2, but the discrepancy used is now

$$\Delta(\theta) = \sum_{i=1}^{I} \frac{(\pi_i - h_i(\theta))^2}{h_i(\theta)}$$

and the empirical discrepancy is

$$\Delta_n(\theta) = \sum_{i=1}^{I} \frac{(p_i^{(n)} - h_i(\theta))^2}{h_i(\theta)}.$$

Pearson's chi-squared, $\chi_P^2(\hat{\theta}_n)$, is then $n\Delta_n(\hat{\theta}_n)$.

One needs the same regularity conditions C_1 and $C_2^{(\chi^2)}$ and the proof of B1–B8' is as in the last section. The matrices Σ and Ω are

$$\Sigma = 4 \sum_{i=1}^{I} \frac{\pi_i^3 h_i'(\theta_0)h_i'(\theta_0)^{\mathsf{T}}}{h_i^4(\theta_0)},$$

$$\Omega = \sum_{i=1}^{I} \frac{\pi_i^2 (2h_i'(\theta_0)h_i'(\theta_0)^{\mathsf{T}} - h_i(\theta_0)h_i''(\theta_0))}{h_i^3(\theta_0)},$$

and again $\Sigma = 2\Omega$ if $\pi_i \equiv h_i(\theta_0)$.

The expectation of Δ_n is now

$$E\Delta_n(\theta) = \Delta(\theta) + \sum_{i=1}^{I} \frac{\pi_i(1 - \pi_i)}{nh_i(\theta)}$$

and again a bias correction is needed. By the same argument which was used in the last section the criterion becomes

$$\chi_P^2(\hat{\theta}_n) + C_P(\hat{\theta}_n) + \operatorname{tr} \Omega_n^{-1}\Sigma_n,$$

where

$$C_\mathrm{P}(\hat{\theta}_n) = \sum_i \frac{p_i(p_i - 1)}{(1 - 1/n)h_i(\hat{\theta}_n)}.$$

The simpler criterion is

$$\chi_\mathrm{P}^2(\hat{\theta}_n) + C_\mathrm{P}(\hat{\theta}_n) + 2p,$$

where one could also omit $C_\mathrm{P}(\hat{\theta}_n)$ (compare the argument at the end of Section A.2.2.) to obtain

$$\chi_\mathrm{P}^2(\hat{\theta}_n) + 2p.$$

A.2.4. The Cramér–von Mises Criterion

This section is a summary of Linhart and Zucchini (1984b). The operating model is characterized by n independently and identically distributed random variables with d.f. F having finite second moments. The approximating models have d.f. G_θ.

The discrepancy is that of Cramér and von Mises:

$$\Delta(G_\theta, F) = \int (F(x) - G_\theta(x))^2 \, dG_\theta(x)$$

with the corresponding empirical discrepancy

$$\Delta_n(\theta) = \Delta(G_\theta, F_n).$$

Apart from C_1 we use the regularity conditions,

$C_2^{(C)}$: $G(x, \theta)$ is continuous in $\mathbb{R}$ for all $\theta \in \Theta$ and twice continuously differentiable in Θ for all $x \in \mathbb{R}$, $G_\theta'(x, \theta)$ and $G_\theta''(x, \theta)$ are of bounded variation,

$C_3^{(C)}$: F has an inverse function F^{-1}, which is continuous on $[0, 1]$,

and show that B1–B7 and B8' hold.

We first transform

$$x = F^{-1}(t),$$

and use the notation

$$\varphi(t, \theta) = G(F^{-1}(t), \theta),$$
$$\psi(t, \theta) = G'_\theta(F^{-1}(t), \theta).$$

Then

$$\Delta(\theta) = \int_0^1 (t - \varphi(t, \theta))^2 d_t \varphi(t, \theta)$$

exists as the Riemann–Stieltjes integral for all θ (e.g., see Bartle, 1976, p. 229). Also

$$\Delta_n(\theta) = \int_{-\infty}^\infty (F_n(x) - F(x) + F(x) - G(x, \theta))^2 d_x G(x, \theta)$$

$$= \int_0^1 [(F_n(F^{-1}(t)) - t)^2 + 2(F_n(F^{-1}(t)) - t)(t - \varphi(t, \theta))] d_t \varphi(t, \theta) + \Delta(\theta).$$

The integrals are "almost sure" Riemann–Stieltjes integrals. Conditions B1, B2, and B3 are assumed under C_1. That $\Delta(\theta)$ and $\Delta_n(\theta)$ are twice continuously differentiable (B6) follows from C_2. Because $F_n(x) \xrightarrow{\text{a.s.}} F(x)$ *uniformly* (e.g., see Serfling, 1980, p. 57) the integrals in the expressions for $\Delta_n(\theta)$, $\Delta'_n(\theta)$, and $\Delta''_n(\theta)$ converge a.s. uniformly on Θ to zero and B5 and B7 hold. Since B6 implies B4, it remains to show that B8 or B8' holds.

If one writes $y_n(t) = \sqrt{n}(F_n(F^{-1}(t)) - t)$, then

$$\sqrt{n}\Delta'_n(\theta_0) = \frac{1}{\sqrt{n}} \int_0^1 y_n^2(t) d_t \psi(t, \theta_0) - 2 \int_0^1 y_n(t)\psi(t, \theta_0) d_t \varphi(t, \theta_0)$$

$$+ 2 \int_0^1 y_n(t)(t - \varphi(t, \theta_0)) d_t \psi(t, \theta_0).$$

It is well known (e.g., see Durbin, 1973, p. 17) that $y_n(t)$ converges in distribution to the tied-down Brownian motion $y(t)$. An argument similar to that of Durbin (1973, p. 31) shows that $\sqrt{n}\Delta'_n(\theta_0)$ converges in distribution to

$$- 2 \int_0^1 y(t)\psi(t, \theta_0) d_t \varphi(t, \theta_0) + 2 \int_0^1 y(t)(t - \varphi(t, \theta_0)) d_t \psi(t, \theta_0)$$

which is normally distributed with zero mean and the covariance matrix

$$\Sigma = 4 \int_0^1 \int_0^1 (\min(s, t) - st)\psi(s, \theta_0)\psi(t, \theta_0)^{\mathsf{T}} d_s\varphi(s, \theta_0)d_t\varphi(t, \theta_0)$$

$$+ 4 \int_0^1 \int_0^1 (\min(s, t) - st)(t - \varphi(t, \theta_0))(s - \varphi(s, \theta_0))d_s\psi(s, \theta_0)d_t\psi(t, \theta_0)^{\mathsf{T}}$$

$$- 8 \int_0^1 \int_0^1 (\min(s, t) - st)(t - \varphi(t, \theta_0))\psi(s, \theta_0)d_s\varphi(s, \theta_0)d_t\psi(t, \theta_0)^{\mathsf{T}}.$$

This establishes the validity of B8′, and Theorems 1 and 2 in Section A.1.1 follow: $\hat{\theta}_n \xrightarrow{\text{a.s.}} \theta_0$, $\Delta_n(\hat{\theta}_n) \xrightarrow{\text{a.s.}} \Delta(\theta_0)$, and $\hat{\theta}_n$ is asymptotically distributed as $N(\theta_0, \Omega^{-1}\Sigma\Omega^{-1}/n)$, where Σ is given above and $\Omega = \Delta''(\theta_0)$.

Estimators Ω_n and Σ_n of Ω and Σ are needed for the criterion. The matrices Σ and Ω are simpler if $F(x) \equiv G(x, \theta_0)$, that is, if the operating model is a member of the approximating family. Experience in similar cases shows that it is advisable to use the criteria based on $F \equiv G_{\theta_0}$, even if one is not prepared to assume that this is the case.

If $F \equiv G_{\theta_0}$, the matrices are

$$\Sigma = 4 \int_0^1 \int_0^1 (\min(s, t) - st)\psi(s, \theta_0)\psi^{\mathsf{T}}(t, \theta_0)ds\, dt$$

and

$$\Omega = 2 \int_0^1 \psi(t, \theta_0)\psi^{\mathsf{T}}(t, \theta_0)dt,$$

and estimators Σ_n and Ω_n are obtained by replacing F by F_n and θ_0 by $\hat{\theta}_n$. For applications it is advisable to transform back by $t = F(x)$. Then, using $x_{(1)}, x_{(2)}, \ldots, x_{(n)}$ to denote the order statistics of $x_1, x_2, \ldots, x_n$, one has

$$\Sigma_n = 4 \iint (\min(F_n(x), F_n(y)) - F_n(x)F_n(y))G'_\theta(x, \hat{\theta}_n)G'_\theta{}^{\mathsf{T}}(y, \hat{\theta}_n)dF_n(x)dF_n(y)$$

$$= \frac{4}{n^2} \Sigma_{ij} \left(\min\left(\frac{i}{n}, \frac{j}{n}\right) - \frac{ij}{n^2} \right) G'_\theta(x_{(i)}, \hat{\theta}_n)G'_\theta{}^{\mathsf{T}}(x_{(j)}, \hat{\theta}_n)$$

and

$$\Omega_n = 2 \int G'_\theta(x, \hat\theta_n) G'^{\mathsf{T}}_\theta(x, \hat\theta_n) dF_n(x)$$

$$= \frac{2}{n} \sum_i G'_\theta(x_{(i)}, \hat\theta_n) G'^{\mathsf{T}}_\theta(x_{(i)}, \hat\theta_n).$$

Also needed (see Proposition 2 in Section A.1.2) is the expectation of $\Delta_n(\theta_0)$:

$$E_F \Delta_n(\theta_0) = \Delta(\theta_0) + \frac{1}{n} \int F(x)(1 - F(x)) dG_{\theta_0}(x)$$

The bias is of order $O(1/n)$ and must be estimated, either by

$$\frac{1}{n-1} \int F_n(x)(1 - F_n(x)) dG_{\hat\theta_n}(x)$$

or, if the approximation $F \equiv G_{\theta_0}$ is used, by $1/6n$. In the latter case no bias correction is necessary because the bias does not depend on the approximating model.

In situations where model selection is based on the expected discrepancy we have the following criterion:

$$n^2 \Delta_n(\hat\theta_n) + C_C + n \operatorname{tr} \Omega_n^{-1} \Sigma_n,$$

whereas in cases where selection is based on the discrepancy due to approximation we have

$$n^2 \Delta_n(\hat\theta_n) + C_C + \frac{n \operatorname{tr} \Omega_n^{-1} \Sigma_n}{2},$$

where

$$C_C = \frac{(n+1) \sum_i G(x_{(i)}, \hat\theta_n) - 2 \sum_i i G(x_{(i)}, \hat\theta_n)}{(n-1)}.$$

The correction C_C which is approximately equal to $-n/6$ can be omitted unless the number of parameters in the approximating family is very small.

The empirical discrepancy which has to be minimized to obtain $\hat{\theta}_n$ is

$$\Delta_n(\theta) = \int F_n^2(x)dG(x, \theta) - 2\int F_n(x)G(x, \theta)dG(x, \theta) + \int G^2(x, \theta)dG(x, \theta)$$

$$= \frac{1}{n^2}\sum_i i^2(G(x_{(i+1)}, \theta) - G(x_{(i)}, \theta))$$

$$- \frac{1}{n}\sum_i i(G^2(x_{(i+1)}, \theta) - G^2(x_{(i)}, \theta)) + \frac{1}{3}$$

[where $G(x_{(n+1)})$ has to be taken as 1],

$$= \frac{1}{n}\sum_i \left(G(x_{(i)}, \theta) - \frac{(2i-1)}{2n}\right)^2 + \frac{1}{12n^2}.$$

A.2.5. The Gauss Criterion

For this section see also Zwanzig (1980) and Linhart and Volkers (1984). The results of this section can also be obtained from the theorems of Chapter 1.1 of Humak (1983) which summarizes the very relevant work on regression under misspecification and on model choice by H. and O. Bunke and other East German authors.

Assume that the operating model is

$$y = \mu + e,$$

where

$$y = (y_{11}, y_{12}, \ldots, y_{1J_1}, \ldots, y_{IJ_I})^\mathsf{T},$$
$$\mu = (\mu_1, \mu_1, \ldots, \mu_1, \ldots, \mu_I)^\mathsf{T},$$
$$e = (e_{11}, e_{12}, \ldots, e_{1J_1}, \ldots, e_{IJ_I})^\mathsf{T},$$

$J_i > 1$ for all i, and the e_{ij} are independently and for each i identically distributed with $Ee_{ij} = 0$ and $\text{Var } e_{ij} = \sigma_i^2$.

Denote the expectation of y under the approximating model by

$$h(\theta) = (h_1(\theta), h_1(\theta), \ldots, h_1(\theta), \ldots, h_I(\theta))^\mathsf{T}.$$

The discrepancy is based on $(\mu - h(\theta))^{\mathsf{T}}(\mu - h(\theta))$ and since $\mu^{\mathsf{T}}\mu$ is inessential we use

$$\Delta(\theta) = -2h(\theta)^{\mathsf{T}}\mu + h(\theta)^{\mathsf{T}}h(\theta)$$

and

$$\Delta_n(\theta) = -2h(\theta)^{\mathsf{T}}\bar{y}^{(n)} + h(\theta)^{\mathsf{T}}h(\theta),$$

where

$$\bar{y} = (\bar{y}_{1.}, \bar{y}_{1.}, \ldots, \bar{y}_{1.}, \ldots, \bar{y}_{I.})^{\mathsf{T}}$$

and $\bar{y}^{(n)}$ is the average of n such vectors obtained from n independent replications of y.

Under the regularity conditions C_1 and

$C_2^{(G)}$: $h_i(\theta)$, $i = 1, 2, \ldots, I$, is twice continuously differentiable in Θ,

the conditions B1, B2, B3, and B6 are satisfied since

$$\Delta_n(\theta) = -2h(\theta)^{\mathsf{T}}\bar{e}^{(n)} + \Delta(\theta),$$
$$\Delta_n'(\theta) = -2h'(\theta)^{\mathsf{T}}\bar{e}^{(n)} + \Delta'(\theta),$$
$$\Delta_n''(\theta) = -2\sum_i J_i h_i''(\theta)\bar{e}_{i.}^{(n)} + \Delta''(\theta),$$

where $\bar{e}^{(n)} = \bar{y}^{(n)} - \mu$.

The sequence $\bar{e}^{(n)}$ obeys the strong law of large numbers and the conditions B5 and B7 follow. From the central limit theorem follows that

$$\sqrt{n}\Delta_n'(\theta_0) \xrightarrow{d} N(0, \Sigma),$$

with

$$\Sigma = 4\sum_i J_i \sigma_i^2 h_i'(\theta_0)h_i'(\theta_0)^{\mathsf{T}}.$$

Theorems 1 and 2 of Section A.1.1 hold with the given Σ and

$$\Omega = \Delta''(\theta_0) = 2\sum_i J_i [h_i''(\theta_0)(h_i(\theta_0) - \mu_i) + h_i'(\theta_0)h_i'(\theta_0)^{\mathsf{T}}].$$

As estimators of Σ and Ω one could use

$$4 \sum_i J_i \hat{\sigma}_i^2 h_i'(\hat{\theta}) h_i'(\hat{\theta})^\mathsf{T}$$

and

$$2 \sum_i J_i [h_i''(\hat{\theta})(h_i(\hat{\theta}) - \bar{y}_{i\cdot}) + h_i'(\hat{\theta}) h_i'(\hat{\theta})^\mathsf{T}],$$

where

$$\hat{\sigma}_i^2 = \frac{\sum_j (y_{ij} - \bar{y}_{i\cdot})^2}{(J_i - 1)}.$$

If the operating model belongs to the approximating family *and* $\sigma_i^2 = \sigma^2$ for all i, then $\operatorname{tr} \Omega^{-1}\Sigma = 2p\sigma^2$ and a suitable criterion is

$$\sum_{ij} (\bar{y}_{i\cdot} - h_i(\hat{\theta}))^2 + 2p\hat{\sigma}^2,$$

where $\sum_{ij} \bar{y}_{i\cdot}^2$ was added to Δ_n for convenience, and

$$\hat{\sigma}^2 = \frac{\sum_{ij} (y_{ij} - \bar{y}_{i\cdot})^2}{(J_\cdot - I)}.$$

A.2.6.   The Criterion for Stationary Time Series

The key reference to this section is Linhart and Volkers (1985) and Volkers (1983).

In addition to C_1 we will assume:

$C_2^{(T)}$:   (a) $k^{-1}(\omega, \theta)$ is continuous on $[-\pi, \pi] \times \Theta$ and is as function of ω in Lip ζ, $\zeta > \frac{1}{2}$,

(b) The derivatives $\phi(\omega, \theta) = \partial k^{-1}(\omega, \theta)/\partial\theta$ and $\psi(\omega, \theta) = \partial^2 k^{-1}(\omega, \theta)/\partial\theta^2$ exist for all $\theta \in U(\theta_0)$, where $U(\theta_0)$ is an open neighborhood of θ_0 in int Θ, are continuous on $[-\pi, \pi] \times U(\theta_0)$, and $\phi(\omega, \theta_0)$ and $\psi(\omega, \theta_0)$ are Lip ζ, $\zeta > \frac{1}{2}$.

and

$C_3^{(T)}$: $x_t,\ t = \cdots, -1, 0, 1, \ldots$, is a stationary Gaussian process with zero mean, covariance function γ satisfying $\sum_{v=-\infty}^{\infty} |v\gamma(v)| < \infty$, and spectral density $f \in \text{Lip } \zeta$, $\zeta > \frac{1}{2}$.

The discrepancy and the empirical discrepancy are

$$\Delta(\theta) = \int_{-\pi}^{\pi} (f(\omega)k^{-1}(\omega, \theta))d\omega,$$

$$\Delta_T(\theta) = \frac{2\pi}{T} \sum_{t=0}^{T-1} I_T(\omega_t)k^{-1}(\omega_t, \theta),$$

where

$$\omega_t = \frac{2\pi t}{T}$$

and

$$I_T(\omega_t) = \frac{1}{2\pi T} \left| \sum_{s=0}^{T-1} x_s e^{i\omega_t s} \right|^2, \qquad t = 0, 1, \ldots, T-1.$$

Conditions B1, B2, and B3 of Section A.1.1 are assumed in C_1; B4 and B6 follow from $C_2^{(T)}$. Lemmas 1 and 2 (to be proved below) imply B5, B7, and B8'.

Lemma 1.   Under C_1, $C_2^{(T)}$ (a), and $C_3^{(T)}$ it follows that

$$\lim_{T \to \infty} \Delta_T(\theta) = \Delta(\theta) \quad \text{a.s.}$$

uniformly for $\theta \in \Theta$. Under C_1, $C_2^{(T)}$ (b), and $C_3^{(T)}$ it follows that

$$\lim_{T \to \infty} \Delta_T'(\theta) = \Delta'(\theta) \quad \text{a.s.}$$

and

$$\lim_{T \to \infty} \Delta_T''(\theta) = \Delta''(\theta) \quad \text{a.s.}$$

uniformly for $\theta \in C$ where $C \subset U(\theta_0)$ is a compact set.

Proof.   We prove the first statement; the proofs of the second and the third are similar. Let $c_n = (1/T) \sum_{t=1}^{T-|n|} x_t x_{t+|n|}$ and q_M be the Cesaro sum of the first M terms of the Fourier series of $k^{-1}(\omega, \theta)$. Since the Cesaro sum converges uniformly in (ω, θ) we may choose M so that $|k^{-1}(\omega, \theta) - q_M(\omega, \theta)|$

$< \varepsilon$. Then, a.s., for T large enough

$$\left| \frac{2\pi}{T} \sum_t I_T(\omega_t)(q_M(\omega_t, \theta) - k^{-1}(\omega_t, \theta)) \right| \leqslant \varepsilon c_0 < \varepsilon(\gamma(0) + 1).$$

But

$$\frac{2\pi}{T} \sum_t I_T(\omega_t) q_M(\omega_t, \theta) = \sum_{n=-M}^{M} q(n) \left(1 - \frac{|n|}{M} \right) c_n + o(1)$$

converges a.s. to

$$\sum_{n=-M}^{M} q(n) \left(1 - \frac{|n|}{M} \right) \gamma_n = \int_{-\pi}^{\pi} \sum_{n=-M}^{M} q(n) \left(1 - \frac{|n|}{M} \right) e^{in\omega} f(\omega) d\omega$$

$$= \int_{-\pi}^{\pi} q_M(\omega, \theta) f(\omega) d\omega$$

and the (a.s.) convergence of $\Delta_T(\theta)$ to $\Delta(\theta)$ follows. The convergence is *uniform*, since by $C_2^{(T)}$ for all $\theta \in \Theta$ and T large enough

$$|\Delta_T(\theta)| \leqslant K \left(\frac{2\pi}{T} \right) \sum_t I_T(\omega_t) = K c_0 \leqslant K(\gamma(0) + 1) \quad \text{a.s.} \qquad \blacksquare$$

Lemma 2. Under C_1, $C_2^{(T)}$, and $C_3^{(T)}$

$$T^{1/2} \Delta'_T(\omega, \theta_0) \xrightarrow{d} N(0, \Sigma),$$

where

$$\Sigma = 4 \int_{-\pi}^{\pi} \phi(\omega, \theta_0) \phi^{\mathsf{T}}(\omega, \theta_0) f^2(\omega) d\omega.$$

Proof. The lemma follows from Theorem 4 of Robinson (1978) because $(2\pi/\sqrt{T}) \sum_t f(\omega_t) \phi(\omega_t, \theta_0)$ converges to $\Delta'(\theta_0) = 0$. [Robinson needs an additional assumption, his A.10, which is satisfied, however, for Gaussian processes with properties $C_3^{(T)}$ by a theorem of Hannan's (1976).] $\qquad \blacksquare$

Theorems 1 and 2 of Section A.1.1 follow with

$$\Omega = \Delta''(\theta_0) = \int_{-\pi}^{\pi} \psi(\omega, \theta_0) f(\omega) d\omega.$$

Proposition 1 holds, but Proposition 2 requires that the bias of $\Delta_T(\theta)$ as estimator of $\Delta(\theta)$ is of the order $o(1/T)$, which is *not* the case here. From Theorem 5.2.2 of Brillinger (1975) and (3.6) of Robinson (1978) follows that

$$E\Delta_T(\theta) = \Delta(\theta) + O\left(\frac{1}{T^{\min(1,\zeta)}}\right).$$

A correction of $\Delta_T(\theta)$ is not possible. The bias is essentially the difference between the Riemann sum $(2\pi/T)\sum_t f(\omega_t)k^{-1}(\omega_t, \theta)$ and the integral $\int_{-\pi}^{\pi} f(\omega)k^{-1}(\omega, \theta)d\omega$. One could remove the difficulty of this bias by switching from the discrepancy $\int f(\omega)k^{-1}(\omega, \theta)d\omega$ to the discrepancy $(2\pi/T)\sum_t f(\omega_t)k^{-1}(\omega_t, \theta)$. We prefer not to do this because, as it stands, $\Delta(\theta)$ is the mean squared error of prediction—a familiar and meaningful quantity.

There is then no option: the bias has to be neglected and the approximation of Proposition 2 can be obtained by omitting a $Eo_p(1/T) + O(1/T^{\min(1,\zeta)})$ term. It seems hard, however, to understand why the $O(1/T)$ term tr $\Omega^{-1}\Sigma/2T$ should be retained. One could argue that corrections should be made, even if they are small, as long as they improve the result. Also a few Monte Carlo experiments (Volkers, 1983) seem to indicate that the omitted term is small in comparison with the trace term (for moderate T). One would also hope that the term does not change its value appreciably if another approximating family is considered. The argument which leads to the criterion runs as before. In general we recommend the simple criterion

$$\Delta_T(\hat{\theta}_T)\left(1 + \frac{2p}{T}\right).$$

A.2.7. The Criteria for Nonstationary Time Series†

In addition to C_1 and $C_3^{(T)}$ we need now $C_2^{(T)'}$, a modified version of $C_2^{(T)}$, and the conditions $C_4^{(T)}$ and $C_5^{(T)}$. We use the notation:

$$Z = (z_1, z_2, \ldots, z_T)^T, \qquad z_t = (z_{t1}, z_{t2}, \ldots, z_{tk})^T,$$

$$d_j(T) = (\textstyle\sum_t z_{tj}^2)^{1/2}, \qquad D_T = \mathrm{Diag}\{d_j(T)\},$$

$$R_T(v) = D_T^{-1}\left(\sum_{t=1}^{T-|v|} z_t z_{t+|v|}^T\right)D_T^{-1}, \qquad |v| = 0, 1, 2, \ldots.$$

†Written by P. Volkers. See also Volkers (1983).

$I_Z(\omega_t)$ is the matrix of the cross-periodograms of the columns of Z, and $I_{Zy}(\omega_t)$ the cross-periodogram of the columns of Z with y.

$C_2^{(T)'}$:   Condition $C_2^{(T)}$ with $U(\theta_0)$ replaced by Θ.

$C_4^{(T)}$:   (a) $\lim_{T \to \infty} d_j(T) = \infty$, $j = 1, 2, \ldots, k$

       (b) $\lim_{T \to \infty} \max_{1 \leqslant t \leqslant T} |z_{tj}|/d_j(T) = 0$, $j = 1, 2, \ldots, k$

       (c) $\lim_{T \to \infty} R_T(v) = R < \infty$, $|v| = 0, 1, 2, \ldots$,

          R positive definite

$C_5^{(T)}$: $(Z^\mathsf{T}Z)^{-1}$ exists for all T.

The condition $C_4^{(T)}$ is a special case of Grenander's conditions. It is satisfied, for example, for $z_{tj} = t^{j-1}$ (Anderson, 1971, p. 581).

We now derive the criterion of Section 12.5. We need the asymptotic normality of the least-squares estimator $\hat{\alpha}_T$ and of the minimum discrepancy estimator $\hat{\theta}_T$ (which exists under C_1 and $C_2^{(T)'}$ by the lemma of Section A.1.1).

The Asymptotic Distributions of $\hat{\alpha}_T$ and $\hat{\theta}_T$

The asymptotic normality of $\hat{\alpha}_T$ follows as a special case from Theorem 2 of Hannan (1973a). For the proof of the asymptotic normality of $\hat{\theta}_T$ we need the following two lemmas:

Lemma 1.   Under $C_3^{(T)}$ and $C_4^{(T)}$, for all continuous, periodic, and symmetric functions $h(\omega)$, $-\pi \leqslant \omega < \pi$,

(a) $2\pi D_T^{-1} \sum_t I_Z(\omega_t)h(\omega_t)D_T^{-1} \to h(0)R$

and

(b) $T^{-1/2}D_T^{-1} \sum_t I_{Zx}(\omega_t)h(\omega_t) \overset{P}{\to} 0$.

     Proof.   Let h_M be the Cesaro sum of the first M terms in the Fourier series of h. There is then, for all $\varepsilon > 0$, an M_ε such that $\sup_\omega |h_M(\omega) - h(\omega)| < \varepsilon$ for all $M > M_\varepsilon$. Let $a_M(\omega) = h_M(\omega) - h(\omega)$, then for $M > M_\varepsilon$,

$$\left\| D_T^{-1} \sum_t I_Z(\omega_t)a_M(\omega_t)D_T^{-1} \right\| \leqslant \frac{\|R_T(0)\|\varepsilon}{2\pi} \leqslant \frac{k^2\varepsilon}{2\pi}.$$

For a proof of (a) it is therefore sufficient to use h_M in place of h. One gets, where h_v are the Fourier coefficients of h,

$$2\pi D_T^{-1} \sum_t I_Z(\omega_t) h_M(\omega_t) D_T^{-1} = \sum_{v=-M}^{M} R_T(v) h_v \left(1 - \frac{|v|}{M}\right) + o(1)$$

and by $C_4^{(T)}$ this converges to

$$R \sum_{v=-M}^{M} h_v \left(1 - \frac{|v|}{M}\right) = R h_M(0).$$

The proof of (b) follows similar lines:

$$\left\| D_T^{-1} T^{-1/2} \sum_t I_{Zx}(\omega_t) h(\omega_t) \right\| \leqslant \varepsilon \sum_t \left\| D_T^{-1} T^{-1/2} I_{Zx}(\omega_t) \right\|$$

$$+ \left\| \sum_{v=-M}^{M} h_v \left(1 - \frac{|v|}{M}\right) T^{-1/2} D_T^{-1} c_{Zx}(v) \right\| + o_p(T^{-1/2}),$$

where $c_{Zx}(v) = \sum_{t=1}^{T-|v|} z_t x_{t+|v|}$. By Schwarz' inequality the first term on the r.h.s. is smaller than

$$\varepsilon \| R_T(0) \|^{1/2} |T^{-1} \sum_t x_t^2|^{1/2}.$$

Now

$$R_T(0) = D_T^{-1} \sum_t z_t z_t^{\mathsf{T}} D_T^{-1} = \left\{ \sum_t z_{ti} z_{tj} / (\sum_t z_{ti}^2 \sum_t z_{tj}^2)^{1/2}, \quad i, j = 1, \ldots, k \right\}$$

and by Schwarz' inequality the elements of $R_T(0)$ are smaller than 1 in absolute value. On the other hand, by $C_3^{(T)}$ $(\sum_t x_t^2/T)^{1/2} \xrightarrow{\text{a.s.}} \gamma(0)^{1/2}$. The first term thus converges a.s. to ε times a constant. By $C_4^{(T)}$ and Theorem 2 of Hannan (1973a) $T^{-1/2} D_T^{-1} c_{Zx}(v) \xrightarrow{P} 0$ for $v = 0, 1, 2, \ldots$. This proves (b). ∎

Lemma 2. Let $S_T(\theta) = \Delta_T^{\hat{x}}(\theta) - \Delta_T^x(\theta)$. Then under C_1, $C_2^{(T)\prime}$, $C_3^{(T)}$, $C_4^{(T)}$, and $C_5^{(T)}$, $S_T(\theta)$, $S_T'(\theta)$, and $S_T''(\theta)$ converge in probability to zero, uniformly on Θ, and $T^{1/2} S_T(\theta_0)$, $T^{1/2} S_T'(\theta_0)$, and $T^{1/2} S_T''(\theta_0)$ converge in probability to zero.

Proof. By Hannan (1973a, Theorem 2) $D_T(Z^{\mathsf{T}}Z)^{-1} Z^{\mathsf{T}} x$ is asymptotically normally distributed. By Lemma 1

$$T^{1/2} D_T^{-1} \Delta_T^{(Zx)}(\theta_0) \xrightarrow{P} 0, \quad T^{1/2} D_T^{-1} \Delta_T^{(Z)}(\theta_0) D_T^{-1} \rightarrow 0; \quad D_T^{-1} \Delta_T^{(Zx)}(\theta) \xrightarrow{P} 0$$

for all $\theta \in \Theta$ and $D_T^{-1} \Delta_T^{(Z)}(\theta) D_T^{-1} \rightarrow 0$ for all $\theta \in \Theta$. (Since Θ is compact the convergence is uniform on Θ.)

After some algebra one can show that

$$S_T(\theta) = -(\Delta_T^{(Zx)}(\theta))^{\mathsf{T}}(Z^{\mathsf{T}}Z)^{-1}Z^{\mathsf{T}}x$$

and by the above and Slutsky's theorem (Serfling, 1980, p. 19) $S_T(\theta) \xrightarrow{P} 0$ uniformly on Θ and $T^{1/2}S_T(\theta_0) \xrightarrow{P} 0$.

The proof of the other statements of the lemma is similar. ∎

In Section A.2.6 it was proved that $\Delta_T^{(x)}(\theta) \xrightarrow{\text{a.s.}} \Delta(\theta)$ uniformly on Θ. By Lemma 2 $S_T(\theta) \xrightarrow{P} 0$, uniformly on Θ. Thus $\Delta_T^{\hat{x}}(\theta) \xrightarrow{P} \Delta(\theta)$ uniformly on Θ, that is, a weaker form of B5 holds for $\Delta_T^{(\hat{x})}(\theta)$. Since B4 obviously also holds, Theorem 1 of Section A.1.1 may be invoked in its weaker version (which can be proved analogously as the strong version of Section A.1.1): $\hat{\theta}_T \xrightarrow{P} \theta_0$, $\Delta_T^{(\hat{x})}(\hat{\theta}_T) \xrightarrow{P} \Delta(\theta_0)$.

For a proof of the asymptotic normality of $\hat{\theta}_T$ one develops $T^{1/2}\Delta_T^{(\hat{x})}(\theta)$ into a Taylor series around θ_0. By Lemma 2 the asymptotic distribution of $-T^{1/2}\Delta_T^{(\hat{x})\prime}(\theta_0)$ is the same as that of $-T^{1/2}\Delta_T^{(x)\prime}(\theta_0)$ and $\Delta_T^{(\hat{x})\prime\prime}(\theta) \xrightarrow{P} \Delta''(\theta)$ uniformly on Θ. As in Section A.2.6 it follows that $T^{1/2}(\hat{\theta}_T - \theta_0) \xrightarrow{d} N(0, \Omega^{-1}\Sigma\Omega^{-1})$, where Σ is given in Section A.2.6.

The asymptotic independence of $\hat{\alpha}_T$ and $\hat{\theta}_T$ follows as in Hannan (1973b, p. 142).

The Criterion of Section 12.5

The approximation of the expected discrepancy is obtained as in Section A.2.6:

$$E\Delta(\hat{\theta}_T) \approx \Delta(\theta_0) + \frac{\text{tr}\,\Omega^{-1}\Sigma}{2T}.$$

Furthermore,

$$\Delta_T^{(\hat{x})}(\hat{\theta}_T) = \Delta_T^{(x)}(\theta_0) + \Delta_T^{(x)\prime}(\theta_0)^{\mathsf{T}}(\hat{\theta}_T - \theta_0) + \tfrac{1}{2}(\hat{\theta}_T - \theta_0)^{\mathsf{T}}\Omega(\hat{\theta}_T - \theta_0)$$

$$+ S_T(\theta_0) + S_T'(\theta_0)^{\mathsf{T}}(\hat{\theta}_T - \theta_0) + \tfrac{1}{2}(\hat{\theta}_T - \theta_0)^{\mathsf{T}}S_T''(\theta_0)(\hat{\theta}_T - \theta_0)$$

$$= \Delta_T^{(x)}(\theta_0) + \Delta_T^{(x)\prime}(\theta_0)^{\mathsf{T}}(\hat{\theta}_T - \theta_0) + \tfrac{1}{2}(\hat{\theta}_T - \theta_0)^{\mathsf{T}}\Omega(\hat{\theta}_T - \theta_0)$$

$$- 2\Delta_T^{(Zx)}(\theta_0)^{\mathsf{T}}\Delta_T^{(Z)}(\theta_0)^{-1}D_T k(0, \theta_0)^{-1}RD_T(Z^{\mathsf{T}}Z)^{-1}Z^{\mathsf{T}}x/T$$

$$+ x^{\mathsf{T}}Z(Z^{\mathsf{T}}Z)^{-1}D_T k(0, \theta_0)^{-1}RD_T(Z^{\mathsf{T}}Z)^{-1}Z^{\mathsf{T}}x/T + M_T,$$

where $M_T = o_p(1/T)$.

The order of M_T can be seen as follows:

$$M_T = S'(\theta_0)^\mathsf{T}(\hat{\theta}_T - \theta_0) + \tfrac{1}{2}(\hat{\theta}_T - \theta_0)^\mathsf{T} S_T''(\theta_0)(\hat{\theta}_T - \theta_0)$$

$$- 2[\Delta_T^{(Zx)}(\hat{\theta}_T)^\mathsf{T} \Delta_T^{(Z)}(\hat{\theta}_T)^{-1} - \Delta_T^{(Zx)}(\theta_0)^\mathsf{T} \Delta_T^{(Z)}(\theta_0)^{-1}]$$

$$\times \frac{D_T R D_T (Z^\mathsf{T} Z)^{-1} Z^\mathsf{T} x}{Tk(0, \theta_0)}$$

$$- 2\Delta_T^{(Zx)}(\hat{\theta}_T)^\mathsf{T} \left[\mathscr{I} - \frac{\Delta_T^{(Z)}(\hat{\theta}_T)^{-1} D_T R D_T}{Tk(0, \theta_0)} \right] (Z^\mathsf{T} Z)^{-1} Z^\mathsf{T} x$$

$$+ x^\mathsf{T} Z (Z^\mathsf{T} Z)^{-1} \left[\Delta_T^{(Z)}(\hat{\theta}_T) - \frac{D_T R D_T}{Tk(0, \theta_0)} \right] (Z^\mathsf{T} Z)^{-1} Z^\mathsf{T} x$$

$$+ \tfrac{1}{2}(\hat{\theta}_T - \theta_0)^\mathsf{T} [\Delta_T^{(x)\prime\prime}(\bar{\theta}_T) - \Omega](\hat{\theta}_T - \theta_0),$$

where $\|\bar{\theta}_T - \theta_0\| \leqslant \|\hat{\theta}_T - \theta_0\|$ for all T. The asymptotic normality of $T^{1/2}(\hat{\theta}_T - \theta_0)$ together with Lemma 2 implies that the first two terms are $o_p(1/T)$; together with $C_2^{T\prime}$(b) it implies that the last term is $o_p(1/T)$. The third term is $o_p(1/T)$ by Hannan's Theorem 2 in (1973a) and $C_2^{T\prime}$. Lemma 1(a) with $h(\omega) = k(\omega, \theta_0)^{-1}$ and the asymptotic normality of $D_T(Z^\mathsf{T} Z)^{-1} Z^\mathsf{T} x$ imply that the fourth and the fifth terms are $o_p(1/T)$.

The expectation of the first three terms in the series for $\Delta_T^{(\hat{x})}(\hat{\theta}_T)$ is approximated by $\Delta(\theta_0) - \operatorname{tr} \Omega^{-1}\Sigma/2T$. (This can be shown as in Section A.1.1 using the asymptotic normality of $\hat{\theta}_T$.) By Hannan (1973a) the expectation of the two last terms is approximated by $-2\pi k f(0)/Tk(0, \theta_0)$. [The approximations neglect $Eo_p(1/T)$ terms.] The expected discrepancy can thus be approximated by

$$E\Delta_T^{(\hat{x})}(\hat{\theta}_T) + \frac{\operatorname{tr} \Omega^{-1}\Sigma}{T} + \frac{2\pi k f(0)}{Tk(0, \theta_0)}.$$

If the approximating model for x uses white noise, $k(\omega, \theta) \equiv 1$, $\Delta_T^{(x)}(\theta) = \sum_t x_t^2/T$, $\Delta(\theta) = \gamma_0$, $\Delta_T^{(\hat{x})}(\theta) = \Delta_T^{(x)}(\theta) - x^\mathsf{T} Z(Z^\mathsf{T} Z)^{-1} Z^\mathsf{T} x/T$, and $E\Delta_T^{(\hat{x})}(\theta) \approx E\Delta_T^{(x)}(\theta) - 2\pi k f(0)/T$. An approximation of the expected discrepancy in this case is $E\Delta_T^{(\hat{x})}(\hat{\theta}_T) + 2\pi k f(0)/T$.

If the operating model is a member of the approximating family $E\Delta(\hat{\theta}_T) \approx \Delta(\theta_0)(1 + p/T)$, $E\Delta_T^{(\hat{x})}(\hat{\theta}_T) \approx \Delta(\theta_0)(1 - p/T - k/T)$, and the criterion becomes

$$\frac{T + p}{T - p - k} \Delta_T^{(\hat{x})}(\hat{\theta}_T).$$

For white noise this simplifies to

$$\frac{1}{T - k}(y - Z\hat{\alpha})^{\mathsf{T}}(y - Z\hat{\alpha}).$$

A.3. DATA

The data in the following tables are used in the examples of application. References to the origin of these data are given in the appropriate places in the main text.

Table A.3.1. Monthly Gross Evaporation (mm) at Matatiele

SEASON	OCT	NOV	DEC	JAN	FEB	MAR	APR	MAY	JUN	JUL	AUG	SEP
1937/38	185	212	169	191	159	152	120	100	63	88	118	165
1938/39	165	199	192	170	129	140	133	98	69	62	123	122
1939/40	162	154	189	194	175	137	114	103	86	97	122	131
1940/41	172	187	177	189	146	141	126	144	97	96	123	184
1941/42	170	181	246	168	152	139	108	85	95	90	99	122
1942/43	133	151	170	165	157	124	102	70	85	99	96	136
1943/44	172	164	160	197	136	139	123	104	79	100	152	135
1944/45	168	209	242	191	158	133	102	97	91	109	137	147
1945/46	173	214	222	157	156	132	110	73	70	89	124	163
1946/47	171	189	211	185	162	153	91	77	49	84	121	129
1947/48	170	162	164	144	136	106	81	60	66	61	112	164
1948/49	171	181	207	208	136	137	126	91	71	87	132	137
1949/50	157	162	195	178	156	137	88	67	72	74	81	141
1950/51	178	171	187	167	192	144	112	94	80	83	97	134
1951/52	155	220	214	220	150	153	95	92	54	56	107	141
1952/53	172	151	182	176	129	157	99	84	73	66	91	128
1953/54	144	160	165	170	146	138	100	86	52	84	110	163
1954/55	173	153	223	184	141	137	95	70	71	69	120	133
1955/56	140	162	157	218	144	128	109	86	69	76	121	129
1956/57	149	135	150	182	154	128	97	91	67	73	106	128
1957/58	163	213	158	173	163	149	104	97	71	69	125	146
1958/59	188	209	191	177	146	144	141	77	64	58	96	139
1959/60	169	169	175	200	151	146	92	79	68	70	107	131
1960/61	174	158	198	200	154	139	93	73	60	68	102	150
1961/62	186	155	192	207	154	133	104	66	78	78	114	154
1962/63	168	162	192	161	157	125	92	74	57	68	76	79
1963/64	86	143	86	91	137	156	89	64	63	62	101	117
1964/65	115	185	167	166	159	165	96	89	51	65	120	127
1965/66	138	152	213	201	140	144	94	74	59	88	102	132
1966/67	160	172	172	180	135	124	71	51	57	53	91	128

Table A.3.2. Monthly Precipitation (mm) at Matatiele

SEASON	OCT	NOV	DEC	JAN	FEB	MAR	APR	MAY	JUN	JUL	AUG	SEP
1937/38	23	52	127	109	128	33	93	3	19	20	32	19
1938/39	51	44	127	112	122	53	6	21	3	13	34	76
1939/40	36	54	121	117	159	59	36	91	11	6	2	67
1940/41	75	55	78	160	118	52	44	0	5	8	9	8
1941/42	41	39	21	181	161	128	120	15	0	0	14	19
1942/43	53	118	243	85	45	122	76	36	11	28	79	17
1943/44	92	210	149	53	86	84	6	22	46	4	0	78
1944/45	71	74	65	128	218	140	14	8	0	1	0	23
1945/46	37	39	44	156	125	75	42	35	0	5	1	10
1946/47	57	105	112	98	154	108	41	0	36	5	5	35
1947/48	42	106	124	141	99	137	22	7	0	1	1	5
1948/49	39	35	91	194	77	137	46	9	0	3	2	18
1949/50	26	61	125	68	123	161	39	18	1	13	87	6
1950/51	37	50	162	71	112	53	38	2	2	4	32	49
1951/52	67	10	50	195	193	59	28	9	14	11	4	31
1952/53	43	86	140	77	49	35	44	6	0	0	9	51
1953/54	65	89	79	69	121	74	28	74	17	5	2	53
1954/55	84	59	113	237	149	78	45	8	44	2	0	28
1955/56	45	98	87	104	167	105	17	19	4	3	5	18
1956/57	42	93	194	217	164	106	18	8	5	10	35	124
1957/58	93	77	98	131	75	57	45	18	0	3	1	8
1958/59	26	140	216	58	87	64	89	122	0	29	19	16
1959/60	34	109	99	82	72	64	47	31	2	0	26	39
1960/61	53	139	183	99	98	138	88	23	5	0	17	22
1961/62	8	107	119	121	141	124	23	14	0	0	5	11
1962/63	42	128	111	294	173	94	29	6	0	44	0	2
1963/64	63	106	124	157	88	151	60	3	58	3	5	43
1964/65	115	45	71	141	50	29	37	23	57	47	64	15
1965/66	34	78	31	209	108	12	38	58	2	0	34	25
1966/67	48	54	96	205	92	172	75	16	24	25	5	4

Table A.3.3.   Daily Rainfall (mm) in Roma, Lesotho, 1966–1979

1966

Day	J	F	M	A	M	J	J	A	S	O	N	D
1	0	4	0	0	0	0	0	0	0	0	0	0
2	0	23	0	0	0	0	0	0	0	0	0	0
3	9	12	0	0	0	0	0	0	0	0	10	3
4	0	7	1	0	0	0	0	0	0	0	42	13
5	0	17	1	4	0	13	0	0	0	0	0	0
6	0	0	0	0	0	0	0	0	0	0	0	0
7	1	0	0	0	0	4	0	0	0	0	0	0
8	29	2	1	0	0	0	0	0	0	0	4	0
9	3	0	0	0	0	1	0	0	0	0	7	0
10	0	0	0	1	0	0	0	0	0	0	0	0
11	1	0	0	4	8	0	0	0	0	0	0	0
12	0	0	0	13	17	0	0	0	0	0	0	2
13	0	0	0	0	0	0	0	0	0	14	0	0
14	0	10	0	1	0	0	0	0	0	0	0	0
15	0	21	0	0	0	0	0	0	0	0	0	0
16	0	0	0	0	0	0	0	0	0	0	0	7
17	6	0	0	0	0	0	0	0	0	0	1	35
18	17	0	0	3	0	0	0	0	2	13	0	0
19	2	3	0	0	0	0	0	0	0	16	0	4
20	81	0	0	0	0	0	0	0	0	10	0	2
21	108	0	0	0	0	0	0	0	0	0	0	10
22	3	0	0	0	0	0	0	0	0	0	0	0
23	2	0	0	0	0	0	0	0	0	0	0	0
24	0	0	0	0	0	0	0	0	0	0	0	0
25	0	0	0	0	0	0	0	0	0	0	0	0
26	8	0	0	0	0	0	0	0	0	0	0	0
27	0	0	38	0	0	0	0	0	0	0	0	0
28	22	0	1	2	0	0	0	3	0	0	0	7
29	0		0	0	0	0	0	8	0	0	0	8
30	0		0	8	0	0	0	0	0	0	0	0
31	0		0		0		0	0		0		2

1967

Day	J	F	M	A	M	J	J	A	S	O	N	D
1	6	1	19	0	0	0	0	0	0	0	0	3
2	0	3	0	0	10	0	0	0	0	5	0	0
3	13	0	0	0	0	0	0	0	0	0	0	1
4	1	0	0	0	0	0	0	0	0	0	0	6
5	4	1	0	13	0	2	0	0	0	0	5	0
6	0	1	0	2	0	0	0	0	0	0	3	0
7	0	8	0	18	0	0	0	0	0	0	10	0
8	0	8	0	17	0	0	0	0	0	0	0	0
9	0	3	15	1	0	3	0	0	0	0	0	0
10	0	0	3	1	0	4	0	0	1	0	1	0
11	19	0	13	28	0	0	0	0	0	0	0	0
12	17	20	0	4	5	0	0	0	0	0	0	0
13	8	0	0	0	12	0	5	0	0	0	0	0
14	0	3	0	0	0	0	0	1	0	0	10	0
15	3	0	0	0	0	0	0	11	0	0	10	0
16	0	12	0	4	0	0	0	2	0	0	2	0
17	8	0	0	3	0	0	0	0	0	1	6	0
18	2	0	17	1	0	0	0	0	0	1	9	3
19	4	0	0	33	0	0	0	0	0	0	0	11
20	25	0	4	0	0	0	0	0	0	0	0	5
21	0	0	0	0	0	0	0	0	0	0	0	3
22	0	0	0	0	0	0	0	0	0	0	0	12
23	0	0	0	0	0	0	0	0	0	0	7	0
24	0	0	0	0	0	0	0	0	0	0	0	0
25	0	2	9	0	0	0	0	0	0	0	0	0
26	0	9	0	17	9	0	0	0	0	0	0	0
27	0	0	0	0	0	0	0	0	6	0	8	0
28	0	0	0	0	0	0	0	0	0	0	0	0
29	4		0	0	11	0	0	0	0	0	35	0
30	63		1	0	4	0	0	0	0	27	0	12
31	42		4		15		0	0		19		18

1968

Day	J	F	M	A	M	J	J	A	S	O	N	D
1	4	0	0	0	0	0	0	0	8	0	0	43
2	12	0	8	2	0	1	0	0	4	0	0	4
3	0	0	2	0	0	0	15	0	0	0	0	0
4	0	0	4	0	0	0	0	0	0	0	0	0
5	0	0	0	0	0	0	0	0	0	0	0	0
6	0	0	17	2	0	0	0	0	0	0	0	9
7	0	0	1	22	0	0	0	0	3	0	0	27
8	6	0	0	0	0	0	5	0	0	0	0	0
9	5	0	0	8	0	0	1	1	0	9	0	0
10	0	1	0	0	1	0	0	0	0	13	0	7
11	0	0	0	0	14	0	0	0	0	0	14	0
12	0	0	1	0	0	1	0	0	0	0	14	0
13	0	0	0	0	0	6	0	0	0	0	0	0
14	1	0	0	0	0	0	0	0	0	0	4	0
15	0	6	17	0	0	0	0	0	0	0	0	0
16	2	0	15	4	1	0	0	0	0	0	1	0
17	0	0	0	30	53	0	0	0	0	0	1	0
18	0	0	0	3	0	0	0	0	0	0	6	0
19	8	0	0	0	0	0	0	0	0	0	0	0
20	7	0	0	0	0	0	0	0	0	1	7	0
21	0	0	0	0	0	0	0	0	0	3	6	14
22	0	0	0	0	0	0	0	0	0	10	1	8
23	0	0	1	0	0	0	0	0	0	20	10	9
24	0	0	4	0	0	0	0	0	0	0	0	0
25	4	0	24	0	0	0	0	0	0	0	0	0
26	0	0	1	0	0	0	0	0	0	0	0	0
27	0	0	0	20	0	0	0	0	3	0	0	0
28	0	0	0	0	0	0	0	0	0	0	0	0
29	0		0	0	0	0	0	0	0	0	0	0
30	1		0	0	0	0	0	0	1	3	0	0
31	26		0		7		0	5		0		0

1969

Day	J	F	M	A	M	J	J	A	S	O	N	D
1	0	6	5	0	0	0	0	0	0	0	0	0
2	0	0	22	11	0	0	0	0	0	0	0	2
3	0	0	0	22	6	0	0	0	0	0	0	0
4	0	0	0	0	0	0	0	0	0	0	0	1
5	6	0	6	0	0	0	0	0	0	0	0	0
6	0	2	0	0	0	0	0	0	0	0	0	15
7	0	8	0	3	0	0	0	0	0	5	0	6
8	0	0	21	30	0	0	0	22	0	0	2	9
9	0	0	23	5	0	0	0	0	0	0	8	4
10	0	1	2	2	0	0	0	0	0	0	0	0
11	0	0	24	0	0	0	0	0	16	0	0	0
12	0	1	0	0	0	0	0	0	0	0	0	1
13	0	1	0	0	0	0	0	0	0	0	0	0
14	0	0	0	0	0	0	0	0	0	0	0	0
15	1	16	0	0	0	0	0	0	0	4	0	3
16	0	12	0	13	0	0	0	0	0	8	0	0
17	7	0	0	0	8	0	0	0	0	5	0	0
18	0	24	0	0	0	0	0	0	0	8	0	4
19	0	0	0	0	16	0	0	0	0	16	0	0
20	0	0	17	0	13	0	0	0	0	1	11	0
21	0	0	0	0	17	0	0	0	0	7	0	0
22	0	0	0	0	0	0	0	0	0	1	0	0
23	0	0	0	0	0	0	0	0	0	0	1	0
24	0	14	0	0	0	0	0	0	0	32	0	5
25	0	4	0	0	0	0	0	0	0	0	0	0
26	0	14	39	0	0	0	0	0	0	24	0	0
27	6	14	12	0	0	0	0	0	0	18	0	8
28	0	27	0	0	0	3	0	0	0	12	0	0
29	0		2	0	0	2	0	0	0	0	0	0
30	5		1	0	2	0	0	0	1	0	0	0
31	4		0		5		0	0		0		8

1970

Day	J	F	M	A	M	J	J	A	S	O	N	D
1	0	0	2	0	0	0	0	0	0	0	0	2
2	4	0	0	0	0	0	0	0	0	0	5	13
3	3	0	0	0	0	0	0	0	0	0	26	15
4	12	0	0	0	0	0	0	0	0	0	0	50
5	0	11	3	0	0	0	0	0	0	0	0	24
6	0	14	0	0	4	0	0	0	0	0	0	0
7	9	6	9	0	5	0	0	0	0	0	1	0
8	0	0	9	0	0	0	0	0	0	0	3	0
9	0	0	0	0	0	0	0	0	0	0	0	0
10	0	0	0	0	0	0	0	0	0	46	1	0
11	0	0	0	0	0	0	0	0	0	18	6	0
12	0	0	0	7	0	0	0	0	0	3	0	0
13	0	0	0	0	0	0	0	0	0	7	0	0
14	0	6	0	5	0	0	0	0	0	0	0	0
15	0	0	0	0	0	0	0	0	0	0	0	0
16	10	0	0	0	0	0	17	0	0	0	0	0
17	35	0	0	0	0	0	0	0	0	0	0	0
18	0	0	1	9	0	13	0	0	0	0	0	2
19	0	7	2	4	0	0	0	0	0	0	0	22
20	0	0	5	0	0	0	0	0	0	0	0	18
21	0	11	0	0	0	0	0	0	0	0	0	0
22	10	0	0	0	0	0	0	0	0	0	0	0
23	2	0	0	0	0	0	0	0	0	0	1	0
24	2	0	0	0	0	0	0	0	0	0	0	17
25	1	0	0	0	0	0	0	25	0	0	0	4
26	0	0	0	0	0	0	0	0	1	2	0	0
27	6	0	2	0	0	6	0	0	0	1	0	0
28	4	0	0	0	0	0	0	0	48	0	0	0
29	2		0	0	0	0	0	0	7	1	0	0
30	0		0	0	0	0	0	0	0	5	0	0
31	0		0		0		0	7		0		0

1971

Day	J	F	M	A	M	J	J	A	S	O	N	D
1	0	0	0	2	15	0	0	0	0	0	4	0
2	8	0	0	0	12	0	0	0	0	0	6	7
3	1	20	0	28	0	0	0	0	0	0	0	5
4	0	3	0	0	0	11	0	0	0	0	0	1
5	2	4	0	0	0	0	0	0	0	0	1	4
6	9	9	0	2	0	0	0	0	0	8	0	1
7	9	0	14	0	0	0	0	0	0	18	0	0
8	0	0	6	2	0	0	0	0	0	3	7	0
9	1	9	0	0	0	0	0	0	0	1	4	0
10	19	0	0	0	11	0	0	0	0	0	0	0
11	3	0	0	0	0	0	0	0	0	0	0	0
12	0	0	0	0	0	0	0	0	0	0	5	0
13	6	0	2	0	0	0	0	0	0	0	3	0
14	0	0	0	3	0	0	0	0	0	8	0	0
15	0	3	0	0	0	0	0	0	0	0	0	0
16	0	6	0	0	0	0	0	0	0	0	0	0
17	0	0	13	0	0	0	0	0	0	2	0	0
18	0	0	0	0	0	0	0	0	0	0	7	0
19	16	13	0	0	0	0	0	0	0	0	0	7
20	0	0	0	0	0	0	0	0	0	0	2	5
21	0	6	2	0	0	2	0	4	0	0	0	6
22	0	2	12	0	0	0	0	0	0	0	0	0
23	18	0	8	4	5	0	0	0	0	0	0	0
24	7	0	22	0	0	0	3	0	0	0	0	0
25	1	6	0	0	0	0	0	0	0	0	0	0
26	0	1	0	0	0	0	0	0	0	5	0	4
27	0	4	0	0	0	0	0	0	0	0	0	16
28	1	0	1	0	2	0	0	0	0	18	0	6
29	0		7	8	0	0	0	0	0	0	0	0
30	17		6	23	0	0	10	0	0	0	4	0
31	4		2		0		0	0		0		2

1972

Day	J	F	M	A	M	J	J	A	S	O	N	D
1	2	0	0	0	0	0	0	0	14	0	0	0
2	1	0	0	0	0	0	0	0	0	3	12	0
3	4	0	0	0	0	0	0	0	0	12	25	0
4	0	0	0	0	0	20	0	0	0	10	2	0
5	0	0	0	0	13	0	0	0	0	25	1	0
6	0	0	0	0	0	0	0	0	0	0	5	0
7	0	0	9	0	0	0	0	0	0	0	0	0
8	0	0	5	1	0	0	0	0	0	0	12	14
9	22	0	0	0	0	0	0	0	0	0	0	9
10	6	0	0	0	0	0	0	0	0	5	0	0
11	49	0	0	13	0	0	0	0	0	0	0	9
12	3	0	0	0	0	0	0	0	0	0	0	0
13	13	0	17	0	0	0	0	0	0	0	0	0
14	14	0	0	0	0	0	0	0	0	11	0	0
15	1	0	51	0	0	0	0	0	0	0	0	0
16	43	3	0	0	0	0	0	0	0	0	0	0
17	1	5	20	0	0	0	0	0	0	0	0	0
18	9	0	21	0	0	0	0	0	0	0	0	0
19	49	12	9	0	0	0	0	0	0	0	0	0
20	9	9	0	0	0	0	0	0	0	0	0	0
21	0	0	0	0	0	0	0	0	0	0	0	0
22	0	7	16	0	0	0	0	0	0	0	0	0
23	0	0	40	0	0	0	0	0	25	0	0	0
24	0	55	30	5	0	0	0	0	15	0	11	0
25	3	35	0	0	0	0	0	0	0	0	1	0
26	0	0	0	1	0	0	0	0	0	0	0	0
27	0	0	32	7	0	0	0	0	0	0	0	0
28	13	25	0	13	0	0	0	0	0	0	0	0
29	0	0	0	8	0	0	0	0	0	0	0	0
30	0		0	8	6	0	0	0	0	0	0	0
31	0		0		0		0	0		0		0

1973

Day	J	F	M	A	M	J	J	A	S	O	N	D
1	0	15	4	1	3	0	0	3	0	0	14	0
2	9	0	0	2	0	0	0	0	0	0	14	0
3	5	4	0	2	0	0	0	0	0	0	0	18
4	0	21	0	0	0	3	0	30	0	0	4	15
5	2	21	0	0	0	0	0	18	0	0	0	0
6	0	8	0	20	0	0	0	13	0	0	0	0
7	0	68	10	3	0	0	0	4	0	0	0	1
8	0	19	30	0	0	0	0	0	0	0	1	9
9	0	11	0	0	0	0	0	0	0	0	6	1
10	0	0	0	12	0	0	0	0	0	0	0	0
11	11	0	0	0	0	0	0	0	0	0	0	0
12	0	0	0	0	0	0	0	0	0	0	0	0
13	0	13	0	0	0	0	0	0	0	0	0	0
14	0	0	0	1	1	0	0	0	0	0	0	0
15	0	0	0	0	0	0	0	0	0	0	0	1
16	0	0	1	0	0	0	0	0	0	1	0	19
17	0	0	8	0	0	0	0	0	0	0	0	0
18	0	5	4	0	0	0	0	0	0	0	0	1
19	0	15	1	2	0	0	0	0	0	8	0	0
20	0	4	2	6	0	0	0	0	0	0	29	0
21	0	20	0	0	0	0	0	0	0	0	18	0
22	0	6	0	0	0	0	0	0	0	0	4	6
23	0	0	0	0	0	0	0	0	0	4	0	3
24	0	3	5	4	2	0	0	0	0	0	1	0
25	1	0	0	0	0	0	0	0	0	0	1	0
26	4	0	32	0	1	0	11	0	0	3	0	2
27	0	1	5	0	0	0	0	0	12	0	0	3
28	0	1	0	0	2	0	0	0	25	0	0	0
29	0		4	0	0	0	0	0	0	1	0	0
30	0		0	0	0	0	0	0	2	0	8	10
31	11		0		0		0	0		1		0

Table A.3.3. (*Continued*)

1974

Day	J	F	M	A	M	J	J	A	S	O	N	D
1	4	5	0	0	0	0	0	0	0	0	7	4
2	3	3	0	10	0	0	0	0	0	0	0	18
3	4	26	4	6	0	0	0	0	0	0	0	0
4	0	2	0	0	0	0	0	0	0	0	0	0
5	0	0	0	0	14	0	0	0	0	1	0	0
6	1	4	0	0	0	0	0	0	0	0	31	0
7	1	0	0	6	0	1	0	0	0	0	28	0
8	1	5	0	0	0	0	0	0	0	1	27	10
9	4	1	0	0	0	0	0	0	0	0	0	12
10	17	0	27	0	0	0	0	0	0	0	10	2
11	7	0	2	0	0	4	0	0	0	0	3	2
12	0	0	2	0	3	0	8	0	0	0	0	0
13	0	26	0	0	1	4	0	1	1	1	0	0
14	0	23	0	1	0	0	0	0	0	0	0	0
15	0	5	0	0	2	0	0	2	5	2	3	19
16	10	10	15	1	0	0	0	1	0	0	0	0
17	4	0	5	1	0	0	0	0	0	0	10	19
18	0	0	0	0	0	0	0	20	0	0	0	6
19	7	15	0	1	0	0	0	0	8	0	6	0
20	3	4	1	1	0	0	0	0	0	0	0	0
21	12	6	0	0	0	0	0	2	0	0	0	5
22	12	6	0	0	0	0	0	9	0	0	15	0
23	16	0	0	0	0	0	0	2	0	0	10	0
24	1	5	8	11	2	0	0	0	0	4	20	0
25	9	3	0	3	0	0	0	0	0	2	0	16
26	0	24	0	0	0	0	0	0	0	0	0	0
27	27	3	0	0	0	0	0	0	0	0	0	0
28	21	1	0	0	0	0	0	0	0	0	13	0
29	25		0	0	1	0	0	0	0	11	4	0
30	2		0	0	0	0	0	0	0	0	0	0
31	0		18		0		0	0		2		0

1975

Day	J	F	M	A	M	J	J	A	S	O	N	D
1	0	0	48	0	0	0	0	0	4	0	4	2
2	0	0	0	1	0	0	0	0	0	0	9	0
3	0	0	4	0	9	0	0	0	0	0	0	0
4	0	1	26	0	0	0	0	0	0	0	0	9
5	0	2	5	0	0	0	30	0	0	0	0	8
6	0	0	0	0	0	4	0	0	0	0	0	0
7	0	0	14	0	0	0	0	0	0	0	0	0
8	0	0	0	0	0	0	0	0	0	0	0	0
9	0	24	0	2	1	0	0	0	0	0	2	0
10	0	4	0	12	0	0	0	0	0	0	0	1
11	0	0	26	0	0	0	0	0	0	3	26	1
12	0	0	27	0	0	0	0	0	0	5	22	7
13	0	26	0	0	0	0	0	0	6	0	17	3
14	0	1	0	4	0	0	0	0	0	0	0	1
15	1	5	1	1	0	0	0	0	0	0	0	16
16	13	36	12	0	0	0	0	0	0	0	2	3
17	4	4	6	0	0	0	0	0	0	2	0	2
18	1	0	2	0	0	0	0	0	4	8	1	2
19	0	22	0	0	0	0	0	0	0	1	0	0
20	0	2	0	1	0	0	0	0	0	0	0	1
21	0	0	1	6	0	0	0	0	0	0	0	0
22	4	0	2	0	0	0	0	0	2	0	0	2
23	22	13	5	4	0	0	0	1	0	8	17	0
24	2	13	0	0	0	0	0	0	0	19	12	0
25	25	0	0	0	0	0	0	0	0	0	8	17
26	19	0	0	1	0	8	0	0	14	0	22	1
27	4	0	0	0	0	9	0	0	36	0	2	0
28	0	0	0	9	0	0	0	0	6	0	0	1
29	15		0	0	0	0	0	0	0	0	0	0
30	0		0	0	0	0	0	0	0	0	0	0
31	0		0		0		0	1		2		1

1976

Day	J	F	M	A	M	J	J	A	S	O	N	D
1	1	7	0	4	0	3	0	0	0	14	1	1
2	0	35	2	6	0	3	0	0	0	47	11	1
3	2	3	0	16	0	0	0	0	0	5	0	0
4	14	9	11	0	1	0	0	0	0	8	9	1
5	37	0	1	0	0	6	0	0	0	2	0	0
6	6	5	0	0	0	2	0	0	0	1	0	0
7	0	23	15	0	0	0	0	0	0	0	0	0
8	15	2	2	0	0	0	0	0	0	3	0	0
9	11	26	0	0	0	0	0	0	0	0	0	0
10	8	31	0	0	0	0	0	0	0	0	0	0
11	31	1	11	0	0	0	0	2	0	0	0	0
12	0	0	0	2	0	0	0	0	0	1	0	6
13	0	0	0	0	0	0	0	0	0	0	0	0
14	18	0	6	0	0	0	0	0	0	11	0	16
15	0	0	0	31	0	8	0	0	0	5	0	21
16	0	0	0	19	0	0	0	0	0	5	0	0
17	20	9	5	0	0	0	0	0	0	0	0	0
18	11	0	50	7	0	0	0	0	0	0	0	0
19	9	0	3	0	0	6	1	0	0	0	0	0
20	6	0	19	0	0	0	0	0	6	0	0	0
21	0	3	0	0	0	0	0	0	0	0	0	0
22	0	0	1	0	0	0	0	0	6	1	0	20
23	0	3	0	0	0	0	0	0	0	2	1	0
24	2	0	1	0	0	0	0	0	0	0	0	1
25	1	0	0	0	0	0	0	0	6	17	0	4
26	28	0	0	0	0	0	0	0	37	0	0	1
27	6	0	5	0	2	0	0	0	11	0	1	0
28	0	1	17	0	27	0	0	0	3	19	0	0
29	0		0	11	0	0	0	0	9	0	0	0
30	0		0	0	0	0	0	0	19	36	0	0
31	0		0		0		0	0		0		1

1977

Day	J	F	M	A	M	J	J	A	S	O	N	D
1	0	0	1	6	0	0	0	0	0	0	0	12
2	0	3	9	0	14	0	0	0	0	0	0	0
3	0	5	9	0	0	0	0	0	0	0	0	0
4	0	15	67	0	0	0	0	0	0	0	0	0
5	0	0	15	0	0	0	0	0	0	0	0	0
6	0	0	0	2	0	0	0	0	0	0	0	0
7	0	0	0	10	0	0	0	0	0	0	0	4
8	13	0	0	0	0	0	0	0	0	0	0	0
9	11	0	4	0	0	0	0	0	0	13	0	0
10	8	10	0	0	0	0	0	0	0	1	0	0
11	0	0	0	0	0	0	0	0	0	0	0	0
12	13	3	1	0	0	0	0	0	0	2	0	0
13	0	6	5	0	0	0	0	0	0	0	0	1
14	0	6	6	0	0	0	0	0	0	0	0	0
15	0	3	14	0	0	0	0	0	0	0	0	2
16	0	0	3	0	0	0	0	0	0	24	0	0
17	0	0	1	0	0	0	0	0	0	13	0	6
18	1	26	0	9	0	0	0	0	23	0	0	0
19	2	0	0	0	0	0	0	0	0	0	0	3
20	0	0	0	0	0	0	0	0	0	0	0	0
21	1	1	0	0	0	0	0	0	29	9	16	14
22	3	15	7	0	0	0	0	0	7	9	0	1
23	2	0	9	10	0	0	0	0	0	0	3	0
24	9	54	3	0	0	0	0	0	12	0	3	6
25	11	11	1	0	0	0	0	5	28	0	0	2
26	0	0	0	0	0	0	0	0	0	0	0	0
27	0	0	0	0	1	0	0	0	4	5	8	0
28	53	0	0	1	0	0	0	0	0	12	7	0
29	22		0	0	5	0	0	0	0	2	28	4
30	0		1	0	0	0	0	0	0	0	11	42
31	11		0		0		0	1		13		16

1978

	J	F	M	A	M	J	J	A	S	O	N	D
1	0	0	0	5	1	0	0	0	0	0	2	10
2	5	0	0	0	0	0	0	0	0	0	0	23
3	0	0	5	0	0	0	0	0	0	0	1	0
4	29	3	5	0	0	0	0	0	0	0	0	4
5	1	0	0	0	0	0	0	0	0	0	0	9
6	12	0	0	6	0	0	0	0	12	0	1	0
7	5	0	2	23	1	0	0	0	0	0	0	1
8	0	0	2	15	0	0	0	0	11	0	0	14
9	0	23	3	10	0	0	0	0	0	0	0	62
10	0	5	0	7	0	0	0	0	0	0	1	16
11	19	0	1	0	0	10	0	0	1	0	9	0
12	0	1	0	0	0	0	0	0	0	0	0	0
13	0	0	5	0	0	0	0	0	0	0	0	0
14	0	0	1	0	0	0	0	0	0	0	0	1
15	2	0	7	0	0	0	0	0	0	0	0	0
16	13	0	17	4	0	0	4	0	0	0	0	1
17	0	11	0	11	0	0	0	0	2	3	0	3
18	0	0	0	12	0	0	0	0	3	20	0	0
19	1	0	0	9	0	0	0	0	8	0	0	0
20	0	10	0	0	0	0	0	0	22	0	0	9
21	1	0	0	0	0	0	0	0	4	2	0	2
22	0	0	0	0	0	0	0	0	0	0	0	13
23	35	0	0	0	0	0	0	2	0	3	7	0
24	9	0	0	7	0	0	0	0	5	0	0	1
25	38	0	0	0	0	0	0	0	0	0	0	0
26	5	0	23	0	0	0	0	17	3	0	0	3
27	1	0	7	0	0	0	0	4	0	0	0	0
28	53	0	11	0	0	0	0	0	0	6	11	5
29	0		0	0	0	0	0	0	0	0	0	0
30	0		13	5	0	0	0	0	0	0	0	4
31	0		42		0		0	0		5		0

1979

	J	F	M	A	M	J	J	A	S	O	N	D
1	0	0	0	0	0	7	0	0	0	0	0	0
2	0	0	0	0	1	0	0	0	0	0	0	10
3	0	0	1	0	0	0	0	0	0	0	0	0
4	0	2	0	0	0	0	0	0	0	0	2	0
5	4	0	0	1	25	1	0	0	0	0	0	1
6	0	4	0	0	0	0	0	0	1	0	0	0
7	0	4	0	0	0	0	0	0	0	0	6	0
8	0	23	0	0	0	0	2	0	0	0	0	0
9	1	12	0	0	0	0	8	0	0	0	0	0
10	0	0	0	0	0	0	0	0	0	5	0	0
11	1	0	0	0	0	0	0	0	0	0	12	1
12	0	4	0	0	0	0	0	0	0	0	0	37
13	0	2	3	0	0	0	0	0	0	0	3	0
14	0	3	0	0	0	0	10	0	1	0	0	3
15	8	37	1	0	0	0	10	0	1	0	0	1
16	9	3	0	0	0	0	0	10	0	0	0	0
17	5	4	0	4	0	0	0	5	0	14	2	0
18	6	0	0	8	0	0	0	5	0	37	2	2
19	16	0	0	4	0	0	3	10	0	40	10	17
20	0	0	0	1	0	0	5	3	17	0	2	0
21	0	0	0	0	0	0	6	3	0	7	0	2
22	0	22	2	7	0	0	0	0	0	0	11	0
23	0	5	0	2	13	0	0	0	0	0	26	23
24	0	6	6	0	0	0	0	0	0	0	2	1
25	0	29	2	8	0	0	0	5	0	0	6	11
26	0	0	0	0	0	0	0	2	0	2	7	5
27	7	0	0	0	0	0	0	0	0	0	0	0
28	1	0	1	0	0	0	0	0	7	0	0	0
29	11		0	0	0	0	0	0	1	0	0	0
30	2		0	1	0	0	0	0	0	0	0	0
31	0		0		0		0	0		0		0

Table A.3.4. **Moisture Content (%) of Pinus Pinaster and Climatic Variables in the Stellenbosch Area, Cape Province**

y(1)	y(2)	x(1)	x(2)	x(3)	x(4)	x(5)	x(6)	x(7)	x(8)	x(9)
11.70	10.48	0.0	5.7	10.9	108.0	96.0	40.3	17.4	25.6	9.1
30.45	12.31	1.8	1.8	4.6	183.0	95.7	59.3	12.6	14.0	11.1
45.10	13.67	2.2	3.4	5.9	156.1	96.0	53.3	13.7	16.3	11.1
9.59	9.99	0.0	11.3	12.1	447.5	91.3	35.7	19.2	25.4	13.1
10.96	9.08	11.5	3.0	1.9	139.1	96.0	46.7	17.1	21.9	12.3
29.36	12.80	7.5	1.4	0.0	131.1	95.0	49.0	14.0	16.0	12.0
31.92	16.35	9.7	2.7	2.5	137.8	94.7	52.7	13.0	16.5	9.5
9.22	10.21	0.0	11.5	12.5	490.1	96.3	36.0	18.4	25.1	11.6
7.71	9.11	0.0	8.3	12.5	135.8	91.0	31.0	19.7	27.7	11.6
9.49	9.94	0.0	8.3	11.7	142.3	96.0	46.7	19.9	26.7	13.0
9.79	9.95	0.0	8.7	11.8	231.8	93.3	47.0	17.1	21.3	12.9
31.07	16.94	0.0	8.2	12.4	147.0	96.7	43.3	17.7	23.2	12.1
28.36	14.48	1.0	6.0	10.8	126.1	95.7	47.0	20.9	25.7	16.1
8.01	8.96	0.4	9.6	12.2	295.3	94.0	38.3	21.6	28.7	14.4
15.59	10.77	0.0	9.0	4.2	150.7	95.0	34.7	17.0	21.8	12.2
7.96	8.65	0.0	9.2	12.3	215.2	95.3	43.0	19.7	26.1	13.2
10.97	11.23	0.0	10.0	12.2	306.1	93.3	42.0	21.9	28.4	15.4
9.10	8.15	0.0	10.0	12.3	307.8	95.0	42.0	20.3	26.7	13.8
9.07	8.33	0.0	5.5	5.5	174.0	92.7	46.0	17.4	24.4	10.4
7.09	6.33	0.1	6.9	8.6	133.0	93.3	50.0	23.1	29.4	16.7
9.23	8.25	0.0	7.7	10.2	193.0	94.3	55.3	21.7	27.1	16.4
6.95	6.85	0.0	15.2	11.4	401.6	49.7	23.0	21.6	28.1	15.0
13.00	7.46	0.0	6.2	9.5	193.1	95.7	50.0	19.0	24.1	13.9
6.45	6.99	0.0	5.8	10.5	114.6	97.3	30.3	19.1	27.4	10.6
43.33	20.98	2.0	2.0	0.3	182.7	99.0	51.3	15.1	18.8	11.4
10.69	10.10	0.0	2.2	5.3	118.2	98.7	56.3	16.4	20.8	12.0
12.62	9.34	0.0	4.5	7.7	172.1	92.0	41.7	14.4	19.0	9.8
3.87	5.55	4.5	5.6	7.4	148.5	72.7	20.0	26.8	36.3	17.2
4.93	6.00	0.0	3.3	9.1	91.5	96.7	27.3	20.7	29.5	11.8
13.95	11.69	0.0	2.8	9.2	77.2	97.3	43.0	14.7	20.3	9.0
6.87	8.13	0.0	5.3	9.0	351.1	80.0	41.0	16.5	21.9	11.0
13.64	10.64	0.0	2.3	8.7	87.8	98.7	43.3	14.3	20.7	7.9
5.41	8.33	0.0	2.5	8.2	62.5	97.0	30.0	18.9	28.2	9.5
17.49	10.94	0.0	3.0	5.3	249.4	96.5	48.5	11.4	15.9	6.9
8.87	8.85	0.0	3.0	4.7	160.8	99.0	53.0	12.0	15.5	8.4
6.88	8.26	0.0	1.8	8.2	75.3	97.0	29.7	13.6	21.1	6.1
7.00	7.51	0.0	3.7	8.1	108.4	88.7	18.0	13.5	22.5	4.4
11.02	11.00	0.1	2.1	2.8	190.4	96.0	63.0	10.4	15.0	5.7
44.00	19.06	3.0	2.2	0.0	50.7	97.3	91.0	12.5	13.6	11.4
16.86	11.89	0.0	2.0	7.1	63.0	94.0	41.0	7.9	13.6	2.1
10.54	11.38	0.0	2.5	8.5	64.9	97.3	38.0	10.4	17.9	2.9
44.07	19.70	8.0	0.2	0.1	134.0	98.0	77.0	10.6	12.2	8.9
22.03	12.18	0.0	3.0	8.8	66.6	98.0	35.3	10.3	16.9	3.6
25.49	13.81	0.0	3.0	8.9	124.5	96.7	31.7	13.7	18.5	8.8
14.53	12.16	0.0	2.5	7.4	106.5	93.7	55.3	12.6	16.5	8.6
6.92	7.31	0.0	5.0	9.6	344.5	85.7	33.3	11.5	19.4	3.6
39.46	10.80	26.8	0.0	0.5	215.2	94.3	82.3	10.6	12.4	8.8
10.40	11.78	0.0	3.7	10.2	99.2	79.7	27.7	11.0	18.3	3.7
16.79	11.91	0.0	3.7	9.1	349.3	90.0	36.0	13.2	17.3	9.0
22.88	12.96	14.2	4.1	1.9	381.1	95.7	51.0	12.4	15.6	9.2
7.23	8.61	0.0	4.5	7.8	112.3	96.0	55.3	15.0	19.8	10.2
7.67	8.71	0.0	5.2	8.5	150.0	96.3	50.3	11.8	18.0	5.6

Note: The moisture content was measured weekly (10/1/80–9/30/81) on 12 cubes each of size 100 mm × 50 mm which were either untreated [y(1)] or treated with a water repellent [y(2)] (Rypstra, 1982). The climatic variables were: x(1) rainfall (mm), x(2) evaporation (mm), x(3) solar radiation (h), x(4) wind velocity (km/D), x(5) maximum relative hyumidity (%), x(6) minimum relative hyuidity (%), x(7) average temperature (°C), x(8) maximum temperature (°C), x(9) minimum temperature (°C).

Table A.3.5. Tree-Ring Indices of Widdingtonia Cedarbergensis in the Cedarberg Area, Cape Province, 1564–1970 (La Marche et al., 1979)

28	47	33	56	48	40	62	47
60	84	95	79	86	85	102	90
108	102	117	171	140	112	155	96
115	169	115	76	72	85	75	88
52	39	41	49	110	124	134	91
96	76	76	71	71	65	75	102
99	88	103	108	76	72	93	108
123	96	91	81	79	90	90	94
104	111	116	94	100	93	106	108
99	116	117	121	112	108	94	97
109	136	115	137	138	135	116	97
98	93	89	88	82	99	94	86
85	97	109	110	113	94	98	83
120	152	124	104	126	111	108	71
82	88	98	89	88	96	98	110
107	112	127	137	132	147	103	135
97	84	136	115	131	113	117	126
99	111	129	128	102	117	85	105
102	117	108	104	83	99	94	88
78	77	81	81	67	101	119	104
95	72	84	73	76	81	80	67
79	87	102	85	85	104	88	71
65	76	95	97	109	119	94	94
100	104	121	121	134	105	98	79
95	93	95	75	104	84	97	115
104	81	105	105	124	107	112	110
104	106	111	132	151	135	113	99
91	91	103	101	108	97	105	215
182	139	127	94	101	87	91	146
93	94	69	89	79	87	88	94
84	99	101	104	103	103	124	110
104	108	107	85	157	97	90	97
78	67	70	81	85	62	73	71
96	76	95	103	100	86	95	103
91	79	92	101	84	85	85	125
85	91	98	93	88	105	113	102
83	76	97	82	100	88	99	103
89	88	96	116	95	92	86	76
89	123	92	85	61	76	107	106
110	132	101	124	86	73	115	133
106	115	95	103	110	115	93	92
119	107	110	132	93	117	107	127
162	113	89	95	92	79	101	88
80	99	80	80	90	86	85	138
125	107	129	127	134	101	106	96
101	116	103	94	74	111	71	109
75	116	115	101	106	99	85	71
83	92	104	117	110	113	112	120
128	139	184	167	131	127	120	123
108	90	59	99	68	68	92	79
76	73	64	73	85	84	86	79
91	89	96	85	145			

Table A.3.6. Ratio of the Price of Gold to the Price of Silver on the Hamburg Stock Exchange, 1687–1832 (Soetbeer, 1879)

14.94	14.94	15.02	15.02	14.98	14.92	14.83	14.87
15.02	15.00	15.20	15.07	14.94	14.81	15.07	15.52
15.17	15.22	15.11	15.27	15.44	15.41	15.31	15.22
15.29	15.31	15.24	15.13	15.11	15.09	15.13	15.11
15.09	15.04	15.05	15.17	15.20	15.11	15.11	15.15
15.24	15.11	14.92	14.81	14.94	15.09	15.18	15.39
15.41	15.18	15.02	14.91	14.91	14.94	14.92	14.85
14.85	14.87	14.98	15.13	15.26	15.11	14.80	14.55
14.39	14.54	14.54	14.48	14.68	14.94	14.87	14.85
14.15	14.14	14.54	15.27	14.99	14.70	14.83	14.80
14.85	14.80	14.72	14.62	14.66	14.52	14.62	14.62
14.72	14.55	14.54	14.68	14.80	14.72	14.78	14.42
14.48	14.70	14.92	14.96	14.92	14.65	14.75	15.04
15.05	15.17	15.00	15.37	15.55	15.65	15.41	15.59
15.74	15.68	15.46	15.26	15.41	15.41	15.79	15.52
15.43	16.08	15.96	15.77	15.53	16.11	16.25	15.04
15.26	15.28	15.11	15.35	15.33	15.62	15.95	15.80
15.84	15.82	15.70	15.76	15.74	15.78	15.78	15.82
15.72	15.73						

Table A.3.7. Productivity ($1000, 1929) in the United States 1890–1953 (U.S. Department of Commerce, 1966)

1.196	1.237	1.292	1.319	1.356	1.404	1.357	1.455
1.419	1.424	1.411	1.518	1.487	1.524	1.492	1.543
1.631	1.627	1.543	1.642	1.628	1.661	1.681	1.708
1.609	1.660	1.803	1.729	1.770	1.822	1.852	1.913
1.899	2.012	2.088	2.070	2.138	2.158	2.170	2.248
2.140	2.152	1.995	1.937	1.991	2.180	2.278	2.352
2.321	2.407	2.518	2.669	2.679	2.681	2.808	2.816
2.821	3.100	2.856	2.842	3.052	3.122	3.184	3.267

Table A.3.8. Price of Sugar (Mills, Coded by Subtracting 40) in the United States, 1875–1936 (Durbin and Watson, 1951)

67.0	65.0	73.0	55.0	48.0	56.0	57.0	52.0
45.0	28.0	24.0	21.0	20.0	30.0	36.0	22.0
6.0	3.0	8.0	1.0	2.0	5.0	5.0	10.0
9.0	13.0	10.0	5.0	6.0	8.0	13.0	5.0
6.0	10.0	8.0	10.0	13.0	10.0	3.0	7.0
16.0	29.0	37.0	38.0	50.0	74.0	22.0	19.0
44.0	35.0	15.0	15.0	18.0	15.0	10.0	6.0
4.0	.0	3.0	1.0	3.0	7.0		

Table A.3.9.   Production of Zinc (10^3 tons) in Spain, 1861–1976 (Schmitz, 1979)

8.2	13.7	16.0	26.7	23.4	24.7	28.9	43.8
37.8	37.9	35.8	29.8	33.7	35.5	33.4	35.7
23.6	23.8	20.3	16.8	14.3	19.1	18.1	16.6
16.5	13.3	23.0	24.8	23.9	27.1	25.8	24.7
20.9	19.7	18.0	21.6	24.6	33.3	39.9	28.7
31.9	34.0	41.1	59.4	58.3	66.9	70.6	62.2
36.0	63.1	64.7	56.8	38.4	33.6	26.2	51.5
37.2	31.5	32.1	29.9	17.6	21.9	38.1	41.9
48.8	53.0	47.1	43.0	52.9	90.0	60.0	50.0
50.0	45.0	45.0	55.0	30.9	46.1	49.1	47.1
44.0	42.7	43.4	35.8	31.7	39.4	42.6	47.2
52.0	62.4	75.5	86.0	83.7	88.3	92.3	87.2
80.8	82.3	85.9	86.1	88.0	78.5	91.7	88.5
39.2	57.2	57.9	74.6	84.3	98.1	87.5	89.4
94.2	93.4	84.2	82.3				

Table A.3.10.   Price of Magnesium (£/ton) in London, 1800–1911 (Schmitz, 1979)

18.86	22.10	24.46	27.36	27.56	27.11	35.09	29.72
29.58	30.66	28.34	23.67	22.83	24.74	26.13	20.47
15.99	17.96	26.87	22.19	21.16	22.14	22.00	21.90
20.67	24.90	18.70	18.06	16.73	14.02	11.96	12.01
11.47	12.40	6.34	17.22	23.77	18.90	18.21	16.88
18.01	18.45	16.24	15.99	17.22	19.19	18.21	18.45
16.49	15.70	17.23	16.90	17.49	23.03	23.29	22.78
23.62	23.46	21.23	21.95	21.96	21.11	20.55	20.47
21.26	19.81	20.30	19.24	18.99	18.77	18.36	17.84
19.83	22.76	21.65	22.14	21.31	20.24	16.44	14.02
16.05	14.73	14.14	12.70	10.95	11.31	13.01	12.34
13.35	12.13	13.00	12.10	10.31	9.54	9.32	10.33
11.12	12.17	12.78	14.70	16.72	12.32	11.08	11.40
11.79	13.50	17.10	18.73	13.23	12.84	12.72	13.75

Table A.3.11.   Average Yield of Wheat (cwt/acre) in Great Britain, 1885–1944 (Ministry of Agriculture, 1968)

17.3	14.9	17.7	15.5	16.5	17.0	17.3	14.6
14.4	17.0	14.5	18.6	16.1	19.2	18.1	15.8
17.0	18.2	16.7	14.8	18.1	18.9	18.7	17.8
18.4	16.6	18.6	15.5	17.6	18.3	17.5	15.8
16.5	18.4	16.1	15.7	19.9	17.1	17.6	17.7
18.3	16.5	17.5	18.3	19.2	16.1	16.2	17.4
19.2	20.0	18.6	16.4	16.4	20.4	18.6	18.1
17.8	20.4	19.9	19.5				

A.4. TABLES

Tables A.4.1–A.4.4 for S_p are based on the calculation of percentage points of multiple R in the multivariate normal case (Volkers, 1985). The series giving the density was integrated term by term. The percentage points of R were computed to an accuracy of four significant figures; the given values of F are rounded.

The percentage points of noncentral F, Tables A.4.5–A.4.8, were computed to an accuracy of four significant figures according to the algorithm of Norton (1983) and then rounded.

Table A.4.1. The 50% Points of F in the Test of S_p

n_2 \ n_1	1	2	3	4	5	6	7	8	9	10
2	3.49	5.61	7.28	8.83	10.34	11.82	13.29	14.74	16.23	17.65
3	2.24	3.46	4.29	5.02	5.71	6.38	7.04	7.69	8.34	8.99
4	1.85	2.78	3.36	3.85	4.30	4.74	5.16	5.58	5.99	6.40
5	1.66	2.46	2.92	3.30	3.64	3.96	4.27	4.58	4.88	5.18
6	1.54	2.27	2.67	2.98	3.25	3.51	3.76	4.00	4.24	4.47
7	1.47	2.15	2.50	2.77	3.00	3.22	3.43	3.63	3.82	4.02
8	1.42	2.06	2.38	2.62	2.82	3.01	3.19	3.36	3.53	3.70
9	1.38	1.99	2.29	2.51	2.69	2.86	3.02	3.17	3.32	3.47
10	1.35	1.94	2.23	2.43	2.59	2.74	2.89	3.02	3.16	3.28
11	1.32	1.90	2.17	2.36	2.51	2.65	2.78	2.91	3.02	3.14
12	1.30	1.87	2.13	2.31	2.45	2.58	2.70	2.81	2.92	3.03
13	1.29	1.84	2.09	2.26	2.40	2.52	2.63	2.73	2.83	2.93
14	1.27	1.82	2.06	2.22	2.35	2.46	2.57	2.66	2.76	2.85
15	1.26	1.80	2.04	2.19	2.31	2.42	2.52	2.61	2.69	2.78
16	1.25	1.78	2.01	2.16	2.28	2.38	2.47	2.56	2.64	2.72
17	1.24	1.77	1.99	2.14	2.25	2.35	2.43	2.52	2.59	2.67
18	1.23	1.76	1.98	2.12	2.22	2.32	2.40	2.48	2.55	2.62
19	1.23	1.74	1.96	2.10	2.20	2.29	2.37	2.44	2.51	2.58
20	1.22	1.73	1.95	2.08	2.18	2.27	2.34	2.41	2.48	2.55
21	1.21	1.73	1.93	2.06	2.16	2.25	2.32	2.39	2.45	2.51
22	1.21	1.72	1.92	2.05	2.15	2.23	2.30	2.36	2.43	2.49
23	1.20	1.71	1.91	2.04	2.13	2.21	2.28	2.34	2.40	2.46
24	1.20	1.70	1.90	2.03	2.12	2.19	2.26	2.32	2.38	2.43
25	1.20	1.70	1.90	2.02	2.11	2.18	2.24	2.30	2.36	2.41
30	1.18	1.67	1.86	1.98	2.06	2.12	2.18	2.23	2.28	2.33
40	1.16	1.64	1.82	1.92	2.00	2.05	2.10	2.14	2.18	2.22
50	1.15	1.62	1.80	1.90	1.96	2.01	2.06	2.09	2.13	2.16
60	1.14	1.61	1.78	1.88	1.94	1.99	2.03	2.06	2.09	2.12
80	1.14	1.60	1.76	1.85	1.91	1.95	1.99	2.02	2.04	2.06
100	1.13	1.59	1.75	1.84	1.89	1.93	1.97	1.99	2.01	2.03
120	1.13	1.58	1.74	1.83	1.88	1.92	1.95	1.97	2.00	2.01
150	1.12	1.57	1.73	1.82	1.87	1.91	1.94	1.96	1.98	1.99
***	1.11	1.56	1.71	1.79	1.83	1.86	1.89	1.90	1.92	1.93

n_2 \ n_1	11	12	13	14	15	16	17	18	19	20
2	19.17	20.57	21.99	23.46	24.86	26.29	27.79	29.42	30.65	31.92
3	9.63	10.26	10.91	11.56	12.19	12.82	13.47	14.09	14.75	15.41
4	6.80	7.21	7.61	8.01	8.41	8.82	9.21	9.62	10.02	10.42
5	5.47	5.77	6.06	6.35	6.65	6.94	7.23	7.52	7.81	8.10
6	4.71	4.94	5.17	5.40	5.63	5.86	6.08	6.31	6.54	6.77
7	4.21	4.40	4.59	4.78	4.97	5.16	5.35	5.53	5.72	5.91
8	3.87	4.03	4.19	4.35	4.51	4.67	4.83	4.99	5.15	5.30
9	3.61	3.75	3.89	4.03	4.17	4.31	4.45	4.59	4.72	4.86
10	3.41	3.54	3.66	3.79	3.91	4.03	4.16	4.28	4.40	4.52
11	3.26	3.37	3.48	3.59	3.70	3.82	3.92	4.03	4.14	4.25
12	3.13	3.23	3.34	3.44	3.54	3.64	3.74	3.84	3.94	4.03
13	3.03	3.12	3.21	3.31	3.40	3.49	3.58	3.67	3.76	3.85
14	2.94	3.03	3.11	3.20	3.28	3.37	3.45	3.54	3.62	3.70
15	2.86	2.94	3.03	3.11	3.18	3.26	3.34	3.42	3.50	3.57
16	2.80	2.88	2.95	3.03	3.10	3.17	3.25	3.32	3.39	3.46
17	2.74	2.81	2.89	2.96	3.02	3.09	3.16	3.23	3.30	3.37
18	2.69	2.76	2.83	2.89	2.96	3.02	3.09	3.15	3.22	3.28
19	2.65	2.71	2.78	2.84	2.90	2.96	3.02	3.08	3.14	3.20
20	2.61	2.67	2.73	2.79	2.85	2.91	2.97	3.02	3.08	3.14
21	2.57	2.63	2.69	2.75	2.80	2.86	2.91	2.97	3.02	3.08
22	2.54	2.60	2.65	2.71	2.76	2.81	2.87	2.92	2.97	3.02
23	2.51	2.57	2.62	2.67	2.72	2.77	2.82	2.87	2.92	2.97
24	2.49	2.54	2.59	2.64	2.69	2.74	2.79	2.83	2.88	2.93
25	2.46	2.51	2.56	2.61	2.66	2.70	2.75	2.80	2.84	2.89
30	2.37	2.41	2.45	2.49	2.53	2.57	2.61	2.65	2.69	2.72
40	2.25	2.29	2.32	2.35	2.38	2.41	2.44	2.47	2.50	2.53
50	2.19	2.21	2.24	2.27	2.29	2.31	2.34	2.36	2.39	2.41
60	2.14	2.16	2.19	2.21	2.23	2.25	2.27	2.29	2.31	2.33
80	2.09	2.11	2.12	2.14	2.16	2.18	2.19	2.21	2.22	2.24
100	2.05	2.07	2.09	2.10	2.12	2.13	2.14	2.16	2.17	2.18
120	2.03	2.05	2.06	2.07	2.09	2.10	2.11	2.12	2.13	2.14
150	2.01	2.02	2.04	2.05	2.06	2.07	2.08	2.09	2.10	2.11
***	1.94	1.94	1.95	1.96	1.96	1.97	1.97	1.98	1.98	1.98

Table A.4.2.   The 90% Points of F in the Test of S_p

n_2 \ n_1	1	2	3	4	5	6	7	8	9	10
2	35.51	45.78	55.52	65.14	75.01	84.69	94.40	104.2	113.9	123.7
3	16.71	18.97	21.40	23.89	26.40	28.93	31.46	34.03	36.61	39.23
4	12.02	12.71	13.70	14.82	15.98	17.16	18.37	19.62	20.84	22.07
5	10.00	10.10	10.57	11.18	11.84	12.55	13.27	14.01	14.75	15.50
6	8.89	8.71	8.92	9.28	9.70	10.16	10.65	11.14	11.65	12.17
7	8.19	7.85	7.91	8.12	8.41	8.73	9.08	9.44	9.81	10.19
8	7.71	7.26	7.23	7.35	7.55	7.78	8.04	8.31	8.60	8.90
9	7.37	6.85	6.75	6.81	6.94	7.11	7.31	7.52	7.75	7.99
10	7.10	6.53	6.38	6.40	6.49	6.62	6.77	6.94	7.12	7.32
11	6.90	6.28	6.10	6.08	6.14	6.23	6.35	6.49	6.64	6.80
12	6.73	6.09	5.88	5.83	5.86	5.93	6.02	6.14	6.26	6.39
13	6.60	5.93	5.70	5.63	5.63	5.68	5.76	5.85	5.95	6.07
14	6.49	5.80	5.55	5.46	5.44	5.48	5.53	5.61	5.70	5.80
15	6.39	5.68	5.42	5.31	5.29	5.30	5.35	5.41	5.49	5.57
16	6.31	5.59	5.31	5.19	5.15	5.16	5.19	5.24	5.30	5.38
17	6.24	5.50	5.21	5.09	5.04	5.03	5.05	5.10	5.15	5.21
18	6.17	5.43	5.13	4.99	4.94	4.92	4.94	4.97	5.01	5.07
19	6.12	5.37	5.06	4.91	4.85	4.83	4.83	4.86	4.89	4.94
20	6.07	5.31	4.99	4.84	4.77	4.74	4.74	4.76	4.79	4.83
21	6.02	5.26	4.93	4.78	4.70	4.67	4.66	4.67	4.69	4.73
22	5.98	5.21	4.88	4.72	4.64	4.60	4.58	4.59	4.61	4.64
23	5.94	5.17	4.84	4.67	4.58	4.54	4.52	4.52	4.54	4.56
24	5.92	5.13	4.79	4.62	4.53	4.48	4.46	4.46	4.47	4.49
25	5.89	5.09	4.75	4.58	4.48	4.43	4.40	4.40	4.41	4.42
30	5.77	4.96	4.60	4.41	4.30	4.23	4.19	4.17	4.16	4.17
40	5.62	4.80	4.42	4.21	4.08	3.99	3.94	3.90	3.87	3.86
50	5.54	4.70	4.32	4.09	3.95	3.86	3.79	3.74	3.71	3.68
60	5.49	4.64	4.25	4.02	3.87	3.77	3.70	3.64	3.60	3.57
80	5.42	4.56	4.16	3.93	3.77	3.66	3.58	3.52	3.47	3.43
100	5.38	4.52	4.11	3.87	3.71	3.60	3.51	3.45	3.39	3.35
120	5.35	4.49	4.08	3.84	3.67	3.55	3.47	3.40	3.34	3.30
150	5.33	4.46	4.05	3.80	3.63	3.51	3.42	3.35	3.29	3.25
***	5.24	4.36	3.94	3.68	3.50	3.37	3.27	3.19	3.13	3.07

n_2 \ n_1	11	12	13	14	15	16	17	18	19	20
2	133.5	143.4	153.3	163.2	173.1	183.0	193.0	202.9	212.8	222.7
3	41.54	44.25	46.82	49.39	51.97	54.54	57.10	59.67	62.23	64.80
4	23.30	24.52	25.80	27.03	28.32	29.49	30.76	32.00	33.16	34.45
5	16.26	17.02	17.78	18.54	19.32	20.05	20.83	21.61	22.37	23.16
6	12.68	13.21	13.74	14.27	14.80	15.34	15.86	16.43	16.96	17.49
7	10.58	10.97	11.36	11.76	12.16	12.56	12.96	13.36	13.77	14.17
8	9.19	9.50	9.81	10.12	10.43	10.75	11.06	11.39	11.69	12.01
9	8.23	8.47	8.72	8.97	9.23	9.48	9.74	10.00	10.27	10.52
10	7.51	7.72	7.92	8.13	8.34	8.56	8.77	8.99	9.21	9.43
11	6.97	7.14	7.31	7.49	7.67	7.85	8.03	8.22	8.41	8.60
12	6.54	6.68	6.83	6.98	7.14	7.30	7.46	7.62	7.78	7.94
13	6.19	6.31	6.44	6.58	6.71	6.85	6.99	7.13	7.27	7.42
14	5.90	6.01	6.12	6.24	6.36	6.48	6.61	6.73	6.86	6.98
15	5.66	5.76	5.86	5.96	6.07	6.18	6.28	6.40	6.51	6.62
16	5.46	5.54	5.63	5.72	5.82	5.91	6.01	6.11	6.22	6.32
17	5.28	5.36	5.43	5.52	5.60	5.69	5.78	5.87	5.97	6.06
18	5.13	5.20	5.27	5.34	5.42	5.50	5.58	5.66	5.75	5.83
19	4.99	5.05	5.12	5.18	5.25	5.33	5.40	5.48	5.55	5.63
20	4.88	4.93	4.99	5.05	5.11	5.18	5.25	5.32	5.39	5.46
21	4.77	4.82	4.87	4.93	4.98	5.04	5.11	5.17	5.24	5.30
22	4.68	4.72	4.77	4.82	4.87	4.93	4.98	5.04	5.10	5.17
23	4.59	4.63	4.67	4.72	4.77	4.82	4.87	4.93	4.98	5.04
24	4.52	4.55	4.59	4.63	4.67	4.72	4.77	4.82	4.87	4.93
25	4.45	4.48	4.51	4.55	4.59	4.63	4.68	4.73	4.78	4.83
30	4.18	4.19	4.21	4.24	4.26	4.29	4.32	4.36	4.39	4.43
40	3.85	3.85	3.86	3.86	3.88	3.89	3.91	3.92	3.94	3.96
50	3.67	3.66	3.65	3.65	3.65	3.66	3.66	3.67	3.68	3.70
60	3.55	3.53	3.52	3.51	3.51	3.51	3.51	3.51	3.52	3.52
80	3.40	3.38	3.36	3.35	3.34	3.34	3.33	3.32	3.32	3.32
100	3.32	3.29	3.27	3.25	3.23	3.22	3.21	3.20	3.20	3.19
120	3.26	3.23	3.21	3.19	3.17	3.15	3.14	3.13	3.12	3.11
150	3.21	3.17	3.15	3.12	3.10	3.08	3.07	3.06	3.04	3.04
***	3.02	2.98	2.94	2.91	2.88	2.86	2.83	2.81	2.79	2.77

Table A.4.3.   The 95% Points of F in the Test of S_p

n_2 \ n_1	1	2	3	4	5	6	7	8	9	10
2	75.50	95.86	115.7	135.3	154.8	174.3	193.6	213.1	232.5	251.9
3	29.65	32.71	36.50	40.46	44.57	48.73	52.93	57.02	61.62	65.77
4	19.66	20.01	21.27	22.81	24.50	26.23	28.01	29.81	31.64	33.47
5	15.63	15.11	15.55	16.29	17.16	18.10	19.08	20.09	21.12	22.17
6	13.49	12.60	12.68	13.03	13.54	14.12	14.73	15.39	16.05	16.73
7	12.19	11.10	10.96	11.13	11.43	11.81	12.23	12.68	13.14	13.63
8	11.32	10.10	9.84	9.89	10.06	10.32	10.62	10.94	11.29	11.65
9	10.69	9.40	9.06	9.02	9.11	9.29	9.51	9.74	10.01	10.29
10	10.23	8.88	8.48	8.38	8.42	8.53	8.69	8.87	9.08	9.30
11	9.86	8.48	8.04	7.89	7.89	7.95	8.07	8.21	8.38	8.55
12	9.57	8.16	7.68	7.51	7.47	7.50	7.58	7.70	7.83	7.97
13	9.33	7.90	7.39	7.20	7.13	7.14	7.20	7.28	7.39	7.50
14	9.14	7.68	7.16	6.94	6.85	6.85	6.88	6.94	7.02	7.12
15	8.98	7.50	6.96	6.73	6.62	6.59	6.61	6.66	6.72	6.80
16	8.83	7.35	6.80	6.54	6.43	6.38	6.38	6.42	6.47	6.53
17	8.71	7.21	6.65	6.39	6.26	6.20	6.19	6.21	6.25	6.31
18	8.60	7.10	6.53	6.25	6.11	6.04	6.02	6.03	6.06	6.11
19	8.50	7.00	6.41	6.13	5.98	5.91	5.88	5.88	5.90	5.94
20	8.43	6.91	6.32	6.03	5.87	5.79	5.75	5.74	5.75	5.78
21	8.35	6.83	6.23	5.93	5.77	5.68	5.64	5.62	5.62	5.65
22	8.28	6.75	6.15	5.85	5.68	5.58	5.53	5.51	5.51	5.52
23	8.21	6.69	6.08	5.77	5.59	5.49	5.44	5.41	5.40	5.42
24	8.16	6.63	6.02	5.70	5.52	5.42	5.35	5.32	5.31	5.32
25	8.11	6.58	5.96	5.64	5.46	5.35	5.28	5.24	5.23	5.23
30	7.91	6.37	5.73	5.40	5.20	5.07	4.99	4.93	4.90	4.88
40	7.67	6.11	5.47	5.11	4.89	4.74	4.64	4.57	4.51	4.48
50	7.53	5.97	5.31	4.94	4.71	4.55	4.44	4.36	4.29	4.25
60	7.44	5.87	5.21	4.84	4.60	4.43	4.31	4.22	4.15	4.10
80	7.33	5.76	5.09	4.71	4.46	4.28	4.16	4.06	3.98	3.92
100	7.26	5.69	5.01	4.63	4.38	4.20	4.06	3.96	3.88	3.81
120	7.22	5.64	4.97	4.58	4.32	4.14	4.00	3.90	3.81	3.74
150	7.18	5.60	4.92	4.53	4.27	4.08	3.94	3.84	3.75	3.68
***	7.03	5.45	4.76	4.36	4.09	3.90	3.75	3.63	3.53	3.45

n_2 \ n_1	11	12	13	14	15	16	17	18	19	20
2	271.2	290.5	309.8	329.1	348.3	367.4	386.6	405.7	424.7	443.6
3	69.92	74.09	78.32	82.59	86.95	91.40	95.93	100.6	104.0	108.3
4	35.31	37.16	39.01	40.86	42.78	44.75	46.27	48.19	50.10	51.99
5	23.19	24.27	25.33	26.40	27.48	28.56	29.64	30.72	31.79	32.87
6	17.42	18.12	18.81	19.53	20.24	20.97	21.70	22.37	23.11	23.85
7	14.12	14.62	15.12	15.63	16.15	16.66	17.17	17.72	18.22	18.76
8	12.02	12.40	12.79	13.17	13.56	13.96	14.36	14.76	15.16	15.56
9	10.58	10.87	11.18	11.48	11.80	12.11	12.42	12.76	13.08	13.40
10	9.53	9.77	10.02	10.27	10.53	10.78	11.04	11.31	11.57	11.84
11	8.74	8.94	9.14	9.35	9.56	9.78	10.00	10.22	10.44	10.67
12	8.13	8.29	8.46	8.64	8.82	9.00	9.19	9.38	9.57	9.77
13	7.63	7.77	7.92	8.07	8.22	8.38	8.54	8.71	8.87	9.04
14	7.23	7.35	7.47	7.60	7.74	7.88	8.02	8.16	8.31	8.45
15	6.90	7.00	7.10	7.22	7.34	7.46	7.58	7.71	7.84	7.97
16	6.61	6.70	6.80	6.89	7.00	7.10	7.22	7.33	7.44	7.56
17	6.37	6.45	6.53	6.62	6.71	6.81	6.90	7.00	7.11	7.21
18	6.16	6.23	6.30	6.38	6.46	6.54	6.63	6.72	6.82	6.91
19	5.98	6.04	6.10	6.17	6.24	6.32	6.40	6.48	6.56	6.65
20	5.82	5.87	5.92	5.98	6.05	6.12	6.19	6.26	6.34	6.42
21	5.68	5.72	5.77	5.82	5.88	5.94	6.00	6.08	6.14	6.22
22	5.55	5.59	5.63	5.68	5.73	5.78	5.84	5.91	5.97	6.04
23	5.44	5.47	5.50	5.54	5.59	5.64	5.70	5.76	5.81	5.87
24	5.33	5.36	5.39	5.43	5.47	5.52	5.57	5.62	5.67	5.73
25	5.24	5.26	5.29	5.32	5.36	5.40	5.45	5.50	5.54	5.60
30	4.88	4.88	4.90	4.91	4.93	4.96	4.99	5.02	5.05	5.09
40	4.45	4.44	4.43	4.43	4.44	4.44	4.45	4.46	4.48	4.49
50	4.21	4.19	4.17	4.16	4.15	4.15	4.15	4.15	4.15	4.16
60	4.06	4.03	4.00	3.98	3.97	3.96	3.95	3.95	3.95	3.95
80	3.87	3.83	3.80	3.77	3.75	3.73	3.72	3.71	3.70	3.69
100	3.76	3.72	3.68	3.65	3.62	3.60	3.58	3.56	3.55	3.54
120	3.69	3.64	3.60	3.57	3.54	3.51	3.49	3.47	3.46	3.44
150	3.62	3.57	3.52	3.49	3.46	3.43	3.40	3.38	3.36	3.35
***	3.38	3.32	3.27	3.22	3.18	3.15	3.11	3.08	3.06	3.03

Table A.4.4. The 99% Points of F in the Test of S_p

n_2 \ n_1	1	2	3	4	5	6	7	8	9	10
2	393.1	495.8	595.4	694.9	794.2	893.5	992.7	1091.	1191.	1289.
3	96.76	103.9	114.9	126.9	138.5	151.6	164.5	177.4	190.5	203.6
4	51.81	50.66	53.04	56.43	60.17	64.26	68.40	72.74	76.75	81.37
5	36.64	33.67	33.97	35.20	36.83	38.73	40.58	42.62	44.61	46.89
6	29.41	25.93	25.43	25.84	26.61	27.61	28.62	29.79	31.03	32.21
7	25.09	21.52	20.79	20.78	21.11	21.70	22.30	23.01	23.81	24.59
8	22.52	18.82	17.83	17.64	17.74	18.07	18.47	18.96	19.50	20.03
9	20.65	17.02	15.90	15.54	15.52	15.67	15.90	16.27	16.63	17.03
10	19.32	15.64	14.47	14.05	13.95	14.00	14.18	14.37	14.68	14.96
11	18.30	14.65	13.45	12.94	12.77	12.78	12.87	13.00	13.20	13.42
12	17.54	13.82	12.61	12.07	11.86	11.81	11.84	11.94	12.08	12.26
13	16.89	13.25	11.98	11.42	11.15	11.05	11.05	11.05	11.22	11.34
14	16.35	12.75	11.43	10.86	10.57	10.45	10.40	10.44	10.50	10.62
15	15.90	12.30	11.04	10.43	10.10	9.94	9.89	9.88	9.95	10.00
16	15.55	11.97	10.66	10.03	9.70	9.53	9.45	9.43	9.43	9.51
17	15.22	11.64	10.35	9.71	9.36	9.18	9.09	9.04	9.05	9.08
18	14.94	11.40	10.07	9.43	9.07	8.87	8.74	8.70	8.70	8.73
19	14.73	11.15	9.84	9.19	8.82	8.60	8.46	8.42	8.39	8.41
20	14.49	10.92	9.64	8.98	8.60	8.38	8.24	8.16	8.13	8.13
21	14.29	10.79	9.47	8.78	8.41	8.17	8.02	7.94	7.90	7.90
22	14.11	10.61	9.30	8.62	8.24	7.99	7.85	7.75	7.70	7.67
23	13.96	10.47	9.15	8.49	8.07	7.82	7.67	7.57	7.51	7.48
24	13.83	10.34	9.02	8.33	7.93	7.68	7.51	7.41	7.34	7.30
25	13.72	10.23	8.88	8.21	7.80	7.54	7.38	7.27	7.20	7.16
30	13.23	9.76	8.45	7.74	7.33	7.04	6.85	6.71	6.64	6.57
40	12.62	9.20	7.90	7.19	6.75	6.45	6.24	6.08	5.97	5.88
50	12.31	8.91	7.59	6.89	6.43	6.12	5.90	5.73	5.60	5.50
60	12.09	8.72	7.40	6.68	6.22	5.91	5.68	5.50	5.37	5.26
80	11.83	8.46	7.16	6.44	5.98	5.65	5.41	5.24	5.09	4.97
100	11.69	8.34	7.03	6.30	5.83	5.51	5.26	5.07	4.93	4.81
120	11.58	8.24	6.94	6.20	5.73	5.41	5.17	4.97	4.82	4.69
150	11.47	8.15	6.84	6.12	5.65	5.31	5.06	4.87	4.72	4.59
***	11.16	7.86	6.55	5.82	5.34	5.00	4.74	4.54	4.38	4.24

n_2 \ n_1	11	12	13	14	15	16	17	18	19	20
2	1388.	1486.	1585.	1683.	1781.	1878.	1976.	2073.	2170.	2267.
3	216.7	229.9	241.1	254.0	266.9	278.6	289.2	299.3	308.9	318.1
4	85.78	90.17	94.74	99.44	104.4	107.1	111.9	116.9	121.5	126.1
5	48.91	51.01	53.25	55.49	57.83	59.76	62.03	64.52	66.23	68.77
6	33.50	34.86	36.06	37.34	38.71	40.02	41.26	42.69	43.95	45.36
7	25.43	26.33	27.14	28.03	28.89	29.80	30.76	31.59	32.45	33.47
8	20.66	21.27	21.86	22.47	23.15	23.74	24.43	25.04	25.72	26.36
9	17.47	17.92	18.39	18.87	19.36	19.87	20.35	20.84	21.32	21.87
10	15.28	15.61	15.98	16.35	16.75	17.10	17.50	17.91	18.33	18.69
11	13.65	13.94	14.26	14.51	14.80	15.15	15.44	15.76	16.11	16.42
12	12.47	12.67	12.89	13.13	13.39	13.63	13.88	14.18	14.43	14.71
13	11.52	11.69	11.85	12.07	12.26	12.47	12.69	12.91	13.13	13.37
14	10.72	10.86	11.03	11.19	11.36	11.55	11.74	11.92	12.11	12.33
15	10.11	10.20	10.32	10.47	10.62	10.79	10.93	11.10	11.27	11.44
16	9.57	9.67	9.77	9.89	10.00	10.14	10.28	10.44	10.57	10.72
17	9.13	9.21	9.30	9.41	9.50	9.62	9.72	9.86	10.00	10.12
18	8.76	8.82	8.89	8.97	9.07	9.16	9.27	9.38	9.48	9.61
19	8.44	8.48	8.54	8.61	8.69	8.77	8.86	8.96	9.06	9.17
20	8.15	8.18	8.23	8.29	8.36	8.42	8.51	8.61	8.69	8.78
21	7.89	7.92	7.96	8.02	8.06	8.13	8.20	8.29	8.36	8.44
22	7.67	7.69	7.72	7.76	7.82	7.86	7.93	7.99	8.06	8.15
23	7.47	7.48	7.51	7.54	7.58	7.63	7.69	7.74	7.81	7.88
24	7.30	7.30	7.31	7.34	7.38	7.42	7.47	7.53	7.59	7.64
25	7.14	7.14	7.14	7.16	7.20	7.23	7.28	7.32	7.38	7.43
30	6.53	6.50	6.49	6.48	6.49	6.51	6.53	6.55	6.58	6.62
40	5.82	5.77	5.73	5.71	5.69	5.68	5.68	5.69	5.69	5.70
50	5.43	5.36	5.32	5.28	5.25	5.23	5.21	5.21	5.19	5.19
60	5.18	5.11	5.06	5.01	4.97	4.94	4.92	4.90	4.88	4.87
80	4.88	4.80	4.73	4.69	4.64	4.60	4.56	4.53	4.51	4.49
100	4.71	4.62	4.55	4.49	4.44	4.40	4.36	4.32	4.30	4.27
120	4.60	4.51	4.44	4.37	4.32	4.27	4.23	4.19	4.16	4.13
150	4.48	4.39	4.32	4.25	4.19	4.14	4.10	4.06	4.02	3.99
***	4.13	4.03	3.94	3.86	3.79	3.74	3.68	3.63	3.59	3.54

Table A.4.5. The 50% Points of $F(n_1, n_2; \lambda = n_1)$

n_2 \ n_1	1	2	3	4	5	6	7	8	9	10
2	1.57	2.18	2.41	2.52	2.60	2.64	2.68	2.70	2.73	2.74
3	1.39	1.93	2.13	2.23	2.29	2.33	2.36	2.38	2.40	2.41
4	1.31	1.82	2.00	2.10	2.15	2.19	2.22	2.24	2.26	2.27
5	1.27	1.76	1.94	2.03	2.08	2.12	2.14	2.16	2.18	2.19
6	1.24	1.72	1.89	1.98	2.03	2.07	2.09	2.11	2.13	2.14
7	1.22	1.70	1.86	1.95	2.00	2.03	2.06	2.08	2.09	2.10
8	1.20	1.68	1.84	1.93	1.98	2.01	2.03	2.05	2.07	2.08
9	1.19	1.66	1.83	1.91	1.96	1.99	2.01	2.03	2.05	2.06
10	1.18	1.65	1.81	1.89	1.94	1.98	2.00	2.02	2.03	2.04
11	1.18	1.64	1.80	1.88	1.93	1.96	1.99	2.01	2.02	2.03
12	1.17	1.63	1.79	1.87	1.92	1.95	1.98	2.00	2.01	2.02
13	1.16	1.63	1.79	1.87	1.91	1.95	1.97	1.99	2.00	2.01
14	1.16	1.62	1.78	1.86	1.91	1.94	1.96	1.98	1.99	2.00
15	1.16	1.62	1.77	1.85	1.90	1.93	1.96	1.97	1.99	2.00
16	1.15	1.61	1.77	1.85	1.90	1.93	1.95	1.97	1.98	1.99
17	1.15	1.61	1.77	1.85	1.89	1.92	1.95	1.96	1.98	1.99
18	1.15	1.60	1.76	1.84	1.89	1.92	1.94	1.96	1.97	1.98
19	1.15	1.60	1.76	1.84	1.89	1.92	1.94	1.96	1.97	1.98
20	1.14	1.60	1.76	1.83	1.88	1.91	1.94	1.95	1.97	1.98
21	1.14	1.60	1.75	1.83	1.88	1.91	1.93	1.95	1.96	1.97
22	1.14	1.59	1.75	1.83	1.88	1.91	1.93	1.95	1.96	1.97
23	1.14	1.59	1.75	1.83	1.87	1.91	1.93	1.94	1.96	1.97
24	1.14	1.59	1.75	1.83	1.87	1.90	1.93	1.94	1.95	1.97
25	1.14	1.59	1.75	1.82	1.87	1.90	1.92	1.94	1.95	1.96
30	1.13	1.58	1.74	1.82	1.86	1.89	1.92	1.93	1.94	1.95
40	1.13	1.57	1.73	1.81	1.85	1.88	1.91	1.92	1.93	1.94
60	1.12	1.57	1.72	1.80	1.84	1.87	1.90	1.91	1.92	1.93
80	1.12	1.56	1.72	1.79	1.84	1.87	1.89	1.91	1.92	1.93
100	1.12	1.56	1.71	1.79	1.84	1.87	1.89	1.90	1.92	1.93
150	1.11	1.56	1.71	1.79	1.83	1.86	1.88	1.90	1.91	1.92
200	1.11	1.56	1.71	1.79	1.83	1.86	1.88	1.90	1.91	1.92

n_2 \ n_1	12	15	20	25	30	50	80	100	150	200
2	2.77	2.79	2.82	2.83	2.84	2.86	2.87	2.87	2.87	2.88
3	2.43	2.45	2.48	2.49	2.50	2.51	2.52	2.53	2.53	2.53
4	2.29	2.31	2.33	2.34	2.35	2.36	2.37	2.37	2.38	2.38
5	2.21	2.23	2.25	2.26	2.26	2.28	2.29	2.29	2.29	2.30
6	2.16	2.17	2.19	2.20	2.21	2.23	2.23	2.24	2.24	2.25
7	2.12	2.14	2.16	2.17	2.17	2.19	2.20	2.20	2.20	2.21
8	2.09	2.11	2.13	2.14	2.15	2.16	2.17	2.17	2.18	2.18
9	2.07	2.09	2.11	2.12	2.13	2.14	2.15	2.15	2.16	2.16
10	2.06	2.08	2.09	2.10	2.11	2.12	2.13	2.13	2.14	2.14
11	2.05	2.06	2.08	2.09	2.10	2.11	2.12	2.12	2.13	2.13
12	2.04	2.05	2.07	2.08	2.09	2.10	2.11	2.11	2.11	2.12
13	2.03	2.04	2.06	2.07	2.08	2.09	2.10	2.10	2.10	2.11
14	2.02	2.04	2.05	2.06	2.07	2.08	2.09	2.09	2.10	2.10
15	2.01	2.03	2.05	2.06	2.06	2.08	2.08	2.09	2.09	2.09
16	2.01	2.02	2.04	2.05	2.06	2.07	2.08	2.08	2.08	2.09
17	2.00	2.02	2.04	2.05	2.05	2.07	2.07	2.08	2.08	2.08
18	2.00	2.01	2.03	2.04	2.05	2.06	2.07	2.07	2.07	2.08
19	2.00	2.01	2.03	2.04	2.04	2.06	2.06	2.07	2.07	2.07
20	1.99	2.01	2.02	2.03	2.04	2.05	2.06	2.06	2.07	2.07
21	1.99	2.00	2.02	2.03	2.04	2.05	2.06	2.06	2.06	2.07
22	1.99	2.00	2.02	2.03	2.03	2.05	2.05	2.06	2.06	2.06
23	1.98	2.00	2.01	2.02	2.03	2.04	2.05	2.05	2.06	2.06
24	1.98	2.00	2.01	2.02	2.03	2.04	2.05	2.05	2.05	2.06
25	1.98	1.99	2.01	2.02	2.03	2.04	2.05	2.05	2.05	2.05
30	1.97	1.99	2.00	2.01	2.02	2.03	2.04	2.04	2.04	2.05
40	1.96	1.98	1.99	2.00	2.01	2.02	2.03	2.03	2.03	2.03
60	1.95	1.96	1.98	1.99	2.00	2.01	2.01	2.02	2.02	2.02
80	1.94	1.96	1.97	1.98	1.99	2.00	2.01	2.01	2.01	2.02
100	1.94	1.96	1.97	1.98	1.99	2.00	2.01	2.01	2.01	2.01
150	1.94	1.95	1.97	1.98	1.98	1.99	2.00	2.00	2.01	2.01
200	1.94	1.95	1.97	1.97	1.98	1.99	2.00	2.00	2.00	2.01

Table A.4.6. The 90% Points of $F(n_1, n_2; \lambda = n_1)$

n_2 \ n_1	1	2	3	4	5	6	7	8	9	10
2	17.51	18.25	18.49	18.61	18.69	18.74	18.77	18.80	18.82	18.84
3	11.16	10.88	10.72	10.62	10.56	10.52	10.48	10.46	10.44	10.42
4	9.07	8.52	8.25	8.10	8.00	7.92	7.87	7.83	7.80	7.77
5	8.05	7.39	7.08	6.90	6.78	6.69	6.63	6.58	6.55	6.52
6	7.45	6.74	6.40	6.20	6.07	5.98	5.91	5.86	5.82	5.79
7	7.06	6.31	5.96	5.75	5.61	5.52	5.45	5.39	5.35	5.31
8	6.79	6.02	5.65	5.43	5.29	5.19	5.12	5.06	5.01	4.98
9	6.58	5.80	5.42	5.20	5.06	4.95	4.88	4.82	4.77	4.73
10	6.43	5.63	5.25	5.02	4.87	4.77	4.69	4.63	4.58	4.54
11	6.30	5.49	5.11	4.88	4.73	4.62	4.54	4.48	4.43	4.38
12	6.20	5.38	4.99	4.76	4.61	4.50	4.42	4.36	4.30	4.26
13	6.12	5.29	4.90	4.67	4.52	4.40	4.32	4.25	4.20	4.16
14	6.05	5.22	4.82	4.59	4.43	4.32	4.24	4.17	4.12	4.07
15	5.99	5.15	4.76	4.52	4.36	4.25	4.16	4.10	4.04	4.00
16	5.93	5.10	4.70	4.46	4.30	4.19	4.10	4.03	3.98	3.93
17	5.89	5.05	4.65	4.41	4.25	4.14	4.05	3.98	3.92	3.88
18	5.85	5.01	4.60	4.36	4.20	4.09	4.00	3.93	3.88	3.83
19	5.81	4.97	4.56	4.32	4.16	4.05	3.96	3.89	3.83	3.78
20	5.78	4.93	4.53	4.29	4.13	4.01	3.92	3.85	3.79	3.75
21	5.75	4.90	4.50	4.26	4.09	3.98	3.89	3.82	3.76	3.71
22	5.73	4.88	4.47	4.23	4.06	3.95	3.86	3.78	3.73	3.68
23	5.70	4.85	4.44	4.20	4.04	3.92	3.83	3.76	3.70	3.65
24	5.68	4.83	4.42	4.18	4.01	3.89	3.80	3.73	3.67	3.62
25	5.66	4.81	4.40	4.16	3.99	3.87	3.78	3.71	3.65	3.60
30	5.59	4.73	4.32	4.07	3.90	3.78	3.69	3.61	3.55	3.50
40	5.49	4.63	4.21	3.96	3.79	3.67	3.57	3.50	3.44	3.38
60	5.40	4.53	4.11	3.86	3.69	3.56	3.46	3.39	3.32	3.27
80	5.36	4.48	4.06	3.81	3.63	3.51	3.41	3.33	3.26	3.21
100	5.33	4.45	4.03	3.78	3.60	3.47	3.38	3.30	3.23	3.18
150	5.29	4.42	3.99	3.74	3.56	3.43	3.33	3.25	3.19	3.13
200	5.28	4.40	3.97	3.72	3.54	3.41	3.31	3.23	3.16	3.11

n_2 \ n_1	12	15	20	25	30	40	50	80	100	150
2	18.86	18.89	18.91	18.93	18.94	18.96	18.97	18.97	18.98	18.98
3	10.40	10.38	10.35	10.34	10.32	10.31	10.29	10.28	10.29	10.27
4	7.74	7.70	7.65	7.63	7.61	7.58	7.56	7.56	7.54	7.54
5	6.47	6.42	6.37	6.34	6.32	6.28	6.26	6.25	6.24	6.23
6	5.73	5.68	5.62	5.59	5.57	5.52	5.49	5.49	5.47	5.46
7	5.25	5.20	5.14	5.10	5.08	5.02	5.00	4.99	4.97	4.96
8	4.92	4.86	4.79	4.75	4.73	4.67	4.64	4.63	4.62	4.61
9	4.67	4.60	4.54	4.50	4.47	4.41	4.38	4.37	4.35	4.34
10	4.47	4.41	4.34	4.30	4.27	4.21	4.17	4.16	4.15	4.14
11	4.32	4.25	4.18	4.14	4.11	4.05	4.01	4.00	3.98	3.97
12	4.19	4.13	4.05	4.01	3.98	3.91	3.88	3.86	3.85	3.84
13	4.09	4.02	3.95	3.90	3.87	3.80	3.76	3.75	3.73	3.72
14	4.00	3.93	3.86	3.81	3.78	3.71	3.67	3.65	3.64	3.63
15	3.93	3.85	3.78	3.73	3.70	3.63	3.59	3.57	3.55	3.54
16	3.86	3.79	3.71	3.66	3.63	3.56	3.51	3.50	3.48	3.47
17	3.81	3.73	3.65	3.60	3.57	3.49	3.45	3.44	3.42	3.41
18	3.76	3.68	3.60	3.55	3.51	3.44	3.39	3.38	3.36	3.35
19	3.71	3.63	3.55	3.50	3.46	3.39	3.34	3.33	3.31	3.30
20	3.67	3.59	3.51	3.46	3.42	3.35	3.30	3.28	3.26	3.25
21	3.64	3.56	3.47	3.42	3.38	3.31	3.26	3.24	3.22	3.21
22	3.60	3.52	3.44	3.39	3.35	3.27	3.22	3.21	3.18	3.17
23	3.57	3.49	3.41	3.35	3.32	3.24	3.19	3.17	3.15	3.14
24	3.55	3.47	3.38	3.32	3.29	3.21	3.16	3.14	3.12	3.11
25	3.52	3.44	3.35	3.30	3.26	3.18	3.13	3.11	3.09	3.08
30	3.42	3.34	3.25	3.19	3.15	3.06	3.01	2.99	2.97	2.96
40	3.30	3.22	3.12	3.06	3.02	2.92	2.87	2.84	2.82	2.80
60	3.18	3.09	2.99	2.93	2.88	2.78	2.71	2.69	2.66	2.64
80	3.12	3.03	2.93	2.86	2.81	2.70	2.63	2.61	2.57	2.55
100	3.09	2.99	2.89	2.82	2.77	2.66	2.58	2.56	2.52	2.50
150	3.04	2.94	2.84	2.76	2.71	2.59	2.51	2.48	2.44	2.42
200	3.02	2.92	2.81	2.74	2.68	2.56	2.48	2.45	2.40	2.38

Table A.4.7. The 95% Points of $F(n_1, n_2; \lambda = n_1)$

n_2 \ n_1	1	2	3	4	5	6	7	8	9	10
2	37.51	38.26	38.50	38.62	38.70	38.75	38.78	38.81	38.83	38.85
3	19.94	18.78	18.28	18.00	17.83	17.71	17.62	17.55	17.50	17.46
4	14.93	13.45	12.82	12.47	12.25	12.10	11.99	11.90	11.83	11.78
5	12.66	11.09	10.42	10.05	9.82	9.65	9.53	9.44	9.36	9.31
6	11.38	9.78	9.10	8.72	8.48	8.31	8.18	8.08	8.01	7.95
7	10.57	8.96	8.27	7.88	7.64	7.46	7.33	7.24	7.16	7.09
8	10.01	8.39	7.70	7.31	7.06	6.89	6.76	6.65	6.57	6.51
9	9.60	7.98	7.29	6.90	6.64	6.47	6.34	6.23	6.15	6.09
10	9.29	7.67	6.98	6.58	6.33	6.15	6.02	5.91	5.83	5.76
11	9.05	7.42	6.73	6.34	6.08	5.90	5.77	5.66	5.58	5.51
12	8.85	7.23	6.53	6.14	5.88	5.70	5.57	5.46	5.38	5.31
13	8.68	7.07	6.37	5.98	5.72	5.54	5.40	5.30	5.21	5.14
14	8.55	6.93	6.24	5.84	5.58	5.40	5.26	5.16	5.07	5.00
15	8.43	6.82	6.12	5.73	5.47	5.28	5.15	5.04	4.95	4.88
16	8.33	6.72	6.02	5.63	5.37	5.18	5.05	4.94	4.85	4.78
17	8.24	6.63	5.94	5.54	5.28	5.10	4.96	4.85	4.76	4.69
18	8.17	6.56	5.86	5.47	5.21	5.02	4.88	4.77	4.69	4.61
19	8.10	6.49	5.80	5.40	5.14	4.95	4.81	4.71	4.62	4.54
20	8.04	6.43	5.74	5.34	5.08	4.89	4.75	4.64	4.56	4.48
21	7.99	6.38	5.69	5.29	5.03	4.84	4.70	4.59	4.50	4.43
22	7.94	6.33	5.64	5.24	4.98	4.79	4.65	4.54	4.45	4.38
23	7.90	6.29	5.60	5.20	4.94	4.75	4.61	4.50	4.41	4.33
24	7.86	6.25	5.56	5.16	4.90	4.71	4.57	4.46	4.37	4.29
25	7.82	6.21	5.52	5.12	4.86	4.67	4.53	4.42	4.33	4.26
30	7.68	6.07	5.38	4.98	4.72	4.53	4.39	4.27	4.18	4.11
40	7.50	5.90	5.21	4.81	4.55	4.36	4.21	4.10	4.00	3.93
60	7.33	5.74	5.05	4.65	4.38	4.19	4.04	3.93	3.83	3.75
80	7.25	5.66	4.97	4.57	4.30	4.11	3.96	3.84	3.75	3.67
100	7.20	5.61	4.92	4.52	4.25	4.06	3.91	3.79	3.70	3.61
150	7.13	5.55	4.86	4.46	4.19	3.99	3.84	3.73	3.63	3.55
200	7.10	5.52	4.83	4.42	4.16	3.96	3.81	3.69	3.59	3.51

n_2 \ n_1	12	15	20	25	30	50	80	100	150	200
2	38.87	38.90	38.92	38.94	38.95	38.92	38.90	38.88	38.82	38.99
3	17.40	17.32	17.27	17.22	17.20	17.14	17.10	17.10	17.06	17.08
4	11.70	11.61	11.53	11.47	11.44	11.37	11.33	11.31	11.31	11.29
5	9.21	9.12	9.03	8.97	8.93	8.85	8.81	8.80	8.78	8.76
6	7.85	7.75	7.66	7.59	7.55	7.47	7.42	7.41	7.39	7.35
7	7.00	6.89	6.79	6.73	6.69	6.60	6.55	6.53	6.51	6.49
8	6.41	6.31	6.20	6.13	6.09	6.00	5.95	5.93	5.91	5.89
9	5.98	5.88	5.77	5.70	5.66	5.56	5.51	5.49	5.47	5.45
10	5.66	5.55	5.44	5.37	5.33	5.23	5.18	5.16	5.13	5.11
11	5.41	5.30	5.18	5.11	5.07	4.97	4.91	4.89	4.87	4.85
12	5.20	5.09	4.98	4.90	4.86	4.76	4.70	4.68	4.65	4.64
13	5.03	4.92	4.80	4.73	4.68	4.58	4.52	4.50	4.47	4.46
14	4.89	4.78	4.66	4.59	4.54	4.43	4.37	4.35	4.32	4.31
15	4.77	4.66	4.54	4.46	4.41	4.31	4.24	4.22	4.20	4.18
16	4.67	4.55	4.43	4.36	4.31	4.20	4.13	4.11	4.08	4.07
17	4.58	4.46	4.34	4.26	4.21	4.10	4.04	4.02	3.99	3.97
18	4.50	4.38	4.26	4.18	4.13	4.02	3.95	3.93	3.90	3.89
19	4.43	4.31	4.19	4.11	4.06	3.94	3.88	3.85	3.82	3.81
20	4.37	4.25	4.13	4.05	3.99	3.88	3.81	3.79	3.76	3.74
21	4.31	4.19	4.07	3.99	3.93	3.82	3.75	3.73	3.69	3.68
22	4.26	4.14	4.02	3.94	3.88	3.76	3.69	3.67	3.64	3.62
23	4.22	4.10	3.97	3.89	3.83	3.71	3.64	3.62	3.59	3.57
24	4.18	4.06	3.93	3.84	3.79	3.67	3.60	3.57	3.54	3.52
25	4.14	4.02	3.89	3.80	3.75	3.63	3.55	3.53	3.50	3.48
30	3.99	3.86	3.73	3.65	3.59	3.46	3.39	3.36	3.32	3.31
40	3.81	3.68	3.54	3.45	3.39	3.25	3.17	3.14	3.11	3.09
60	3.63	3.50	3.35	3.26	3.19	3.05	2.96	2.92	2.88	2.86
80	3.54	3.40	3.26	3.16	3.09	2.94	2.84	2.81	2.76	2.74
100	3.49	3.35	3.20	3.10	3.03	2.88	2.78	2.74	2.69	2.66
150	3.42	3.28	3.13	3.02	2.95	2.79	2.68	2.64	2.58	2.55
200	3.38	3.24	3.09	2.99	2.91	2.74	2.63	2.59	2.53	2.50

Table A.4.8.   The 99% Points of $F(n_1, n_2; \lambda = n_1)$

n_2 \ n_1	1	2	3	4	5	6	7	8	9	10
2	197.5	198.3	198.5	198.6	198.7	198.8	198.8	198.8	198.8	198.9
3	65.60	59.81	57.53	56.31	55.57	55.02	54.64	54.34	54.12	53.95
4	39.65	34.16	32.02	30.84	30.13	29.62	29.25	28.99	28.76	28.58
5	29.84	24.79	22.80	21.73	21.05	20.58	20.25	19.99	19.78	19.62
6	24.89	20.15	18.29	17.28	16.64	16.20	15.88	15.64	15.44	15.28
7	21.95	17.44	15.67	14.70	14.09	13.67	13.37	13.12	12.94	12.78
8	20.01	15.68	13.97	13.04	12.45	12.04	11.74	11.51	11.33	11.18
9	18.65	14.45	12.80	11.89	11.32	10.92	10.62	10.40	10.21	10.07
10	17.65	13.55	11.93	11.04	10.49	10.09	9.80	9.58	9.40	9.26
11	16.87	12.86	11.27	10.40	9.85	9.46	9.18	8.96	8.78	8.64
12	16.27	12.31	10.76	9.90	9.35	8.97	8.69	8.47	8.30	8.15
13	15.77	11.87	10.34	9.49	8.95	8.57	8.30	8.08	7.90	7.77
14	15.35	11.51	9.99	9.15	8.62	8.25	7.97	7.75	7.58	7.44
15	15.01	11.21	9.70	8.88	8.35	7.97	7.70	7.48	7.31	7.17
16	14.72	10.95	9.46	8.64	8.11	7.74	7.47	7.26	7.08	6.95
17	14.46	10.73	9.25	8.43	7.91	7.54	7.27	7.06	6.89	6.75
18	14.24	10.54	9.07	8.26	7.74	7.37	7.10	6.89	6.72	6.58
19	14.05	10.37	8.91	8.10	7.58	7.22	6.95	6.74	6.57	6.43
20	13.88	10.22	8.77	7.96	7.45	7.09	6.82	6.61	6.44	6.30
21	13.72	10.08	8.64	7.84	7.33	6.97	6.70	6.49	6.32	6.18
22	13.59	9.96	8.53	7.73	7.22	6.86	6.59	6.38	6.22	6.08
23	13.46	9.86	8.42	7.63	7.12	6.77	6.50	6.29	6.12	5.98
24	13.35	9.76	8.33	7.54	7.04	6.68	6.41	6.20	6.04	5.90
25	13.25	9.67	8.25	7.46	6.96	6.60	6.33	6.13	5.96	5.82
30	12.85	9.32	7.92	7.15	6.65	6.29	6.03	5.82	5.66	5.52
40	12.38	8.91	7.54	6.77	6.28	5.93	5.67	5.46	5.30	5.16
60	11.92	8.52	7.17	6.42	5.93	5.58	5.32	5.12	4.95	4.82
80	11.70	8.34	7.00	6.25	5.76	5.42	5.16	4.95	4.79	4.65
100	11.57	8.23	6.89	6.15	5.67	5.32	5.06	4.86	4.69	4.56
150	11.40	8.08	6.76	6.02	5.54	5.19	4.93	4.73	4.57	4.43
200	11.32	8.01	6.69	5.96	5.47	5.13	4.87	4.67	4.50	4.37

n_2 \ n_1	12	15	20	25	30	50	80	100	150	200
2	198.8	198.8	198.8	198.8	198.9	199.0	199.0	199.0	199.0	199.0
3	53.68	53.41	53.07	52.92	52.83	52.68	52.63	52.22	52.21	52.04
4	28.31	28.05	27.77	27.61	27.48	27.28	27.14	27.10	27.05	27.02
5	19.36	19.11	18.85	18.69	18.58	18.37	18.25	18.22	18.16	18.06
6	15.04	14.79	14.55	14.39	14.29	14.08	13.96	13.93	13.87	13.85
7	12.55	12.31	12.07	11.92	11.82	11.62	11.50	11.46	11.41	11.39
8	10.95	10.72	10.48	10.33	10.23	10.03	9.92	9.88	9.83	9.80
9	9.85	9.61	9.38	9.24	9.14	8.94	8.82	8.78	8.73	8.71
10	9.03	8.81	8.57	8.43	8.33	8.13	8.02	7.98	7.93	7.90
11	8.42	8.19	7.96	7.82	7.72	7.52	7.41	7.37	7.32	7.29
12	7.94	7.71	7.48	7.34	7.24	7.04	6.92	6.89	6.83	6.81
13	7.55	7.32	7.09	6.95	6.85	6.65	6.54	6.50	6.44	6.42
14	7.23	7.00	6.77	6.63	6.53	6.33	6.21	6.17	6.12	6.10
15	6.96	6.74	6.51	6.36	6.26	6.06	5.95	5.91	5.85	5.82
16	6.73	6.51	6.28	6.14	6.04	5.83	5.72	5.68	5.62	5.59
17	6.54	6.32	6.08	5.94	5.84	5.64	5.52	5.48	5.42	5.40
18	6.37	6.15	5.91	5.77	5.67	5.47	5.35	5.31	5.25	5.22
19	6.22	6.00	5.77	5.62	5.52	5.32	5.19	5.15	5.10	5.07
20	6.09	5.87	5.64	5.49	5.39	5.18	5.06	5.02	4.96	4.93
21	5.97	5.75	5.52	5.37	5.27	5.06	4.94	4.90	4.84	4.81
22	5.87	5.64	5.41	5.27	5.17	4.96	4.83	4.79	4.73	4.70
23	5.77	5.55	5.32	5.17	5.07	4.86	4.74	4.69	4.64	4.61
24	5.69	5.47	5.23	5.09	4.98	4.77	4.65	4.60	4.55	4.52
25	5.61	5.39	5.15	5.01	4.91	4.69	4.57	4.52	4.46	4.44
30	5.31	5.08	4.85	4.70	4.60	4.38	4.25	4.21	4.14	4.11
40	4.95	4.72	4.48	4.33	4.23	4.00	3.86	3.82	3.75	3.72
60	4.60	4.38	4.13	3.98	3.87	3.63	3.48	3.43	3.36	3.32
80	4.44	4.21	3.96	3.80	3.69	3.45	3.29	3.24	3.16	3.12
100	4.34	4.11	3.86	3.70	3.59	3.34	3.18	3.12	3.04	3.00
150	4.21	3.98	3.73	3.57	3.45	3.19	3.02	2.96	2.87	2.83
200	4.15	3.92	3.66	3.50	3.38	3.12	2.94	2.88	2.79	2.74

References

Adamson, P. T. and W. Zucchini (1984). On the application of a censored log-normal distribution to partial duration series of storms. *Water S. A.*, **10**, 136–146.

Ahlborn, M. (1982). *Modellauswahl mit Diskrepanzkriterien in der Kovarianzanalyse*. Thesis; Institut für Statistik und Ökonometrie, University of Göttingen.

Akaike, H. (1969). Fitting autoregressive models for prediction. *Ann. Inst. Statist. Math.*, **21**, 243–247.

Akaike, H. (1970). Statistical predictor identification. *Ann. Inst. Statist. Math.*, **22**, 203–217.

Akaike, H. (1973). Information theory and an extension of the maximum likelihood principle. In Petrov and Czaki, eds., *Proceedings of the 2nd International Symposium on Information Theory*, pp. 267–281.

Allen, D. M. (1971). *The Prediction Sum of Squares as a Criterion for Selecting Prediction Variables*. Technical Report No. 23, Department of Statistics, University of Kentucky.

Allen, D. M. (1974). The relationship between variable selection and data augmentation and a method for prediction. *Technometrics*, **16**, 125–127.

Anděl, J. (1982). Fitting models in time series analysis. *Math. Operat.-forschung Statist.*, Ser. *Statist.*, **13**, 121–143.

Anderson, T. W. (1971). *The Statistical Analysis of Time Series*. Wiley, New York.

Bartle, R. G. (1976). *The Elements of Real Analysis*, 2nd ed. Wiley, New York.

Bishop, Y. M. M., S. E. Fienberg, and P. W. Holland (1975). *Discrete Multivariate Analysis*. MIT Press, Cambridge, MA.

Bliss, C. I. (1953). Fitting the negative binomial distribution to biological data. *Biometrics*, **9**, 176–196.

Bliss, C. I. (1970). *Statistics in Biology. Statistical Methods for Research in the Natural Sciences*, *Vol. 2*. McGraw-Hill, New York.

Bloomfield, P. (1973). An exponential model for the spectrum of a scalar time series. *Biometrika*, **60**, 217–226.

Boos, D. D. (1981). Minimum distance estimators for location and goodness of fit. *J. Am. Statist. Assoc.*, **76**, 663–670.

Boos, D. D. (1982). Minimum Anderson–Darling estimation. *Commun. Statist.-Theor. Meth.*, **11**, 2747–2774.

Bowman, A. W. (1984). An alternative method of cross-validation for the smoothing of density estimates. *Biometrika*, **71**, 353–360.

Bowman, A. W., P. Hall, and D. M. Titterington (1984). Cross-validation in nonparametric estimation of probabilities and probability densities. *Biometrika*, **71**, 341–351.

Box, G. E. P. and G. M. Jenkins (1970). *Time Series Analysis: Forecasting and Control*. Holden-Day, San Francisco.

Brillinger, D. R. (1975). *Time Series. Data Analysis and Theory*. Holt, Rinehart and Winston, New York.

Brown, M. B. (1976). Screening effects in multidimensional contingency tables. *Appl. Statist.*, **25**, 37–46.

Browne, M. W. (1969). *Factor Analysis Models and Their Application to Prediction Problems*. Ph.D. thesis, University of South Africa.

Chow, Y.-S., S. Geman, and L.-D. Wu (1983). Consistent cross-validated density estimation. *Ann. Statist.*, **11**, 25–38.

Cogburn, R. and H. T. Davis (1974). Periodic splines and spectral estimation. *Ann. Statist.*, **2**, 1108–1126.

Cox, D. R. (1961). Tests of separate families of hypotheses. *Proc. 4th Berkeley Symp.*, **1**, 105–123.

Cox, D. R. (1962). Further results on tests of separate families of hypotheses. *J. R. Statist. Soc. B*, **24**, 406–424.

Cramér, H. (1946). *Mathematical Methods of Statistics*. University Press, Princeton, NJ.

Don, C. E. (1975). *An Investigation of the Calorific Value and Some Other Properties of Bagasse*. M. Sc. Thesis, University of Natal, Durban.

Draper, N. R. and H. Smith (1981). *Applied Regression Analysis*, 2nd ed. Wiley, New York.

Durbin, J. (1973). *Distribution Theory of Tests Based on the Sample Distribution Function*. SIAM, Philadelphia.

Durbin, J. and M. Knott (1972). Components of Cramér–von Mises statistics. I. *J. R. Statist. Soc. B*, **34**, 290–307.

Durbin, J. and G. S. Watson (1951). Testing for serial correlation in least squares regression. *Biometrika*, **38**, 159–178.

Efron, B. (1979). Bootstrap methods: another look at the jackknife. *Ann. Statist.*, **7**, 1–26.

Efron, B. (1982). *The Jackknife, the Bootstrap and Other Resampling Plans*. SIAM, Philadelphia.

Efron, B. (1985). *How Biased is the Apparent Error Rate of a Logistic Regression*. Technical Report No. 102, Division of Biostatistics, Stanford University.

Ekhart, E. (1949). Über Inversionen in den Alpen. *Meteorol. Rundschau*, **2**, 153–157.

Emerson, P. L. (1968). Numerical construction of orthogonal polynomials from a general recurrence formula. *Biometrics*, **24**, 695–701.

Fienberg, S. E. (1980). *The Analysis of Cross-Classified Categorical Data*. MIT Press, Cambridge, MA.

Fisher, R. A. (1928). The general sampling distribution of the multiple correlation coefficient. *Proc. R. Statist. Soc. A*, **121**, 654–673.

Fisher, R. A. and F. Yates (1963). *Statistical Tables for Biological, Agricultural and Medical Research*, 6th ed. Oliver and Boyd, Edinburgh.

Frauen, M. (1979). Personal communication.

Fuller, W. A. (1976). *Introduction to Statistical Time Series*. Wiley, New York.

Geisser, S. (1974). A predictive approach to the random effect model. *Biometrika*, **61**, 101–107.

Geisser, S. and W. F. Eddy (1979). A predictive approach to model selection. *J. Am. Statist. Assoc.*, **74**, 153–160.

Goodman, L. A. (1971). The analysis of multidimensional contingency tables: stepwise procedures and direct estimation methods for building models for multiple classifications. *Technometrics*, **13**, 33–61.

Grenander, U. and M. Rosenblatt (1957). *Statistical Analysis of Stationary Time Series*. Wiley, New York.

Guttman, L. (1954). A new approach to factor analysis: the radex. In Lazarsfeld, ed., *Mathematical Thinking in the Social Sciences*. Free Press, Glencoe.

Hald, A. (1948). *The Decomposition of a Series of Observations*. G. E. C. Gads, Copenhagen.

Haldane, J. B. S. (1951). A class of efficient estimates of a parameter. *Bull. Int. Statist. Inst.*, **33**, 231–248.

Hall, P. (1982). Cross-validation in density estimation. *Biometrika*, **69**, 383–390.

Hall, P. (1983a). Orthogonal series methods for both qualitative and quantitative data. *Ann. Statist.*, **11**, 1004–1007.

Hall, P. (1983b). Large sample optimality of least squares cross-validation in density estimation. *Ann. Statist.*, **11**, 1156–1174.

Hannan, E. J. (1973a). Central limit theorems for time series regression. *Zeit. Wahrscheinl.-theorie verw. Gebiete*, **26**, 157–170.

Hannan, E. J. (1973b). The asymptotic theory of linear time series models. *J. Appl. Prob.*, **10**, 130–145.

Hannan, E. J. (1976). The asymptotic distribution of serial covariances. *Ann. Statist.*, **4**, 396–399.

Heise, B. (1981). *Diskrepanzkriterien zur Auswahl linearer Modelle*. Thesis, Institut für Statistik und Ökonometrie, University of Göttingen.

Hocking, R. R. (1972). Criteria for selection of a subset regression: which one should be used? *Technometrics*, **14**, 967–970.

Hocking, R. R. (1976). The analysis and selection of variables in linear regression. *Biometrics*, **32**, 1–49.

Hotelling, H. (1940). The selection of variates for use in prediction with some comments on the general problem of nuisance parameters. *Ann. Math. Statist.*, **11**, 271–283.

Humak, K. M. S. (1983). *Statistische Methoden der Modellbildung, Band II*. Akademie-Verlag, Berlin.

Jenkins, G. M. (1961). General considerations in the analysis of spectra. *Technometrics*, **3**, 133–166.

Jenkins, G. M. and D. G. Watts (1968). *Spectral Analysis and Its Applications*. Holden Day, San Francisco.

Jennrich, R. I. (1969). Asymptotic properties of non-linear least squares estimators. *Ann. Math. Statist.*, **40**, 633–643.

Jones, H. L. (1946). Linear regression functions with neglected variables. *J. Am. Statist. Assoc.*, **41**, 356–369.

Kendall, M. G. and A. Stuart (1979). *The Advanced Theory of Statistics. Vol. 1: Distribution Theory*, 4th edition. Griffin, London.

Kitagawa, T. (ed.) (1983). *Johoryo-tokeigaku*. (Information criterion.) Kyoritsushuppan Kabushikigaisha, Tokyo.

Knoch, K. (1947). Die "Normalperiode" 1901–1930 und ihr Verhältnis zu längeren Perioden. *Meteorol. Rundschau*, **1**, 10–23.

Knüppel, H., A. Stumpf, and B. Wiezorke (1958). Mathematische Statistik in Eisenhüttenwerken. Teil I. Regressionsanalyse. *Archiv f. d. Eisenhüttenwesen*, **29**, 521–533.

Krahl, A. (1980). *Methoden zur Auswahl von Familien eindimensionaler Modelle nach Diskrepanzkriterien*. Thesis, Institut für Statistik und Ökonometrie, University of Göttingen.

Kronmal, R. and M. Tarter (1968). The estimation of probability densities and cumulatives by Fourier series methods. *J. Am. Statist. Assoc.*, **63**, 925–952.

Kullback, S. and R. A. Leibler (1951). On information and sufficiency. *Ann. Math. Statist.*, **22**, 79–86.

La Marche (Jr.), V. C., R. L. Holmes, P. W. Dunnwippie, and L. G. Drew (1979). *Tree-Ring Chronologies of the Southern Hemisphere 5. South Africa*. Chronology Series V, Laboratory of Tree-Ring Research, University of Arizona, Tucson, AZ.

Linder, A. and W. Berchtold (1976). *Statistische Auswertung von Prozentzahlen*. Birkhäuser, Basel.

Linhart, H. (1958). Critère de sélection pour le choix des variables dans l'analyse de régression. *Revue suisse d'Economie politique et de Statistique*, **94**, 202–232.

Linhart, H. and A. W. G. van der Westhuyzen (1963). The fibre diameter distribution of raw wool. *J. Textile Inst.*, **54**, T123–T127.

Linhart, H. and P. Volkers (1984). Asymptotic criteria for model selection. *OR Spektrum*, **6**, 161–165.

Linhart, H. and P. Volkers (1985). On a criterion for the selection of models for stationary time series. *Metrika*, **32**, 181–196.

Linhart, H. and W. Zucchini (1980). *Statistik Eins*. Birkhäuser, Basel.

Linhart, H. and W. Zucchini (1982a). *Statistik Zwei*. Birkhäuser, Basel.

Linhart, H. and W. Zucchini (1982b). On model selection in analysis of variance. In B. Fleischmann et al., eds., *Operations Research Proceedings 1981*, pp. 483–493.

Linhart, H. and W. Zucchini (1982c). A method for selecting the covariates in analysis of covariance. *S. Afr. Statist. J.*, **16**, 97–112.

Linhart, H. and W. Zucchini (1984a). An orthogonal class of models for time series. *S. Afr. Statist. J.*, **18**, 59–67.

Linhart, H. and W. Zucchini (1984b). Model selection by the discrepancy of Cramér and von Mises. *Statist. & Dec.*, Suppl. Issue No. 1, 247–306.

Linhart, H. and W. Zucchini (1985). Selection of approximating models by chi-squared discrepancies if the operating model is multinomial. *Statist. & Dec.*, Suppl. Issue No. 2, 375–380.

Lukatis, W. (1972). *Akademiker in Wirtschaftsunternehmen. Ergebnis einer quantitativen Analyse*. Akademische Verlagsgesellschaft, Frankfurt.

Mallows, C. L. (1973). Some comments on C_P. *Technometrics*, **15**, 661–675.

Martin, L. (1960). *Adjustement d'un Faisceau de Régressions Curvilignes au Moyen d'un Système de Polynomes Orthogonaux*. Bureau de Biométrie, Bruxelles.

Ministry of Agriculture, Fisheries and Food (1968). *A Century of Agricultural Statistics. Great Britain 1866–1966*. Her Majesty's Stationery Office, London.

Mosteller, F. and J. W. Tukey (1968). Data analysis, including statistics. In G. Lindzey and E. Aronson, eds., *Handbook of Social Psychology, Vol. 2*. Addison-Wesley, Reading, MA.

Norton, V. (1983). A simple algorithm for computing the non-central F-distribution. *Appl. Statist.*, **32**, 84–85.

Ogata, Y. (1980). Maximum likelihood estimates of incorrect Markov models for time series and the derivation of AIC. *J. Appl. Prob.*, **17**, 59–72.

Oman, S. D. and Y. Wax (1984). Estimating fetal age by ultrasound measurements: an example of multivariate calibration. *Biometrics*, **40**, 947–960.

Parr, W. C. (1981). Minimum distance estimation: a bibliography. *Commun. Statist.-Theor. Meth.*, **A10**, 1205–1224.

Parr, W. C. and W. R. Schucany (1980). Minimum distance and robust estimation. *J. Am. Statist. Assoc.*, **75**, 616–624.

Parr, W. C. and T. De Wet (1981). On minimum Cramér–von Mises norm parameter estimation. *Commun. Statist.-Theor. Meth.*, **A10**, 1149–1166.

Parzen, E. (1974). Some recent advances in time series modelling. *Trans. Auto. Control*, **AC-19**, 723–730.

Parzen, E. (1977). Multiple time series: Determining the order of approximating autoregressive schemes. In P. R. Krishnaiah, ed., *Multivariate Analysis IV*. North-Holland, Amsterdam, pp. 283–295.

Pearson, E. S. and H. O. Hartley (1972). *Biometrika Tables for Statisticians, Vol. II*. Cambridge University Press, Cambridge.

Pearson, E. S. and H. O. Hartley (1976). *Biometrika Tables for Statisticians, Vol. I*, 3rd ed. Cambridge University Press, Cambridge.

Pereira, B. de B. (1977). Discriminating among separate models: a bibliography. *Int. Statist. Rev.*, **45**, 163–172.

Plackett, R. L. (1960). *Principles of Regression Analysis*. Clarendon, Oxford.

Plackett, R. L. (1974). *The Analysis of Categorical Data*. Griffin, London.

Ponnapalli, R. (1976). Deficiencies of minimum discrepancy estimators. *Can. J. Statist.*, **4**, 33–50.

Press, S. J. (1972). *Applied Multivariate Analysis*. Holt, Rinehart and Winston, New York.

Prosch, R. (1985). Sea Fisheries Research Institute, Department of Environment Affairs, Cape Town, South Africa, Personal communication.

Proschan, F. (1963). Theoretical explanation of observed decreasing failure rate. *Technometrics*, **5**, 375–383.

Rayner, A. A. (1967). *A First Course in Biometry for Agriculture Students*. University of Natal Press, Pietermaritzburg.

Robertson, C. A. (1972). On minimum discrepancy estimators. *Sankhyā A*, **34**, 133–144.

Robinson, P. M. (1978). Alternative models for stationary stochastic processes. *Stoch. Processes Appl.*, **8**, 144–152.

Rudemo, M. (1982). Empirical choice of histograms and kernel density estimators. *Scand. J. Statist.*, **9**, 65–78.

Rypstra, T. (1982). Department of Forestry, University of Stellenbosch, Personal communication.

Sahler, W. (1970). Estimation by minimum-discrepancy methods. *Metrika*, **16**, 85–106.

Sakamoto, Y. and H. Akaike (1978a). Analysis of cross-classified data by AIC. *Ann. Inst. Statist. Math.*, **30**, 185–197.

Sakamoto, Y. and H. Akaike (1978b). Robot data screening of cross-classified data by an information criterion. *Proceedings of the International Conference on Cybernetics and Society*. IEEE, New York, pp. 398–403.

Sawa, T. (1978). Information criteria for discriminating among alternative models. *Econometrica*, **46**, 1273–1291.

Schäfer, B. (1985). *Eine statistische Erhebung und Analyse in Zusammenhang mit einem Entscheidungsproblem der Stadtverwaltung Göttingen*. Thesis, Institut für Statistik und Ökonometrie, University of Göttingen.

Schmitz, C. J. (1979). *World Non-Ferrous Metal Production and Prices 1700–1976*. Frank Cass, London.

Scott, D. W. (1979). On optimal data-based histograms. *Biometrika*, **66**, 605–610.

Serfling, R. J. (1980). *Approximation Theorems of Mathematical Statistics*. Wiley, New York.

Snedecor, G. and W. Cochran (1980). *Statistical Methods*, 7th ed. Iowa State University Press, Ames.

Soetbeer, A. (1879). *Edelmetallproduktion*. Justus Perthes, Gotha.

South African Department of Water Affairs (1967). *Hydrographic Survey Publication Number 9*.

Sparks, R. S. (1984). *Selection of Variables in Multivariate and Generalised Multivariate Regression Models*. Ph.D. thesis, University of Natal, Durban.

Sparks, R. S., W. Zucchini, and D. Coutsourides (1985). On variable selection in multivariate regression. *Comm. Statist.-Theor. Meth.*, **A14**, 1569–1587.

Statistisches Bundesamt (1976). *Statistisches Jahrbuch 1976 für die Bundesrepublik Deutschland*. Kohlhammer, Stuttgart.

Stone, M. (1974a). Crossvalidatory choice and assessment of statistical predictions. *J. R. Statist. Soc. B*, **36**, 111–133.

Stone, M. (1974b). Cross-validation and multinomial prediction. *Biometrika*, **61**, 509–515.

Stone, M. (1978). Cross-validation: a review. *Math. Operat.-forschung Statist.*, *Ser. Statist.*, **9**, 127–139.

Thompson, M. L. (1978). Selection of variables in multiple regression, Part I and Part II. *Int. Statist. Rev.*, **46**, 1–19, 129–146.

Tingle, A. C. (1965). *Silviculture, Volume and Yield of Populus Deltoides*. M.Sc. Thesis, University of Stellenbosch.

Titterington, D. M. (1985). Common structure of smoothing techniques in statistics. *Int. Statist. Rev.*, **53**, 141–170.

Toro-Vizcarrondo, C. and T. C. Wallace (1968). A test of the mean square error criterion for restrictions in linear regression. *J. Am. Statist. Assoc.*, **63**, 558–572.

Tukey, J. W. (1961). Discussion, emphasizing the connection between analysis of variance and spectrum analysis. *Technometrics*, **3**, 191–219.

U.S. Department of Commerce, Bureau of Census (1966). *Long Term Economic Growth*. Washington, DC.

Van der Merwe, A. H., P. C. N. Groenewald, D. J. De Waal, and C. A. Van der Merwe (1983). Model selection for future data in the case of multivariate regression analysis. *S. Afr. Statist. J.*, **17**, 147–164.

Verwey, F. A. (1957). *Prediction of Work Performance of Artisan Apprentices During Training by Means of Psychological Tests*. Ph.D. Thesis, University of the Witwatersrand, Johannesburg.

Volkers, P. (1983). *Modellauswahl bei Gauß'schen Zeitreihen*. Ph.D. Thesis, Institut für Statistik und Ökonometrie, University of Göttingen.

Volkers, P. (1985). An algorithm for the distribution of the multiple correlation coefficient. *Appl. Statist.*, in press.

von Mises, R. (1947). On the asymptotic distribution of differentiable statistical functions. *Ann. Math. Statist.*, **18**, 309–348.

Wahba, G. (1980). Automatic smoothing of the log periodogram. *J. Am. Statist. Assoc.*, **75**, 122–132.

White, H. (1982). Maximum likelihood estimation of misspecified models. *Econometrica*, **50**, 1–25.

Wolfowitz, J. (1953). Estimation by the minimum distance method. *Ann. Inst. Statist. Math.*, **5**, 9–23.

Zucchini, W. and P. T. Adamson (1984a). *The occurrence and severity of droughts in South Africa*. WRC Report No. 91/1/84; No. 91/1/84(A), Water Research Commission, South Africa.

Zucchini, W. and P. T. Adamson (1984b). *Assessing the risk of deficiencies in streamflow*. WRC Report No. 91/2/84, Water Research Commission, South Africa.

Zucchini, W. and L. A. V. Hiemstra (1983). A note on the relationship between annual rainfall and tree-ring indices for one site in South Africa. *Water S. A.*, **9**, 153–154.

Zucchini, W. and R. S. Sparks (1984). *Estimating the missing values in rainfall records*. WRC Report No. 91/3/84, Water Research Commission, South Africa.

Zwanzig, S. (1980). The choice of approximative models in nonlinear regression. *Math. Operat.-forschung Statist., Ser. Statist.*, **11**, 23–47.

Author Index

291

Subject Index

Index of Criteria